Proceedings of the 30th International Geological Congress
Volume 23

Engineering Geology

Proceedings of the 30th International Geological Congress

PROCEEDINGS OF THE
30TH INTERNATIONAL GEOLOGICAL CONGRESS

BEIJING, CHINA, 4 - 14 AUGUST 1996

VOLUME 23

ENGINEERING GEOLOGY

EDITORS:
WANG SIJING
INSTITUTE OF GEOLOGY, CHINESE ACADEMY OF SCIENCES, BEIJING, CHINA
P. MARINOS
CIVIL ENGINEERING FACULTY, NATIONAL TECHNICAL UNIVERSITY OF ATHENS, GREECE

CRC Press
Taylor & Francis Group
Boca Raton London New York

CRC Press is an imprint of the
Taylor & Francis Group, an **informa** business

First published 1997 by VSP BV Publishing

Published 2019 by CRC Press
4 Park Square, Milton Park, Abingdon, Oxon, OX14 4RN

and by CRC Press
2385 NW Executive Center Drive, Suite 320, Boca Raton FL 33431

© 1997 by Taylor & Francis Group, LLC

First issued in paperback 2019

CRC Press is an imprint of Informa UK Limited

ISBN 13: 978-0-367-44805-9 (pbk)
ISBN 13: 978-90-6764-240-8 (hbk)

**Visit the Taylor & Francis Web site at
http://www.taylorandfrancis.com**

**and the CRC Press Web site at
http://www.crcpress.com**

CONTENTS

Rock and Soil Properties

Geoenvironment and Geohazards

New Techniques and Application

Construction Material

PREFACE

This volume documents the contributions of engineering geologists from various parts of the world, who attended the 30th International Geological Congress (IGC) held in Beijing on 4-14 August, 1996. Thanks to the meticulous efforts of the Organizing Committee of the 30th IGC, an ideal forum was provided to geologists from different continents to review the recent progress made in geological sciences and their role in the development and well being of our modern society.

The International Association of Engineering Geology (IAEG) was one of the sponsors of Engineering Geology Symposium No. 17 at the IGC. A tremendous amount of organisation work was done by leaders of the IAEG to ensure the success of the symposium and to promote cooperation and scientific exchange among engineering geologists and geoscientists from the various fields.

Engineering geology is one of the boundary branches of geoscience which entails the application of geological theories and principles to solve problems in engineering construction. Any large-scale engineering project inevitably constitutes a complex system of two interconnected elements, i.e. the surface building and geological subsurface. An ingenious coupling of these two elements would be one of the basic requirements for system optimization. The geologic formation represents a structure of natural origin, in contrast to the surface building which was designed and constructed artificially. Hence, the main task in engineering geology is to investigate and understand the engineering properties of the geologic formation. Engineering geology, of course, has its roots in geology. However, as a result of the advance of engineering geology, new information and theories are being added to the realm of geosciences and to the study of Earth as a planet.

Engineering geologists should not only be concerned about the safety and economy of the projects under construction. To fulfill their duty, they should also address the environmental impacts of their engineering activities. There is little doubt that engineering geological investigations and studies will become increasingly significant in the protection of the environment and the reduction of natural disasters in the years ahead.

This volume is divided into several parts, such as the engineering geological and environmental problems of major engineering works, rock and soil properties, protection of the geoenvironment and reduction of geohazards, as well as the application of innovative new techniques in engineering geological investigation and construction materials. Together, this collection of papers reflects the major achievements and advancement of engineering geological science and technology in recent years.

The editors are grateful to their colleagues for their efforts towards the 30th IGC and to all contributors of this volume. Special thanks are due to Mr. Jin Qiusheng, Prof. Du Yonglian, Mr. Du Donghai and Ms. Wang Lingjuan, for their able assistance in the preparation of this volume.

Wang Sijing   Paul Marinos
Institute of Geology   Civil Engineering Faculty
Chinese Academy of Sciences   National Technical University of Athens

ENGINEERING GEOLOGICAL PROBLEMS OF MAJOR ENGINEERING WORKS

Proc. 30th Int'l. Geol. Congr., Vol. 23, pp. 3-15
Wang Sijing and P. Marinos (Eds)
© VSP 1997

Principal Experience And Technical Methodology In Geological Investigation And Research For The Three Gorges Project

CHEN DEJI

Bureau of Investigation & Survey, Changjiang Water Resource Commission, Wuhan ,430010, China.

Abstract

The geological investigation and studies for the Three Gorges Project (TGP) have been lasted for more than 40 years, centering around such five engineering issues as dam site (dam region) comparison, regional stability and seismic risks, reservoir and environmental geology, damsite and structure foundation geology, natural construction materials. Through coordinated muti-discipline research including all the geoscience disciplines relevant to the engineering construction, and employment of state-of-the-art techniques in different periods, a comprehensive yet in-depth geological investigation and study has been carried out . As a result a wealth of reliable data have been obtained and great a many valuable experiences are summarized, which has made a great contribution to the development of relevant disciplines .

The TGP formally commenced to construction on Dec. 14, 1994 and the construction of major structures are now in progress smoothly. The understandings and conclusions reached in investigation phases are now under examination and the results will further enrich engineering geology in both theory and practice with the progress of the project construction.

Keywords: Three Gorges Project, Engineering geological investigation, Techniques, Experience.

A BRIEF ACCOUNT OF THE THREE GORGES PROJECT

Through almost half a century of repeated studies, justification and argumentation, the TGP was eventually approved at the Fifth Plenary Session of the Seventh National People's Congress in March 1992 and included in the ten-year plan of national economic and social development . The main characteristic indices are shown in table 1. With such a construction scale and technical scheme, the TGP has well dealt with the contradiction between the need and the possibility, and achieved the coordination and unity in efficiency, investment environment, resident relocation .

The TGP will have tremendous social and economic benefits . It can effectively control the upstream flood, as a result significantly improving the flood control capacities in the middle and lower reaches of the Changjiang River. According to the adopted technical scheme, the TGP will have a total installed capacity of 18200 MW, yielding an annual energy output of 84. 68 billion kw.h . With such a high capacity and tremendous energy output as well as close-in to the load center, it is an optimum program to alleviate the electric shortage in such areas as the central China, eastern China and eastern Sichun

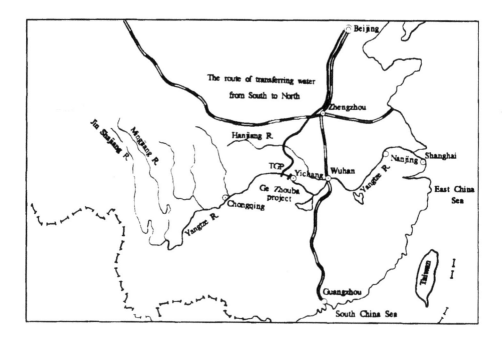

Fig.1   The geographical location map of TGP

province. After completion of the TGP, the navigation capacities of the Changjiang River
the Golden Waterway will be increased greatly and the navigation conditions in the
middle reach will be improved enormously. In addition, the TGP has other tremendous
comprehensive benefits in development of aquatic product, irrigation and tourism
resources; speeding up implementation of the program of shaking off poverty and
building up a fortune in the Three Gorges area; and supply of water resources for the
South-North water transfer project .(Fig.1)

HISTORY OF GEOLOGICAL AND SEISMOLOGICAL INVESTIGATIONS AND STUDIES

Over 70 years have been passed since the earliest conception on the Yangtze (Changjiang)
Gorges project was formally advanced by the pioneer, Dr. Sun Yat-Sen in his industrial
plan "polices of national reconstruction," issued in 1919 . In the past decades, in order to
propose an advanced, economic and reliable construction scheme and blueprint of design,
and to answer various concerns with the construction of the TGP raised by personages of
various circles both in China and abroad, several generations of scientific and
technological workers have made unprecedent studies centering around the scientific and
technical problems relating to the TGP . And a wealth of results have been obtained,
which as a result have not only satisfied the needs of project planning and design, and
also played an important role in advancing the relevant disciplines . Engineering

geological investigation is one important component.

Since 1950s Changjiang Water Resources Commission (formerly Yangtze Valley Planning Office) with the collaboration of 40 plus units of production, scientific research institutes and universities, has extensively done investigations and studies centering around the geological and seismological problems relating to the project. Thousands of earth science workers including one hundred plus veteran Chinese experts in different disciplines have involved in the studies. Nationwide research on the vital geological and seismological problems was organized twice by the State Science and Technology Commission. In addition, many well known experts and scholars from the former Soviet Union, the United States, Sweden, Canada, Italy, France, Austria, Japan, and the World Bank investigated the TGP for cooperation, exchange and consultation .

The geological research work has covered all the geological and seismological aspects relating to the TGP, including regional geology, regional geomorphology, Quaternary geology and neotectonics, seismology and seismogeology, reservoir and damsite engineering geology and hydrogeology, geotechnics, mineral geology, environmental geology, and natural construction materials, etc. And a variety of techniques and methods have been employed, including surface geology, remote sensing geology, aerogeophysical prospecting and deep seismic sounding, borehole drilling, adit and shaft exploration, engineering geophysics, high precision deformation observation, lab and in situ rock and soil mechanical test and research, in situ measurements, physical modeling, numerical analysis, local micro seismic station networks, and other advanced analysis and examination methods and techniques. These research works have been constantly extended and deepened, and thus provided rich and reliable geological data for the plan and design of the TGP in various phases. It can be said that the geological investigation for the TGP is unprecedented in the world engineering investigation history in both its scope and depth.

The Sandouping damsite chosen for the TGP was selected from 15 alternative damsites in two dam regions through repeated comparisons. The study results of many years indicate that the TGP damsite offers excellent geological conditions which are rare to find and the regional and environmental geological conditions are relatively superior, too. Thus it is suitable for building such a large scale hydro project. At present time, the TGP is commencing to all round construction. The findings and conclusions reached in the investigation phases are now under examination and as a result will be constantly summarized and upgraded through the project practice.

PRINCIPAL ASPECT AND TECHNICAL METHODOLOGY IN GEOLOGICAL RESEARCH FOR THREE GORGES PROJECT

Practice demonstrates that in order to make a in-depth study and properly solve the vital geological problems relating to the TGP, it is a key measure to fully, timely and appropriately apply the up to date technical tools and methods. In this respect we have a

Table 1.   Main Characteristic Idices for the TGP

Item	Unit	Indices
drainage area above the dam	$\times 10^4 Km^2$	100
mean annual runoff	$\times 10^9 m^3$	4510
long term average discharge	m^3/s	14300
design flood (p=0.1%)	m^3/s	98800
normal pool level	m	175
limiting level during flood season	m	145
draw down level during dry season	m	155
reservoir area	Km^2	1084
total capacity of reservoir (NPL)	$\times 10^9 m^3$	393
flood control storage (NPL)	$\times 10^9 m^3$	221.5
dam type		concrete gravity
dam crest elevation	m	185
max dam height	m	175
installed capacity	MW	18200
guaranteed output	MW	4990
mean annual energy output	$\times 10^9 Kw.h$	846.8
permanent navigation shiplock		double-line,continous 5-step
effective dimension of ship chamber	m	280 34 5
submerged cultivated land (including citrus land)	$\times 10^4 mu$	38.86
population in submerged area	$\times 10^4$ person	84.46
earth-rock excavation	$\times 10^4 m^3$	10282.9
concrete placement	$\times 10^4 m^3$	2793.5
total construction period	year	17
construction period needed for the first power generation	year	11

wealth of successful experiences relating to the TGP. For example, in the study of the regional stability and seismicity, such techniques have been adopted as deformation survey of the forebay region of reservoir and major faults, special-purposed micro-seismic station network operated for nearly 40 years, artificial deep seismic sounding with a length of 1000km, etc.. Aero and space remote sensing technologies and geography information system covering all of the reservoir area play an important role in the investigation of regional tectonic background and major faults, land use and land classification and relocation environmental capacities as well as study of reservoir bank slope stabilities and rockfalls and landslides. Instrumentation monitoring systems have been set up in some large landslides. In the study of the damsite geology, such techniques have been implemented as deep borehole earth stress measurement devices (by stress relief method in depth of 300m), computer tomography (CT), fissure-networks modeling, borehole color TV, directional boring, geomechanical model, etc. And they are very useful for effectively solving various engineering problems with quantitve or semi-quantitve assessment.

Regional Stability and Seismicities.
This research theme covers:

(1) Studies of the regional geology and geotectonics, including stratigraphy, lithology, historical geology ,lithofacies and tectonic settings.
(2) Studies of the crustal structure thickness of each layer, Moho discontinuity, extended depth of major faults, deep structures and the state of isostatic gravity field. Such means were used in the study as large area airborne gravity and aero-magnetic measurement, deep seismic sounding(Fig.2), etc.

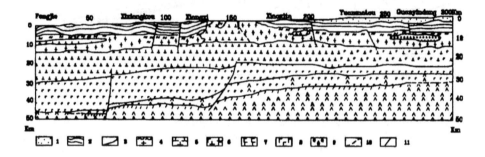

Fig.2   Profile of deep seismic sounding result from Feng Jie to Guan Yingdang

(3) Studies of the regional geomorphology and neotectonic movement.
(4) Studies of the damsite and regional faults in terms of their features, (scale, properties evolution history etc.) and distribution laws, esp. on the several major faults around the periphery of Huangling anticline, where the damsite is located .
(5) Studies on the activities of some major faults, the methods used including: measurement and monitoring of displacement; examination of active age ; measurement of microgravity and Hg gas; analysis of the relationship between microseismic activities and the faults
(6) Studies on the current crustal movement
(7) Studies and analysis of seismicity regularity using the historical data collected for more 2000 years and the monitoring data of local micro seismic station network.
(8) Studies on the seismicities and seismic risks, computation of seismic dynamic parameters under different exceeding probabilities and assessment of their impacts on the project.

Reservoir geology and bank stability
(1) Repeated studies and verification of such problems as reservoir leakage, immersion, bank slump, debris flow, mineral resources submergence, etc.
(2) Special-purposed investigation and stability classification of bank slopes of the stem

stream and its 173 tributaries in terms of the geological conditions, structure types and stability conditions

(3) Special-purposed investigation and stability assessment of the large scale rockfalls, landslides closely relevant to the dam and major towns, and computation and modeling of their possible volumes of slip mass entering the river and the generating surges.

(4) Using various methods to assess the stability of landslides, such as :geological synthetical analysis, fuzzy cluster analysis, sensitivity analysis(Fig.3), failure probability analysis etc.

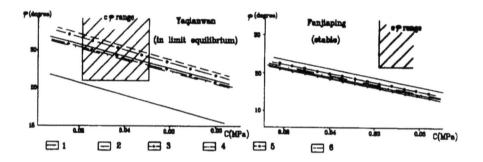

Fig. 3 Diagram of sensitivity analysis of landslide stability (two examples)
1. natural river level, half of slope to be saturated, earthquake intensity 6°; 2. natural river level, no water in slope, earthquake intensity 6°; 3. half of slope saturated, reservoir level 145m , earthquake 6°; 4. half of slope saturated, reservoir level 175m , earthquake 6°; 5. half of slope saturated, reservoir level from 175m suddenly down to 145m, earthquake 6°; 6. natural river level , half of slope saturated, earthquake 7°.
A. Yanqianwan Landslide .   B. Fanjiapin   Landslide

(5)Instrumentation in some key large landslides for displacement monitoring and safty prediction (Fig4. Fig5.).

(6)Hazard impact assessment of slope and landslide failure on dam structures, navigation,and cities,towns, as well as the local people located in the instability site.

(7)Studies on rock slopes stability, esp. on high consequent slopes, in terms of their structure patterns, deformation. mechanism, failure type and assessment of the stability at present situation and prediction after impoundment

(8) Quaternary geological mapping, bankcaving prediction of Quaternary bank slopes of different types.

(9) Prediction of the variation of hydrogeological environment and its consequent impacts on the areas along the reservoir banks (including hydrodynamic field, hydrochemical field and temperature field)after impoundment of the reservoir.

Justification of geological conditions of new sites for relocated towns in the reservoir area

Relocation of townships in the TGP reservoir involves 19 counties (or cities), 140 towns with a population of 480, 000 people. The investigation and study of the geological environments and environmental geological problems with the new sites are carried out in three phases, namely planning, preliminary as well as detailed phases. Following aspects are studied:

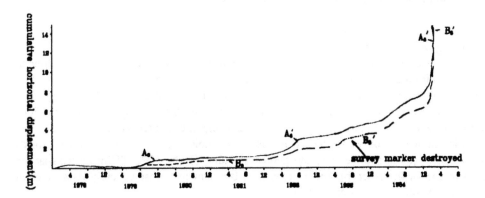

Fig 4. Cumulative curve of horizontal displacement of survey point A3 (A3') B3 (B3') at Xintan slope

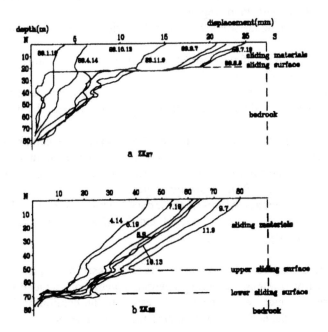

Fig5. Curves of depth-displacement from borehole inclinometer

(1) Identify the geological environments for site selection of relocated cities and towns.
(2) Catastrophic assessment of major environmental geological problems, including rockfalls, landslides, unstable rockmass and debris flows.
(3) Construction suitability zoning in accordance with the engineering geology,

environmental geology and construction planning, which will be used as a geological basis for the planning and construction of newly relocated towns .

(4) Engineering geological assessment of building foundations, and geological supervision and information feedback during construction .

Reservoir induced earthquakes

Following works have been carried out:

(1) Analysis of 102 cases of reservoir induced earthquakes all over the world in order to find out their unique features and common laws so as to make a comparison study.

(2) An all around study on the lithology, structural geology , hydrogeology and watertightness conditions in the reservoir region, esp. on the hydrogeological conditions and fault activities at the damsite and in the nearly region of the reservoir .

(3) Measurements of earth stress in deep boreholes, pore water pressure, rock's permeability, joints and fissures, and geotemperatures (300m 500m 800m in depth)at the damsite and in the reservoir region closed to the damsite. Most valuably data of earth stress in the above area   were obtained .

(4) Intensified observation of earthquakes at several important locations in the same area above mentioned   by means of small diameter seismic station networks in order to find out the characteristics of micro seismicities and their laws and to determine the background conditions of seismicitis in the region .

(5) Numerical simulation of regional stress field and its variation after impoundent of reservoir .

(6) Through fuzzy cluster analysis of the correlation factor between the large-size and deep reservoirs that have already induced earthquakes and those that haven't induced any earthquakes, a prediction has been made to the maximum induced earthquake and the probabilities of earthquakes with different magnitudes possibly induced by the TGP reservoir.

(7) In view of the seismological background in the TGP reservoir , a comprehensive assessment has been made of the possibility of TGP reservoir induced earthquakes based on the lithlogy, tectonics, permenbility , earth stress state and seismicities as well as various numerical analytical results of each reservoir stretch.

(8) According to the characteristics of reservoir induced earthquakes, the dynamic parameters very close to damsite have been studied and the affecting intensities at the damsite are obtained in the forbay region of the TGP.

According to the seismogeological background in the TGP reservoir region and the seismic assessment standards commonly used in the world, a prediction and assessment is made to the TGP reservoir induced earthquakes with respect to the possibility, location, intensity and their impacts on the engineering structures and environments.

Damsite Geology

Following special themed studies have been carried out in accordance with the designing conditions and work features of each hydraulic structure .

(1) Studies of the weathering mantle . Special attention is paid to the engineering geological properties of weathered rocks so as to determine the dam foundation and to

probe into the possibility to chose the lower part of the weakly weathered zone as the dam foundation (Fig6.), in addition to evaluate the impacts of locally anomalous weathering along some discontinuities and of the interlayered weathering rock in the foundation rockmass and to propose corresponding treatment measures .

(2) Studies of the faults and fractures, including detailed investigation of faults within the foundation of hydraulic structures, simulation of fissure networks, studies of the engineering geological and hydrogeological conditions of faults with different directions produced in various periods, assessment of the possible effects of various types of faults and fissures on the project .

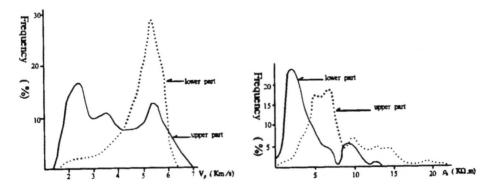

Fig6. Comparasion of Vp ρ, parameters between the upper part and lower part of weakly weathered zones in dam foundation

(3) Studies of low-angle discontinuities in terms of their geneses, distribution laws, properties, dominant directions, connecting ratio, mechanic characteristics, etc. Through zoning of the low-angle discontinuities within the dam foundation according to their development level, a conceptual geological model is proposed, based on which an analysis of the deep-seated stability against sliding of dam foundation and geomechanical modeling tests are made.

(4) Studies of load-release features of rockmass. As a result the impacts of load-release process in different geomorphological units are determined on the engineering geological properties of dam foundation rockmass.

(5) Studies of the hydrogeological features of dam foundation rockmass (or soil). Emphasis is put on the study of the heterogeneity and anisotropy of hydrodynamic field in fissured rockmass, identification of permeability characteristics of rockmass, and underline of the locations with relative heavy permeabilities, as well as proposal of the geological basis for seepage control and dainage measures to be taken.(Fig7.)

(6) Studies of the structure and quality of dam foundation rockmass. On the basis of structure classification, quality gradation and zoning of dam foundation rockmass, a assessment is made by means of integral multi-factor analyses.

(7) Studies of stabilities of the excavated high rock slopes. Based on the investigation of natural and artificial slopes, the basic design slope in the completely and strongly weathered rocks is determined . Studies are carried out respectively of the overall stability and local stabilities of design slopes in fresh and slightly weathered rocks according to the

rockmass structures and strength, and fissure water features in it . Finally the basic design slopes and excavation shapes are proposed after comprehensive studies by means of geomechanical model, 2D or 3D finite-element analysis, block theory and limiting equilibrium method, etc..(Fig.8)

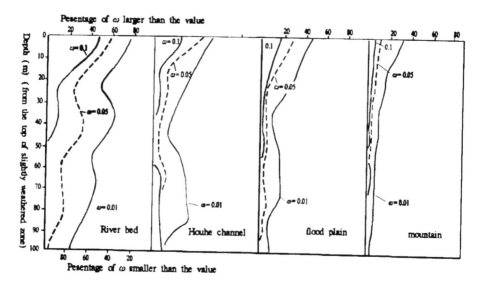

Fig7. Variation of unit water adsorption (ω)with depth

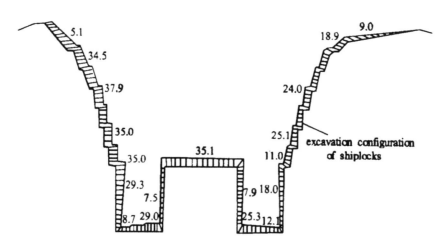

Fig8. A sketch showing the displacement after excavation of shiplock (Unit:mm)

(8) Studies of the tests of physical and mechanical properties of rockmass and soil in order to determine the shear strength and other parameters of the different weathered and fresh rocks and of the discontinuities of various types as well as the interface between concrete and bedrock .

Natural construction materials

Large quantities of and multi-type natural construction materials are needed for the TGP. More than 10 types of material resources and over 30 borrow (quarry) sites have been investigated and compared.

(1) Studies of the concrete aggregates. Two types of resources are investigated. One is the natural sand and gravels distributing in the flood plain and riverbeds of the ChangJiang river and its tributaries, the other is the artificial aggregate processed from crystalline rocks and carbonate rocks.

(2) Investigation of the earth materials for the core of cofferdam and wall-stabalization.

(3) Investigation of rockfill materials and block materials for cofferdam.

MAIN CONCLUSIONS REACHED IN THE INVESTIGIATION PHASES

(1) In terms of tectonics, the TGP is located in a relatively stable region with low level of seismicity. In the Huangling crystalline massif where the damsite is located exist no strong earthquake bearing structures . The basic seismic intensity there-at is of VI degrees. According to seismic risk analysis, the peak horizontal acceleration at the damsite is of 0 . 125g under a given yearly exceeding probability of 10^{-4} which can be used as the basis for anti-seismic design of major hydraulic structures, such as dam, etc.

(2) According to the regional geological setting, reservoir-induced earthquakes would probably occur in some local locations in the reservoir region but it would be a rare case for some strong induced earthquakes to happen.

(3) No leakage problems will be expected in the TGP reservoir. Submerging loss of mineral resources is minimum. Debris flow and bankcaving will be in a very small scale, and the overall stability of bank slopes is good. Some potentially unstable rockfalls, landslides, as well as dangerous rockmass which are fortunately far from the damsite and will pose no dangers to the structures(Fig.9) except the nearby towns and residence areas, thus special attention should be paid in the relocation plan for new town sites. Compared to the natural conditions, the possible interruption to the navigation caused by rockfalls and landslides will be mitigated significantly after impoundment of reservoir and no navigation obstruction will occur(Fig.10).

(4) The geological investigation results obtained in the preliminary and detailed phases for the new town sites provided the basis for new townsite planning . Attention must be paid to the geological conditions in the new townsite planning and construction so as to appropriately utilize the suitable areas for buildings and to avoid those unfavorable conditions, and thus to guarantee the new townsite with favorable geological conditions. Proper planning programme and technical measures should be taken to avoid the occurrence of geological disasters and to reduce the unnecessary losses.

(5) The damsite bedrocks consist of granitic rocks of Presinian Era. With their homogeneous lithology, integral property, high mechanic strength, poor permeability, and mostly well cemented fault tectonites, they offer an excellent foundation for high dams and very competent rockmass for excavation of high slopes and large dimension underground caverns . Even though there exist some geological defects in the dam

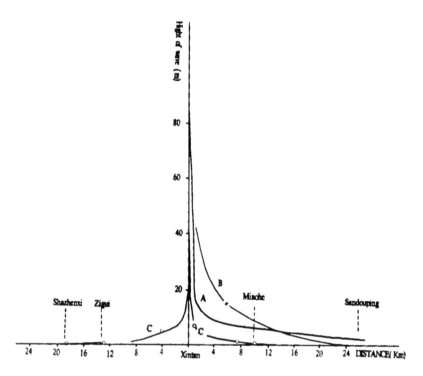

Fig 9. Attenuation curve of the generating wave of Xintan landslide
(A) Calculated curve assuming volume of slip mass entering river with speed of 100m/s, water level 150m
(B) Physical model tested curve assuming volume of slip masses 16 million m³ ,with speed of 67m/s, water level 130m
(C) Observing curve of Xintan landslide occurred on June 12,1985

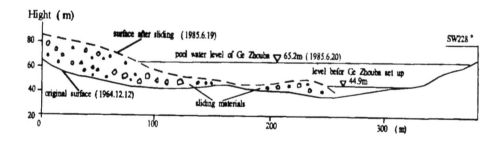

Fig10. Diagram of submarine section before and after Xinatan sliding

foundation such as low-angle discontinuities, individual poorly cemented faults and deep weathered zones as well as local strongly permeable zones, they can be solved through optimum design and some necessary engineering treatments .

(6) The natural construction materials required by the TGP can be obtained within a reasonable distance.

In summary, large quantities of geological investigation and research works have been completed for the TGP and the principal geological problems are well understood and concluded. Now the construction and excavation is underway for various engineering structures at the damsite, and the development relocation in the reservoir area is in full swing . The geological work in construction period provides an invaluable opportunity to examine the analysis and conclusions made in the investigation phases and as a result to verify the understandings previously got and also an important measure to guarantee the safety of construction and operating period

With the obtained and the being obtained principal experiences and technical methodology in construction of the TGP, an important contribution will be made to the progress of engineering geology in hydropower development.

REFERENCES

Chen Deji ,1986, The Geological Study of the Three Gorges Project in China .Proceedings 5th International Congress . IAEG

Chen Deji Xue Guofu ,A Study on the Xintan Landslide 1987.Proceedings International Symposium on Engineering Geologic Environment in Mountainous Area

Chen Deji, 1990, A Brief Account of Research on Reservoir Slope Stability of Three Gorge Project

Chen Deji ,1993, Theory and practice in Geological study of Three Gorge Project. Journal of Engineering Geology Vol.1 NO.1

Proc. 30th Int'l. Geol. Congr., Vol. 23, pp. 17-29
Wang Sijing and P. Marinos (Eds)
© VSP 1997

Engineering Geology and Seismotectonic Investigations for the 4 x 950 MW Darlington Nuclear Power Project

C.F. LEE
Department of Civil and Structural Engineering, the University of Hong Kong, Hong Kong

Abstract

This paper presents the case history of a 4 x 950 MW nuclear power station located in the intraplate seismic environment of Eastern Canada. The pertinent seismotectonic features of such an environment are described, including the prevalence of a state of high horizontal compressive stress, and the absence of surface rupturing during most intraplate seismic events. The design earthquakes for the various potential sources zones are determined, including the Western Quebec Zone, the Niagara Region and the Clarendon-Linden Fault System in western New York State. The design basis seismic ground motion parameters are then deduced for the Darlington site, along with the ground response spectra. Differences between the intraplate and plate-boundary seismic environments are discussed.

INTRODUCTION

In the seismic design of nuclear power plants, each seismogenic fault within the area of interest constitutes a potential source zone. A design earthquake (DE) is deduced for each source zone based on historical seismicity and/or evidence of neotectonic movement. The design basis earthquake (DBE) is then taken as the most severe of all DE's from the various possible source zones, incorporating an appropriate safety margin. A seismogenic fault normally manifests itself by co-locating with a significant number of epicentres, or by showing signs of neotectonic movement. In many cases, particularly at the boundaries of tectonic plates, the identification of seismogenic faults based on their surface expressions is a viable task, given due diligence in geological investigation and a good knowledge of regional seismicity. This may or may not be the case, however, in the intraplate environment where the seismogenic faults are located at great depth, without surface expressions or evidence of surface rupturing due to seismic events. This renders the determination of the design basis seismic ground motion (DBSGM) parameters a rather challenging task. The present paper provides such an example of earthquake - resistant design in the intraplate environment. Specifically, the case history of the 4 x 950 MW Darlington Nuclear Power Station design will be used to illustrate the challenges encountered and the methodologies adopted under such circumstances.

THE INTRAPLATE SEISMIC ENVIRONMENT

In tectonic terms, Eastern Canada is located in the interior of the North American Plate, remote from the plate boundaries on the Pacific Coast and at the Mid-Atlantic Ridge. Seismicity in Eastern Canada is, to a large extent, related to its glacial history and to the rejuvenation of fault activities under the condition of glacio-isostatic uplift. At the peak of the Wisconian Glaciation, the land subsided by as much as several hundred metres, depending on the specific locale under consideration (Paterson 1972; Cathles 1975). Upon deglaciation some 13,000 years B.P., the land rebounded. While the bulk of the glacio-isostatic uplift has since occurred, much of Eastern Canada is still rebounding today at the rate of a few millimetre per year. Crustal deformation during glaciation and deglaciation resulted in significant changes to the state of stress and strain in the rock mass at depth, leading to the rejuvenation of ancient fault systems and hence seismicity in the region.

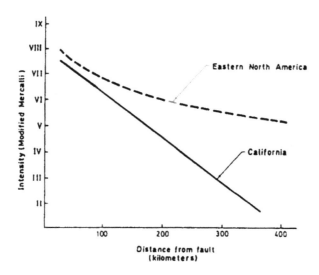

Figure 1. Attenuation of intensity for California (plate boundary environment) and Eastern North America (intraplate environment).

Intraplate seismicity in Eastern Canada differs from plate-margin seismicity (or interplate seismicity) in two fundamental aspects. Firstly, deformation at plate margins often resulted in a highly fractured or disintegrated rock mass around the seismogenic fault (e.g. the San Andreas Fault System in California). This led to a larger damping effect for the seismic motions generated by fault rupture. In contrast, the rock mass conditions in much of Eastern Canada tend to be generally more massive and more competent. This means that the rock formations in Eastern Canada have a better ability to transmit seismic waves and a smaller damping effect for seismic motions. Thus, seismic motions would be attenuated more rapidly at plate margins than in the intraplate environment. Figure 1 illustrates such a comparison, in terms of Modified Mercalli intensity. This difference also explains why intraplate earthquakes are felt over relatively large areas. Comparing a plate-margin earthquake to an intraplate earthquake of the same magnitude, one may notice that the former is often characterized by greater damage near the seismogenic fault zone. For the intraplate earthquake of the same magnitude, the felt area will normally be larger while the damage will generally be less substantial, reflecting the dissipation of seismic energy over a wider area.

The second major difference between plate-margin and intraplate earthquakes lies in the stress field around the seismogenic fault. Field measurements indicate that the magnitude of the normal compressive stress tends to decrease towards a plate boundary fault, such as the San Andreas Fault (Tullis 1977; Engelder et al 1978). In other words, the fault plane itself is not subject to a state of high lateral compression. This makes it easier for the fault to slip under shear loading. In contrast, much of Eastern Canada is subject to a state of high horizontal compression. The value of horizontal compressive stress measured generally ranges from 5 MPa to 10 MPa near the bedrock surface, trending also to increase with depth (Herget 1974; Herget 1980; Haimson and Lee 1980; Lee 1978; Lee 1981; Lee and White 1994). This state of compressive stress often led to the occurrence of "pop-ups" in the bedrock surface, which are folds or buckles in the bedrock surface due to horizontal compression (White et al 1973; Franklin and Hungr 1976; Lee and White 1994). This state of horizontal compression also results in a high degree of tightness in the rock mass. Thus there are numerous tunnels which were found to be "bone dry" during excavation, despite being located under large bodies of surface water and having relatively thin rock covers (Lee et al 1988; Lee and White 1994). In fracture mechanics term, the predominant direction of fracture propagation is parallel to the direction of the major principal stress (Fairhurst and Cook 1966). This means that the seismogenic fault may well propagate in a predominantly horizontal direction at a considerable depth in the crust. Thus intraplate seismicity may or may not be associated with surface rupturing or a surface expression of the seismogenic fault. This is in direct contrast to the case of plate-margin earthquake, where surface rupturing is routinely observed, with most of the damage occurring at the apex of a rupturing fault.

The state of high horizontal compressive stress generally observed in Eastern Canada is believed to be also related to the glacial history of the region. At the peak of the Wisconsian Glaciation, the maximum thickness of the Laurentide ice sheet reached 3000 - 3750 m (Paterson 1972; Cathles 1975). This sustained heavy glacial load resulted in the viscoelastic deformation of crustal rocks in much of Eastern Canada. It has been demonstrated that under the condition of lateral confinement, a significant amount of horizontal residual stress could be inherited by the rock mass upon deglaciation (Lee 1978; Lee and Asmis 1979). Another possible major contributing factor is plate tectonics. The spreading of the seafloor at the Mid-Atlantic Ridge produces a compression effect on the North American Plate, predominantly in a westerly direction. This is generally consistent with the predominant direction of the major principal stress observed in much of Eastern North America (Sbar and Sykes 1973; Sykes 1978).

SITE AND REGIONAL GEOLOGY

The site of the Darlington Nuclear Generating Station is located on the north shore of Lake Ontario, approximately 65 km to the east of the City of Toronto (Figure 2). The station features four 950 MW nuclear power reactors of the CANDU type, which is a Canadian design that uses natural uranium as fuel and heavy water as moderator. Construction on the site began in the 1978 and the four reactor units were commissioned respectively between 1990 and 1993. The overall project cost was in the order of US$10 B, including interest during construction. The station was designed, constructed and operated by Ontario Hydro, a provincial electrical utility that supplies electricity to the Province of Ontario, Canada. Ontario Hydro currently has an installed generating capacity of approximately 32,000 MW, including nuclear, coal-fired and hydroelectric power stations.

The Darlington site is underlain by approximately 20 m of Quaternary glacial deposits of till and granular materials. Beneath the Quaternary deposits are approximately 200 m of flat-lying (or very gently dipping) Ordovician limestones, shaly limestones, shales, siltstones/sandstones, which are in turn underlain by Precambrian basement rocks of massive, migmatitic gneisses. The Paleozoic sediments dip very gently to the south, at an angle of about half a degree, thickening considerably at the Niagara Escarpment to the west of Toronto, where Silurian (and further west, Devonian) sediments overlie those of Ordovician age (Figure 3).

A comprehensive program of state-of-the-art geophysical, geological, seismological and geotechnical investigations was carried out for the Darlington project, beginning in the mid-1970's. These included regional airborne and surface geophysical surveys, extensive drilling, in-situ stress measurement by overcoring and hydrofracturing, high-precision packer testing for hydraulic conductivity and a variety of specialized geomechanical testing. The geophysical work focussed on regional faulting and lineament tracing, including aeromagnetic gradiometer survey, maxi-probe electrical resistivity survey, mini-sosie seismic reflection survey, ground radar survey, etc. The results of geophysical surveys were cross-checked prior to confirmation drilling. There were incidences when innovative geophysical techniques of a research nature hinted at the possibility of faulting in river beds or along creeks and streams that drain into Lake Ontario. Such a possibility, however, was not substantiated by systematic drilling carried out subsequently to validate the results of geophysical surveys. There are, of course, faults in the sedimentary sequence as well as in the Precambrian basement rocks in the vicinity of the site and the general region of interest. With the exception of the Clarendon-Linden Fault System in western New York State (Figure 2), however, there appears to be no spatial correlation between such faults and the epicentres of historical and contemporary seismic events recorded in the region. The faults in the Precambrian basement show no signs of penetrating the overlying sedimentary sequence. This applies also to the Central Metasedimentary Belt Boundary Zone (CMBBZ) of the Grenville basement rocks, which manifests itself as a major shear zone with a width of several kilometres in outcrops at approximately 100 km north of the Lake Ontario shoreline (Easton and Carter 1991; Easton 1992; Lee and White 1994).

The Clarendon-Linden Fault System is a north-northeasterly trending, high-angle reverse fault system in Ordovician sediments in western New York State. Since the late 1920's, a number of seismic events have been recorded and attributed to fluid injection and solution mining activities in the vicinity of the fault system (Van Tyne 1977; Fletcher and Sykes 1977). The largest event, which occurred in 1929, has been given a Richter magnitude rating of M5.8. This fault system has been postulated to extent northeasterly into Lake Ontario, possibly also into Prince Edward County on the north shore of Lake Ontario. While the fault system has been associated with induced seismicity due to fluid injection, it was nonetheless taken as an active fault and a source zone for the seismic design of the Darlington Nuclear Generating Station.

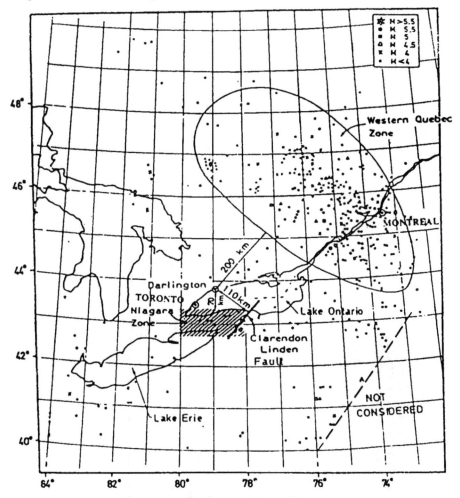

Figure 2. Darlington site location, regional seismicity and design earthquakes

Otherwise, there is no conclusive evidence to date of any major, deep-seated neotectonic movement along any of the reported faults in the sedimentary sequence of southern Ontario. A number of minor faults were reported by White et al (1973) in the sedimentary bedrocksof the region. They are not considered to be deep-seated and are believed to be related to the occurrence of high horizontal stresses in the southern Ontario region (Lee and White 1994), similar to other near-surface, high-angle reverse faults observed by Oliver et al (1970) in New York State and in the Province of Quebec, Canada.

C.F.Lee

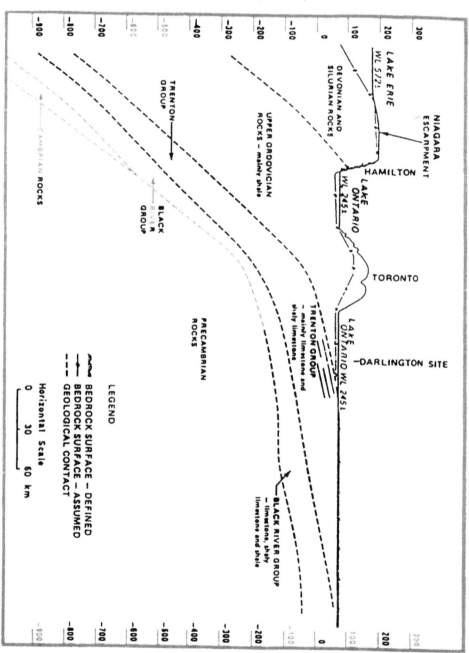

Figure 3. Geological section along north shore of Lake Ontario

In engineering geology terms, both the Ordovician sediments and the Precambrian basement rocks are generally massive and competent rocks. The Ordovician sediments exhibit well defined, virtually horizontal bedding planes. The regional strike is approximately east-west, while the dip is to the south. The thickness of the beds vary from 2.5 to 75 cm. There are two dominant joint sets with orientations of N 70^0 - 90^0 E and N 45^0 W - N 40^0 E respectively, Joint spacing is close at the bedrock surface (0.1 - 0.3 m approximately) and much more widely spaced at depth. In the cooling water intake and discharge tunnels, constructed at the Darlington site at a depth of some 35 m below the Lake Ontario bottom (bedrock) surface, the east-west joints were mapped at a spacing of 40 m apart. The tunnel excavations were found to be "bone dry", requiring water to be brought down from the surface for dust control (Lee and White 1993; Lee and White 1994). The Ordovician sediments as exposed in the walls of the tunnel excavations were obviously very tight and massive, with no signs of faulting or shear zones. The uniaxial compressive strength of the sediments is typically in the range of 50 - 100 MPa. All nuclear structures for the Darlington project are founded on competent limestone bedrock of the Ordovician age.

As noted earlier, the Precambrian basement rock at the Darlington site consist of massive and migmatitic/granitic gneisses of the Gvenville Province. While foliation shear and minor alteration along fractures are not uncommon, the basement rocks are generally tight and massive, with a uniaxial compressive strength which is typically twice of that for the Ordovician limestones. Figure 4 illustrates the in-situ horizontal stresses measured at a 300 m deep test hole drilled at the Darlington site. The solid symbols refer to results obtained at shallow depths using the stress-relief, overcoring technique and the United States Bureau of Mines borehole deformation gauge. The open symbols represent data from hydrofracturing tests carried at greater depths in the same test hole. It is clear from Figure 4 that a state of high horizontal compressive stress prevails at the Darlington site, similar to many other parts of Eastern North America. The horizontal stresses measured also show a trend to increase with depth. For comparison purposes, the pressure due to the weight of the overlying rock is plotted in Figure 4. Also presented in this figure are the values of hydraulic conductivity measured using a high-precision double packer system. The measured values typically range from 10^{-13} to 10^{-12} m/s, with higher values locally in fracture zones. Both the in-situ stresses and the hydraulic conductivity values measured reflect a high degree of compression and watertightness in the rock mass at the Darlington site (Haimson and Lee 1980; Lee 1981; Lee and White 1993; Lee and White 1994).

DESIGN BASIS EARTHQUAKE

Figure 2 illustrates the distribution of epicentres of historical earthquakes in the region of interest to Darlington seismic design (Basham 1975). The highest density of seismic events is found in the Western Quebec Zone (WQZ), located northwest of the site (Figure 2). The zone includes much of the Ottawa - St. Lawrence Valleys. Fault rejuvenation is believed to be the main cause of seismicity within the zone. For the seismic design of the Darlington station, the WQZ was taken as a major far-field source zone. Figure 5 shows the magnitude recurrence relation for the WQZ. A recurrence rate of 1 in 1000 years was adopted for the seismic design of the Darlington station. The corresponding Richter magnitude of Design

24

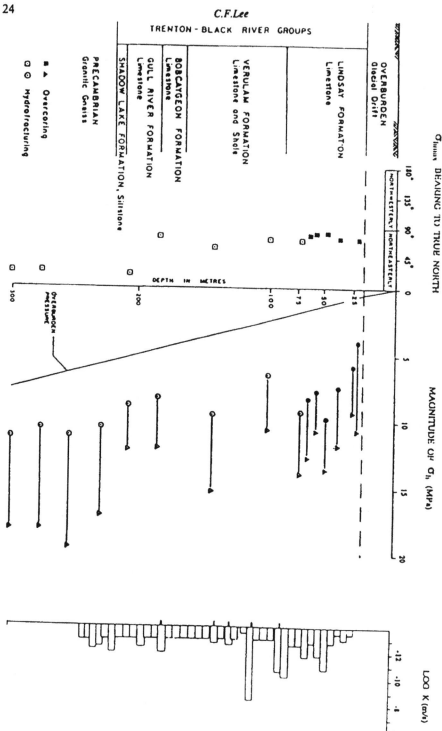

Figure 4. In-situ horizontal stresses and hydraulic conductivity measured at Darlington site

Earthquake #1 (DE 1), due to WQZ, was hence determined as M6.7. The shortest possible distance from the boundary of WQZ to the Darlington site is 200 km.

The second major source zone for the Darlington site is the Clarendon-Linden Fault System in western New York State. As noted earlier, fluid injection and solution mining led to induced seismicity around this fault system since the 1920's, with a maximum event of M5.8 which occurred in 1929. For the Darlington project, Design Earthquake #2 (DE 2) was taken as an M6.5 event, occurring on the fault system and its postulated projection into Lake Ontario, at a point which is closest to the Darlington site. This shortest possible distance to the site is approximately 110 km (Figure 2).

The second major source zone for the Darlington site is the Clarendon-Linden Fault System in western New York State. As noted earlier, fluid injection and solution mining led to induced seismicity around this fault system since the 1920's, with a maximum event of M5.8 which occurred in 1929. For the Darlington project, Design Earthquake #2 (DE 2) was taken as an M6.5 event, occurring on the fault system and its postulated projection into Lake Ontario, at a point which is closest to the Darlington site. This shortest possible distance to the site is approximately 110 km (Figure 2).

Design Earthquake #3 (DE 3) is based in the Niagara Region to the south of the site. The region includes the Clarendon-Linden Fault System as well as the Niagara Peninsula. The magnitude recurrence relation for this region is given in Figure 5. For a recurrence rate of 1 in 1000 years, a Richter magnitude M6 event was determined for DE 3, at a shortest possible distance of 70 km from the site (Figure 2).

With the three design earthquakes thus determined, the peak ground motion parameters for the Darlington site could be deduced using an appropriate set of attenuation equations based on strong ground motion statistics. For the Darlington project, the attenuation equations developed by Milne and Davenport (1969) specifically for Eastern Canada were used:

$$a = 0.06 \, e^{0.92M} R^{-1.38}$$
$$v = 0.43 \, e^{1.31M} R^{-1.36}$$
$$d = 0.18 \, e^{1.11M} R^{-1.0}$$

where $a =$ peak acceleration in %g
$v =$ velocity in cm/sec
$d =$ displacement in cm
$M =$ Richter magnitude (of design earthquake in this case)

$$R = \sqrt{\Delta^2 + h^2}$$

$D =$ epicentral distance in km (i.e. shortest distance to site in this case)
$h =$ focal depth in km

Given the relatively deep-seated nature of intraplate earthquakes in Eastern Canada, a focal depth equivalent to mid-crustal depth was commonly assumed in seismicity studies for Eastern Canada. This assumption was also adopted in the Darlington study.

Table 1 shows the peak ground motion parameters thus calculated for the Darlington site, based on the three DE's and attenuation equations aformentioned. Accordingly, the design basis seismic ground motion (DBSGM) parameters were determined to be:

> a = 0.08 g
> v = 90 mm/s
> d = 60 mm

for the Darlington project. The corresponding design base response spectrum, at 5% damping, is illustrated in Figure 6. These design parameters were used in the seismic qualification of the Darlington nuclear structures and equipment (Tang and Lee 1993).

Table 1. Design earthquakes and peak ground motion parameters

Design Earthquake	Richter Magnitude	Epicentral Distance (km)	Peak Ground Motion Parameters		
			a (%g)	v (mm/s)	d (mm)
DE 1	6.7	200	0.06	42	30
DE 2	6.5	110	0.07	70	44
DE 3	6	70	0.08	66	40

CONCLUDING REMARKS

Seismic design of the Darlington Station as outlined in this paper reflects the seismotectonic framework of Eastern Canada, with particular reference to the intraplate seismotectonic features of the general region. Such features include the prevalence of a state of high horizontal compressive stress in bedrock, largely due to viscoelastic deformation during glaciation. This state of stress facilitates fault propagation in a horizontal rather than vertical direction. Seismicity in Eastern Canada is caused mainly by the rejuvenation of faults at depth in response to glacio-isostatic uplight. The prevalent state of stress thus implies that intraplate seismicity may not necessarily be associated with surface rupturing of the seismogenic faults. In other words, the seimotectonic investigation programs should not focus just on the surface expressions of the faults. Without a proper understanding of the intraplate seismic environment, one might risk misinterpreting surficial stress-relief features (such as pop-ups, bedding planes slips and offset boreholes) as neotectonic deformation of seismological significance. These features are simply the surface manifestation of the state of stress in the bedrock, being consistent with its mechanical behaviour and the mechanics of surficial unloading.

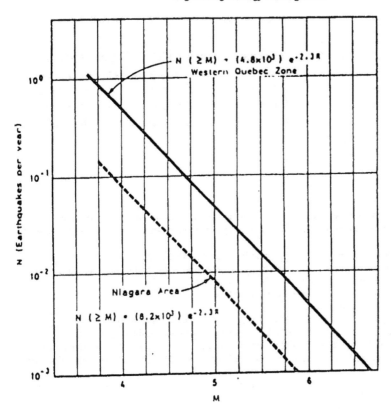

Figure 5. Magnitude recurrence relations for Western Quebec Zone and Niagara Region

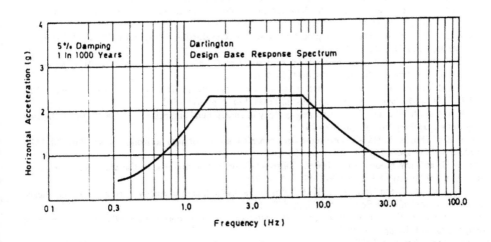

Figure 6. Design base response spectrum for Darlington Nuclear Generating Station

Acknowledgments

The author is indebted to his many colleagues for their collaboration on seismotectonic studies over a period of almost two decades. He is particularly grateful to Harold Asmis of Ontario Hydro; Peter Basham, John Adams, Ann Stevens and David Forsyth of the Geological Survey of Canada; Gail Atkinson of Carlton University; and Owen White of the Ontario Geological Survey, for sharing their expertise and insights throughout this long (and sometimes challenging) period.

REFERENCES

Basham, P.W. (1975). Design basis seismic ground motion for Darlington Nuclear Generating Station A. Seismological Service of Canada, Report 75-16, Earth Physics Branch, Department of Engergy, Mines and Resources, Canada, December 1995.

Cathles, L.M. (1975). The viscosity of the Earth's mantle. Princetion University Press, pp 386.

Easton, R.M. (1992). The Grenville Province and the Proterozoic history of central and southern Ontario. In: Geology of Ontario. Ontario Geological Survey, Spec. Vol. 4(2), pp 715 - 904.

Easton, R.M. and Carter, T.R. (1991). Extension of Grenville basement beneath southwestern Ontario. Ontario Geological Survey, Open File Map, 162.

Engelder, T., Sbar, M.L., Marshak, S. and Plumb, R. (1978). Near surface in situ stress pattern adjacent to the San Andreas Fault, Palmdale, California. Proc. 19th U.S. Rock Mechanics Symposium, pp 95 - 101.

Fairhurst, C. and Cook, N.G.W. (1966). The phenomenon of rock splitting parallel to the direction of maximum compression in the neighbourhood of a surface. Proc. 1st International Congress on Rock Mechanics, Lisbon, Vol. 1, pp 687 - 692.

Fletcher, J.B. and Sykes, L.R. (1977). Earthquakes related to hydraulic mining and natural seismic activity in Western New York State. Journal of Geophysical Research, Vol. 32, No. 26, pp 3767 - 3780.

Franklin, J.A. and Hungr, O. (1976). Rock stresses in Canada. Proc. 25th Geomechanical Colloquy, Salzburg, Austria.

Haimson, B.C. and Lee, C.F. (1980), Hydrofracturing stress determinations at Darlington, Ontairo. Proc. 13th Canadian Rock Mechanics Symposium, Toronto, Canada, pp 42 - 50.

Herget, G. (1974). Ground stress determinations in Canada. Rock Mechanics, Vol. 6, pp 53 - 64.

Herget, G. (1980). Regional stresses in the Canadian Shield. Proc. 13th Canadian Rock Mechanics Symposium, Toronto, Canada, pp 9 - 16.

Lee, C.F. (1978). A rock mechanics approach to seismic risk evaluation. Proc. 19th U.S. Rock Mechanics Symposium, Reno, Nevada, pp 77 - 88.

Lee, C.F. (1981). In-situ stress measurements in southern Ontario. Proc. 22nd U.S. Rock Mechanics Symposium, Cambridge, Massachusetts, pp 435 - 442.

Lee, C.F. and Asmis, H.W. (1979). An interpretation of the crustal stress field in northeast North America. Proc. 20th U.S. Rock Mechanics Symposium, Austin, Texas, Vol. 1, pp 655 - 662.

Lee, C.F. and White, O.L. (1993). Engineering geology in underground aggregate mining and space utilization. Engineering Geology Vol. 35, pp 247 - 257.

Lee, C.F. and White, O.L. (1994). One some goemechanical aspects of high horizontal stress. Proc. 7th IAEG International Congress, Lisbon, Vol. 1, pp 295 - 302.

Milne, W.G. and Davenport, A.G. (1969). Distribution of earthquake risk in Canada. Bulletion of the Seismological Society of America, Vol. 59, pp 729 - 754.

Oliver, J., Johnson, T. and Dorman, J. (1970). Postglacial faulting and seismicity in New York and Quebec. Canadian Journal of Earth Science, Vol. 7, pp 579 - 590.

Paterson, W.S.B. (1972). Laurentide ice sheets: estimated volumes during Late Wisconsin. Review in Geophysics, Vol. 10, pp 885 - 917.

Sbar, M.L. and Sykes, L.R. (1973). Contemporary compressive stress and seismicity in Eastern North America: an example of intraplate tectonics. Bulletin, Geological Society of America, Vol. 84, pp 1861 - 1882.

Sykes, L.N. (1978). Earthquakes and recent tectonics of Eastern North America: Reactivation of old zones of weakness. Symposium on Intraplate Seismicity and Tectontics in Eastern North America, Toronto, 1978.

Tang, J.H.K. and Lee, C.F. (1993). Seismic design of Ontario Hydro nuclear generating stations. Ontario Hydro report, May 1, 1993.

Tullis, T.E. (1977). Stress measurements in shallow overcoring on the Palmdale uplift. EOS, Transactions, American Geophysical Union, Vol. 58, pp 1122.

Van Tyne, A.M. (1975). Subsurface investigation of the Clarendon-Linden structure, western New York. Geological Survey, New York State Museum and Science Service, Open-File Report, May 15, 1975.

White, O.L., Karrow, P.F. and MacDonald, J.R. (1973). Residual stress relief phenomena in southern Ontario. Proc. 9th Canadian Rock Mechanics Symposium, Montreal, pp 323 - 348.

Proc. 30th Int'l. Geol. Congr., Vol. 23, pp. 31-39
Wang Sijing and P. Marinos (Eds)
© VSP 1997

The Engineering Geology Conditions and the Designs Features
---an Example of Dam Construction on Active Fault

PENG DUNFU
Xinjiang institute of Water Conservation and Hydropower Investigation and Design ,People's Republic of China

Abstract

The Kezier reservoir on the Weigan river in Xinjiang is a large scale reservoir. The reservoir site outcrops are weak sandstone and mudstone of Neo-tertiary system. All of the terraces and gravel layers of the river bed have been dislocated by the fault F_2, forming a Square Mountain between the main dam and the auxiliary dam, on which relatively large ground fissure can be found everywhere. By crossing fault observation since 1972, the current active level of the fault is obtained. The auxiliary dam right abutment is located on the scarp of the fault F_2. Fault activity and ground fissure are key engineering geology problems of the auxiliary dam right abutment. On the basis of discussion and assessment of the said problem,the design features of the auxiliary dam right abument are given.

Keyword: main dam, auxiliary dam, abutment, engineering geology condition, active fault.

INTRODUCTION

The Kezier reservoir on the Weigan river in Xinjiang is the first large scale reservoir built on active fault. The reservoir has a capacity of 6.4 million m^3. Its normal storage level is 1149.6 m, the main dam is 44 m high, the auxiliary dam 12.5~32.6 m high. The right abutment of the auxiliary dam is located on the scarp of the active fault F_2. Assessment of the engineering condition of the fault F_2 is the biggest key of engineering geology , and the design of the dam located on active fault is the key of design work.

The kezier active fault(F_2) crosses the Kezier river, and then passes the left bank of the Weigan river along the north edge of the Square Mountain ,at last, The Kezier active fault(F_2) crosses the Kezier river ,and then passes the left bank of the Weigan river along the north edge of the crosses obliquely the Weigan river to the east.(Fig. 1)

The Square Mountain is formed by faulting and rising of III-VI grade terraces lying on the left bank of the confluence of the Weigan river and the Kezier river. High and dip angle ground fissures developed on the top of the mountain. The genesis , scale and distribution of the ground fissures on the Square Mountain and their influence should be considered in the dam design, especially in the design of the right abutment of the auxiliary dam.

SCALE AND FEATURE OF THE FAULT F_2

The scale and feature of the fault F₂ on the whole

The fault F_2 is located on the synclinal axis of the Kezier river mouth where the upstand

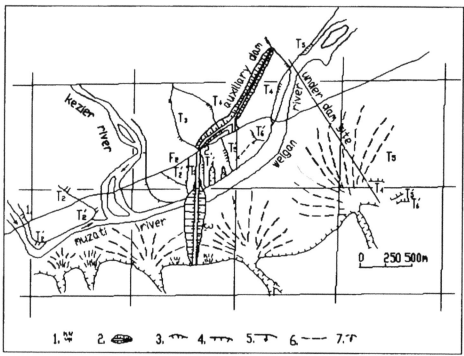

1.small square mountain; 2.square mountain; 3.upper dam site.

Figure 1. The Kezier reservoir dam site geomorphy location of the fault F_2

Strata urn sharply inclination strata. It is 85 km long. The deformation form and intensity are different in the different zone of the fault, some of them is of bending deformation, some is of a group of equal strength imbricate faults. At the dam site, the faults are mainly of rift way. There is a main rift of its, about 10 km long. The main rift is of bending deformation on both end. On the south-west end, bending deformation of the gravel layer of VI grade terrace on the right bank of the Muzati river is the highest, which is of 97 m. On the north-east end, bending deformation height of the gravel bed of IV grade terrace on the opposite bank of the Kezier Qianfudong is 34 m. On the east of the reservoir dam site 40 km (between the Xianshuigou and the Kuche river) there is a group of parallel imbricate faults. On the east of the Square Mountain , the fault F_2 crosses obliquely the Weigan river. Based on the drill data, the vertical throw of the gravel bed of the river bed is 2.68 m. According to above explanation, the fault F_2 has not formed unit fault section or fault zone. By now, it is a active fault and developing fault.

Scale and feature characters of the fault F_2 in the reservoir site

Occurrence of the fault F_2 near the dam site: $65°\angle SE23°$, the fault F_2 is a fault zone, including F_{2-1}, F_{2-2}, F_{2-3} three parallel fault planes and a group of 3~4 upstand interlayer faulted bedding planes. F_{2-1} is a main cross section, being more straight, filled with several centimeter thick of compact fault gouge, or without. The fault zone is about 10 m wide on the north-east end of the Square Mountain and formed into IV terraces, vertical throw is 17 m long, total throw is 44 m long. Fault F_{2-1} is the most active area. On the middle zone of the north slope and south-west end, three active faults are found. Fault F_{2-1} is the most active area, but the fault zone is reduced to 80 m wide, and there are active interbedding displacement in the fault zone.

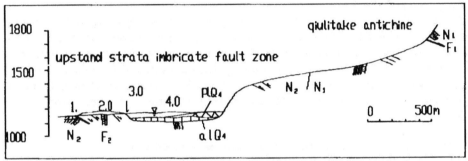

1. kezier syncline; 2. square mountain; 3. weigan river; 4. normal high water level.

Figure 2. Geological section of the reservoir and dam site.

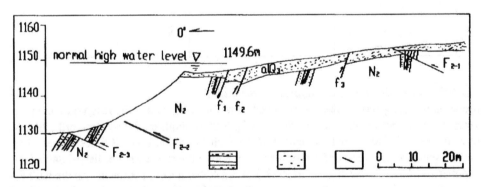

N_2: Pliocene; alQ_3: upper Pleistocene series alluvial layer;
1. sandstone, mudstone interbedded strata; 2. pebble, gravel bed, 3. active fault.

Figure 3. Geological section of the right abutment of the auxiliary dam.

Occurrence of the foundation rock in the F_2 fault zone in the dam site: striking $65°$ upstanding or dipping north-west, dip angle is greater than $80°$ .4~5 active interlayer faults in the fault zone dislocated the gravel bed Q_3 by 0.5~2.7 m, with an occurrence of $65°NW$ $\angle72°~84°$. In the gravel bed, the interlayer fault has a dip angle smaller than about $50°$.

ACTIVE FEATURE OF THE FAULT F_2

Geological relief trace survey of the activity
There are 8 grade terraces on the dam site. The movement of the fault F_2 caused the ground surface deformation. These deformation traces are well defined, and visible on the satellite photograph, aerial photograph and color infrared aerial photograph. These traces are found on the atural section formed by gulleys. Deformation of the fault F_2 in the reservoir and dam site is shown in table 1.

In table 1, deformation of the fault F_2 is bigger than its dislocation. Fault deformation began at arly peroid of Holocene.

Data of cross fault survey
After the fault F_2 was found in 1966, cross fault precise level survey along north edge of the quare Mountain began in 1972, and then base line and seismic survey was done. By now, deformation data has been gathered for 24 years. Duration, consistance and completeness are the No.1 of special trace survey in Xinjiang .The survey data of 16 years from 1988 to now shows: vertical   movement is 0.307 mm/a, horizontal contraction 0.48 mm/a, horizontal back torsion 0.19 mm/a, the tendency of long peroid deformation will become stable.

Table 1 .Vertical deformation of the fault F_2 in the Kezier reservoir and dam site

Geological age	upper Pleistocene (Q_3)					Holocene		
terrace	T_6	T_5	T_4	T_3	T_2	T_1	river bed
vertical bending of terrace plane (m)	97	34						
vertical dislocation of terrace plane (m)	17				6.89	4.3	2.68

The influence of the remote earthquake on the fault F_2
Moderately strong earthquake had occurred eight times on outside the reservoir site in ten years long of period survey for the fault F_2. Five earthquakes in the east-north, 70 km, three of them led abnormal movement of the fault F_2, which was 1.8 times bigger than normal standard deviation,$\pm45u$. North-west 50 Km of the reservoir site, Jierjisi Saike lake experienced 7.2 grade earthquake in 1978, it was felt, but it had no influence on the fault F_2. It was due to action of wave stress site in north-east of the fault F_2, which made resonance action on the left handed stress. 50 km of the reservoir site south, three times moderately strong earthquake had occurred, they all had visible influence on the movement of the fault F_2, abnormal was $\pm60u$. It was the result of north-south principal compressive stress relief after earthquake. All earthquake, in the period of 1974 to 1976 had influence on the fault, abnormal was $-90\sim150u$, total abnormal was 240u. It shows fault F_2 was easily influenced by remote earthquakes, and not subject stress accumulation. At present, it is in extreme equilibrium state of stress.

Type of the fault F_2 movement
According to following reasons. Creeping is main movement of the fault F_2.

(1).In the reservoir site and outer small area of it(circle 20 km), earthquake occurred since 1900 were smaller than 5.0 grade, so it is a zone where seismictivity are relatively stable.

(2).The fault F_2 cut are mudstone and sand stone of Tertiary system. According to indoor simulation test and in-situ test, internal friction angle of the fault gouge is $\phi=11.5^0 \sim 19^0$, cohesive force c=100~250 kPa, shear strength of the fresh mudstone and sand stone is $\phi=34^0$, c=500kPa, $\phi=35^0$, c=580kPa. their compressive strength is 9~24 MPa. The strength of the rock body is lower and plastic, and high, so it has not the condition strong stress.

(3). The plasticity creep deformation is the main movement fault zone F_2. The fault deformation just plays a secondary role. It released energy by different deformation in different zone, the energy released would not be high at each individual movement.

(4).The long axises of pebble and gravel on lower wall of the fault F_2 is arranged along pull direction . It shows that pebble and gravel has enough time to adjust their position, so the fault F_2 was slow movement.

(5) . Long period survey data shows that the fault F_2 has deformed by creeping.

PRESPECT OF THE FAULT F_2 MOVEMENT IN LIFE OF PROJECT

The amount and movement of the fault is a key question in cross fault engineering design.

The deformation of the fault F_2 is creep. The main movement period of the fault F_2 was at early period of Holocene. At present, the movement of the fault is only an extention of the main movement, and it is weak. The active is tending towards the weak . So, the fault F_2 will deform at the present creeping speed in the next 100~200 years. Based on this estimation, cumulative vertical movement will be about 30.7~61.4 mm in next 100~200 years.

If the worst thing occur, such as the fault F_2 creeps, energy gathers to the end or other zone of the fault , where deformation is damped, and which will lead viscosity movement. Based on the result of comparison with the fault scale of south edge of the Tiansan mountain, the creep of fault will be within 1 m.

ENGINEERING GEOLOGICAL CONDITION OF THE SQUARE MOUNTAIN

Mountain is located between the main dam and the auxiliary dam, formed by the fault The Square F_2 dislocating III-VI grade terraces near the river bank and the terraces rising. It is 800 m long from east to west, 150 m(east)~300 m(west)from south to north. The scarp of the fault F2 is on the north of the mountain , the scarp is 22~33 m high. The south-east of the mountain is valley slope land of the Weigan river, the slope land is 35~55 m high. The west of the mountain is valley slope land of the Kezi river, the slope land is 25 m high.

Base rock of the Square Mountain is the second layer (N_2^2) of Pliocene series: sand stone and mud stone interbedded strata, occurrence: 65°NW 80°~90°. There are 14 imbricate faults in the Square Mountain. Their occurrence are 65°SE 23°.

The imbricate fault planes dip to outside of the south slope land of the Square Mountain , and form an istable structure plane. There had occurred 300 m long landslide on the east half of the south slope land. At present, it is unstable. there are small unstable rock bodies on the west half of the south slope land.

On the west half of the top of the Square Mountain, there are NWW direction, high dip angle ground fissures. Most strikes of them are 300°~353°, few are 4°~16°. These fissures can be divided into three groups of fissure. There is a main big fissure in either of the fissure zones. The distance of the fissure zones is 190~195 m. All fissures are wide on the upper and narrow on the lower end, and stretch to base rock through pebble and gravel bed of these terraces. The fissures in pebble gravel bed have been partially filled by diluvial deposit and breaking soil from fissure. Fissures in base rock are not usually filled. All fissures do not occur vertical or horizontal fault. The biggest depth under ground surface is 13.4~14.3 m. The depth under the top of the base rock is 10~11 m. Elevation is 1142.9 m. It is 6.7 m lower than normal water high level.

The fissures of the Square Mountain are intensively abnormal of convex topography. Macroscopic earthquake damage has rather repeating feature in same earthquake structure system in some geological history period. The future earthquake influence on the Square Mountain will occur repeatedly on these original fissures. According to the minimum strength theory, if the square mountain occurs intensity abnormal, it will release energy along the present fissures . The chance occurring new fissures is not big. So if hydraulic structure avoids the fissures zone, earthquake intensity abnormal influence can be avoided or reduced.

So, engineering geological condition of the Square Mountain is rather poor. The scale of the mountain is not big, but it is the supply body of the left dam abutment of the main dam and the right abutment of the auxiliary dam, actually, it is a part of the dam body. It is necessary to select good axis of the dam and strengthen the mountain body, in order to the square mountain works good as dam body in the reservoir service period.

THE INFLUENCE OF THE FAULT F_2 ON DAM OF THE KEZIER RESERVOIR

Based on the weigan river planning of the river basin, the Kezier reservoir is the best project. When the engineering preliminary reconnaissance and design began in 1966, it had been pointed out that the fault F_2 was active fault. At present, it is continuously active. It is a reasonable question and a key question if the Kezier reservoir can be build. The following is the key arguments.

(1). The dam is located on the inner of the east wing of the Baicheng epsilon-type structural system, the qiulitake curved structure, where stress has not gathered. In the reservoir and dam site and a circle of 20 km, earthquakes having occurred since 1900 are smaller than 5.0 grade. It shows the reservoir and dam site are a relatively calm zone of seismic activities. Basic intensity is VIII.

(2). Give up the dam site where fault F_2 passing, select dam site where the fault F_2 passing the right abutment of the auxiliary dam, so that the influence of the fault F_2 on the building is minimum.

(3) .The long period fault survey data has shown that fault F_2 deformation is creep, movement speed is moderate, total movement will be low in the service period.

(4). Some measures can be used in the present earth and rockfill dam design, so that the dam is adaptable to creep of the fault F_2, and even sudden deformation.

Based on movement of the fault F_2, the regional structures where the reservoir and dam is located, and present design level dam can be built.

PLACE OF THE RIGHT ABUTMENT OF THE AUXILIARY DAM SELECTION AND MEASURES AGAINST FAILURES

Influence factor
It can not be avoided that the right abutment of the auxiliary dam will be laid on the scarp of the fault F_2. When selecting location of the dam body , the infulence of the Square Mountain on the stability, anti-seepage of the dam and cracking of the top of the Square Mountain, and measures against failures and etc. , all these are main questions, and the length of the dam is less important.

Siting
In the scheme, but the dam is The right abutment of the auxiliary dam was once put on middle of the north slope of the Square Mountain, the length of the auxiliary dam is shortest just located on east fissures zone. Most of the fissures of the Square Mountain top, middle ground fissure zone are on upstream side of the dam abutment, these fissures will form seepage passage to downstream of the main dam on the south slope of the Square Mountain. Most part of the mountain is in saturation state. It is not good to stability of the mountain, and the engineering work required is great. Considering of the optimization of the dam's engineering design on the Square Mountain and protecting action of the Square Mountain, the right abutment of the auxiliary dam is moved 270 m along upstream to between the middle and west ground fissure zone. With this provision, both of the east and middle ground fissure zones are located behind the dam, bad influence of them in the dam are avoided. It is good to protect the dam by the Square Mountain, when most part of the Square Mountain are behind the dam. Meanwhile, demage from suddently movement is avoided. The second dam will be built on the previous location of the right auxiliary dam.(Fig .1).

Function of the second dam

The second dam is a preventive engineering measure used to prevent emergencies.

The second dam is not used to hold back water. It is dry during reservoir life. The earthquake resistance of the dam is better than the resistance of the right abutment of the auxiliary (the first dam) in operating condition state. Because the second dam is dry earthquake resistance will be high, and breaking strength will be lighter than the first dam. When earthquake occurs, or when the fault F_2 suddenly creeps and if the impervious body of the first dam is damaged ,the second dam will play the role of water holding, so that the pressure of downstream and upstream of the first dam will reach equilibrium quickly and protecting dam from failure.

The cavity between the right abutment of the auxiliary and the second dam will be filled with pebble and gravel. When the dam body of the right abutment of the auxiliary is damaged, seepage caused by the damage will be controlled by sand and gravel, and dam failure will be avoided , and discharge of reservoir will be avoided. If the water level of the reservoir reach the highest water level, 1149.6 m, since elevation of the ground surface behind the second dam is 1140 m, elevation of the surface of the base rock is 1136 m, average seepage gradient will be about 0.05 in the pebble and gravel. It will be stable enough.

ENGINEERING GEOLOGICAL CONDITION OF THE RIGHT ABUTMENT OF THE AUXILIARY DAM(THE FIRST DAM).

The base rocks of the right abutment of the auxiliary dam are soft sandstone and mudstone of the Teritary system. Main composition of the cross section of the dam is sandstone, including mudstone 11%, silty mudstone, 14.5%, sandstone 70.4%, mud siltystone 4.1%.

The fault zone F_2 is formed by the three active fault planes and four interlayer fault planes.

F_{2-1}: occurrence $65^0SE\angle34^0$,the faulted zone is 50 cm wide, it is main active fault plane, outcropping elevation is 1753.7 m.

F_{2-2}: occurrence $65^0SE\angle25^0$, dip angle become higher near ground surface, outcropping elevation is 1140.6 m.

F_{2-3}: occurrence $65^0SE\angle30^0$, the faulted zone is 50 cm wide, outcropping elevation is 1131.5 m.

Four interlayer faults are in the opposite direction of the fault F2. Their occurrence are 72^0 $NW\angle72^0$.The gravel is dislocated by 0.5~2.7 m. dip angle in gravel become lower, about 57^0.Besides those planes ,there are also some small faults ,shorter and closely spaced joints. Total width of the fault zone is 80 m.

Under base rock plane of the fault F_2 30 ~19 m, above 113.7 m elevation, permeation velocity is 40~220 Lu. It is very strong or relatively strong permeable. Longitudinal wave speed is less than 1200 m/s, some are just 600 m/s or smaller than 600 m/s. Seepage velocity under the zone is less than 2 Lu. So impervious body shall extend into strong or relativity strong permeable body by 3~5 m when the right abutment of the auxiliary dam is strengthened. Except grouting material is impervious, it must have some plastic deformation feature in order to suit fault movement. The grouting shall be great.

When the dam body of the right abutment designed, influence of creep of the three active faults and the four interlayer faults on the dam body, should be considered. When stress is put on the abutment for long time, fissures will occur. Most of them are compressive torsional cracks, but tension fissures can ocure too at the contact zone of the dam body and the base rock, so engineering measure should be taken to minimize them.

GUIDELINES OF DESIGN OF THE RIGHT ABUTMENT OF AUXILIARY DAM

The design of the dam body should suit the creeping deformation of fault F_2 and when strong earthquake occur and the fault F_2 dislocate above 1 m, the abutment can be safe. Design guidelines.

(1) . wide dam crest, gentle dam slope, heighten dam body.

(2).make specially wide core wall.

(3) . increase plasticity of the permeable body, fill the core wall with relative high clay content and high plastic soil with high water content to raise plasticity, and lighten damage of the fault F_2 to core wall.

(4) . Lengthen seepage path, lower seepage velocity on upstream bank slope of the core wall, clay pavement along it, in order to lengthen the seepage path through the fault F_2 zone.

(5) .Double filters will be provided on upstream, downstream of the core wall and the downstream slope of upstream clay protection. When the core wall cracks due to dislocating of the fault F_2, sand in the filters will fill up fissures, and make fissures close automatically. The double filters and upstream slope must be thick enough, so that they can not completely disturbed and lose it's function when the fault F_2 dislocating exceeds 1 m.

(6) . Take measures to see to permeable body has good contact with base rock plane and it is not damaged by seepage.

(7) . Thicken curtain wall, consolidation grouting is done in every place of the abutment. Use effective measure to make sure that high quality of grouting is made.

Proc. 30th Int'l. Geol. Congr., Vol. 23, pp. 41-52
Wang Sijing and P. Marinos (Eds)
© VSP 1997

Sediment of Hydroenergetic Run-of-the-river Reservoir of the Djerdap I Hydroelectric Power Plant

-Generation and Problems in Exploration-

MILETA PERISIC*, ANDJELKA MIHAJLOV**, & SLOBODAN KNEZEVIC***
*Geoinstitut, Rovinjska 12, 1000 Belgrade, Yugoslavia.
** Federal Ministry for Develop., Science and Envir. Bul.Lenjina 2, Belgrade,Yugoslavia.
*** Faculty of Mining, University of Belgrade, 11000 Belgrade, Yugoslavia.

Abstract

Construction of the run-of-the-river reservoir of the hydroelectric power plant Djerdap I in the Danube river basin covering 817000 km^2, with the flux from 1950-16000 m^3/s, and retention time from 6-14 days, the conditions for deposition of big masses of sediments have been established. In the project the deposition and removal of suspensions from the reservoir in a ratio 1 : 1, in first 27 years, in an absolute amount of 20 Mt/g, have been foreseen. During exploitation it was found that the total inputs were overrated, but the retention rate was far underestimated, what has particularly been analyzed in this work. During investigations of the water quality changes in the reservoir an exceptionally attractive phenomenon has been ascertained, related to the reduction of primary production and precipitation of planktonic mass from the upstream section into the reservoir. Since the Danube river basin, with tributaries feeding the reservoir, is rich in nutrients and with a high organic production, the total deposited sediments exhibit the high percentage of organic constituent. In the paper have been analyzed conditions of precipitation in the reservoir by use of sedimentological analyses and contents of organic mater, whose participation in the increased efficiency of precipitation was analyzed considering the analysis of changes of the particle's ceta potential in the turbulent flow. The suspended deposit has fixed by itself the numerous toxins, metal ions, organic micro-pollutants (oil derivatives, pesticides, PCBs); that is why the destiny question of these materials in the sediment is of the primordial importance.
In the paper have been analyzed solutions of recovery of influence of sediments with the high toxic substance's contents, whose eluation particularly in the intensified process under the increased temperatures, has limited the development of the living world in the reservoir and in the downstream sector.

Keywords: Sediment, suspended solids, dissolved oxygen, mineralization, heavy metal, plankton

INTRODUCTION

Along the relatively short section of about 250 km. of the Danube in Yugoslavia the water flow greatly increases. By receiving several tributaries the Danube water balance changes (long term mean values):

- The Danube at the inflow profile to Yugoslavia 72 x 109 $m^3/year$,

- The Danube at the outflow profile   183 x 109 $m^3/year$.

By this increase of more than 2.5 times significant changes occur in the ionic composition, specific organic load, micropolutant content etc. Specific flow conditions, especially in the backwater part of the Reservoir Iron Gate I, cause processes and quality changes of the Danube.

Tributaries from Yugoslavia or those formed in neighboring countries have specific water characteristics. Besides differences caused by geological structure of the terrain, differences caused by human action dominate. By constructing of hidroenergetic reservoir in the river flow conditions for a number of physico-chemical and biological processes are changed. This changes are especially pronounced in the processes of sedimentation, organic matter degradation, reaeration and organic production. Intensification of sedimentation processes under the conditions of water retention, occurring in such a reservoir, has lead to the alteration in different processes, effecting as a result water quality of the system. The study of specific changes in the physico-chemical and biochemical processes and in the biological status of the watercourse at varying retention times completes the picture of short-term changes in the investigated system. The examination of settling effects and processes in the sediment of the reservoir provides an insight into some long-term changes due to backwater effects in the investigated Danube section.

Settling of degradable organic matter promotes anaerobic processes in undisturbed bottom sediments. Sediment erosion and transport under the high water flow conditions lead to aerobic decomposition which also effects water quality: oxygen regime, organic and undissolved matter. By permanent accumulation of bottom sediments in the reservoir, the effects on the water quality will be further intensified. By draining the large watershed in the upstream part, the Danube carries a very great load, especially with nutrients. Hydraulic characteristics of the water flow together with a high N and P content promotes high primary production levels. This is the reason that, inspite of prevention measures, the Danube water flow has permanently high organic load. In the Hungarian part of the Danube [9] found chlorophyll "a" content 10-187 ug/l.

Methodology and Study Results

The examined section of the Danube covers the backwater-influenced part of the Djerdap I reservoir from Smederevo to Kladovo (Fig. 1). Changes in the composition of the water mass in this section were analyzed at five profiles. Samples for investigations of the water quality were taken at three points in the profile (on the left bank, in the central flow and on the right bank) under characteristic discharge conditions (spring high-water, summer and autumn low water). Sampling of water was done at a death of 0.5 m. from the water surface. All physico-chemical analyses were made by standard analytical methods [8]. For biological analyses, composite samples of 40 l. were taken from the left and right bank and central flow and filtered through a plankton net N^0 20. The analysis of phytoplankton and zooplankton (qualitative and quantitative composition) was made with regard to species, while the estimate of the water quality was made according to Knopp and Pantle-

Buck. Investigation of the changes in water composition along the section under study, at various discharges, was done with regard to the flow time from profile to profile. Sample for sediment analysis were taken in the autumn low flow conditions.

RESULTS AND DISCUSSION

For the presentation of dominant processes in the studied section which are characteristic for hydropower reservoirs, this work analyses elements of the systems listed in Tables 1 and 2. Changes in the flow conditions, i.e. rate, geometric characteristics (changed ratio

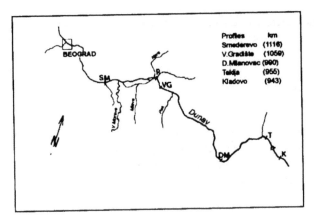

Figure 1. Investigated run-of-the-river section of the Danube

of water surface to the total volume), lead to changes in the deposition of suspended material, reaeration, conditions for plankton development and a number of other processes relevant to the water quality.

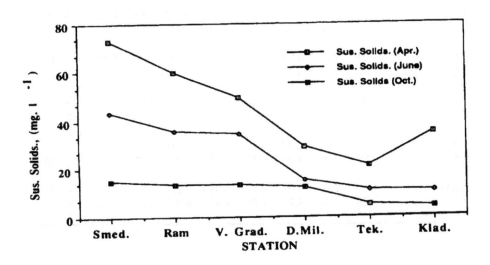

M. Perisic et al.

Figure 2. Removal of suspended material from water along the section

Typical changes of the suspended solids and the quantitative composition of plankton for characteristic conditions in the water course (change of flow rate from 1850 to 11900 m^3 is shown in diagrams Figures 2 and 6.

Transportation processes which are relevant to : sedimentation, reaeration, intensity of exchange on the contact layer water/sediment changed considerably in dependence on the discharge. Some basic data related to the transport processes in the reservoir are shown in Table 2.

The data presented there and the criteria for the evaluation of the stratification based on the value of Froudes number [12].

$$Fd = 10^{-5} \; \frac{L}{zt}$$

where : L- reservoir length (m), z- mean depth (m), t -retention time (yr), enable us to evaluate the potentials of this process. In keeping with this principle, for the values Fd >> 0.32, the reservoir is thermally homogeneous which is the also the case with the analyzed section of the river Danube.

Table 1. Detected changes of some indicators of river Danube water quality the Smederevo-Tekija section

Profile (km.)	Sechi disc, m min max	pH meq/l min max	HCO3 meq/l min max	CO3 mg/l min max	NH4 mg/l min max	NO2 mg/l min max	NO3 mg/l min max	Orto,P mg/l min max	Total,P mg/l min max
Smederevo (1116)	0.20 0.60	7.60 8.50	3.15 3.55	0.00 0.40	0.09 0.44	0.004 0.036	1.85 3.36	0.080 0.255	0.180 0.385
V.Gradiste (1059)	0.25 0.75	7.60 8.45	3.20 4.25	0.00 0.15	0.15 0.52	0.021 0.092	1.80 2.80	0.106 0.213	0.147 0.370
D.Milanovac (990)	0.35 1.20	7.70 8.40	3.05 4.25	0.00 0.05	0.11 0.42	0.026 0.110	1.82 2.97	0.070 0.200	0.170 0.298
Tekija (955)	0.50 1.95	7.70 7.95	3.00 4.20	0.00 0.00	0.06 0.51	0.031 0.070	1.75 2.85	0.070 0.196	0.140 0.245
Kladovo (943)	0.50 1.80	7.70 7.90	3.05 4.25	0.00 0.00	0.115 0.49	0.032 0.062	1.90 2.67	0.077 0.157	0.180 0.254

SEDIMENTATION PROCESSES AND THEIR EFFECTS

One of the basic features of the backwater-influenced reservoirs is a high level of reduction of the entered suspended material. The effects of this process are manifold and reflected in the deposition of organic dissolved and undissolved materials, of a large number of other materials and of specific organic micropollutants. The results of measurements (mean values in the profile) of the suspension composition in the inflow and outflow profile, shown in Figure 2, indicate a considerable degree of correlation between the suspension constituent and discharge. The increase of water transparency under conditions of low water is considerable: from about 0.5 m at the inflow profile to about 2.0 m in the profile of Tekija. The extension of the photic zone to the depth of about 8 m in the outflow part should results in increased bio-production. However, this does not happen to the extent which could be expected in view of this fact and the contents of macro and micro elements.

Table 2. Biochemical degradation and reaeration in the studied section
(characteristic discharges, water temperature about 15° C)

Section	Discharge, m^3/s	BOD5 mg/l	Dissolved O_2, mg/l in/out	Mean velocity, in/out	Retent. time m/s days	Kr 1/day	K2 1/day
	1850	8.17/5.18	7.74/6.56	0.21	3.14	0.145	0.202
Smederevo	3300	5.59/3.44	9.21/8.65	0.47	1.57	0.309	0.358
to	9500	3.39/1.71	7.02/6.90	1.30	0.57	1.23	0.731
V. Gradiste	11900	5.24/3.35	9.77/10.2	1.65	0.44	1.02	0.865
	1850	5.18/3.15	6.56/5.11	0.18	4.44	0.112	0.111
V.Gradiste	3300	3.44/1.87	8.65/6.72	0.22	3.99	0.162	0.128
to	9500	1.71/1.52	6.90/6.86	0.65	1.32	0.189	0.274
D.Milanovac	11900	3.35/2.62	10.2/10.4	0.80	1.10	0.224	0.317
	1850	3.15/2.26	5.10/4.51	0.10	4.03	0.082	0.043
D.Milanovac	3300	1.87/1.28	6.72/6.68	0.16	2.79	0.136	0.060
to	9500	1.52/1.44	6.86/6.80	0.49	0.91	0.060	0.131
Tekija	11900	2.62/2.33	10.4/10.0	0.60	0.74	0.158	0.151

Composition of the sediments, and the effects on water resulting from water retention was predicted while the reservoir system was being planned for construction [1], Fig. 3. Sedimentation in the Iron Gate I, predicted and real effects of sedimentation during life of the reservoir have been much greater than predicted. The greatest are at the stretch from the dam to the mouth of the Nera river (km 943-1074), where results over a period of five years show that the mean yearly deposit was 119 m^3/m' with the maximum at Donji

Milanovac (km. 995), where during the period 1970-1984 bottom sediments of 10 meter thickness were formed.

Figures 4 and 5 represent results of granulometric characteristic of suspended matter before power-plant building and sediment characteristic measured in 1994. on profile Tekija.

The results of our study for the first time point out the significance of this phenomenon, resulting in a very high level of the river water purification, which depending in the water flow and temperature has following values :
- water transparency change from 0.2 m at the profile Smederevo to about 2 m during the low water flow at Tekija and farther downstream,
- reduction of BOD5 values in the range 40-75 % depending on the water flow,
- 90 % decrease of bioseston (plankton),
- pronounced reduction of micropollutants (heavy metals, mineral oil, pesticides).

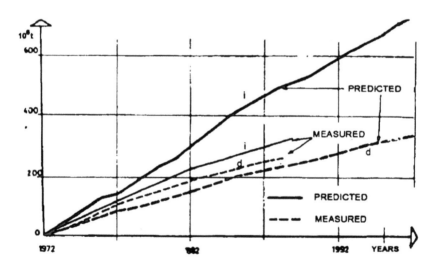

Figure 3. Predicted and measured input (i) and deposition (d) of suspended   solids in Iron Gate I

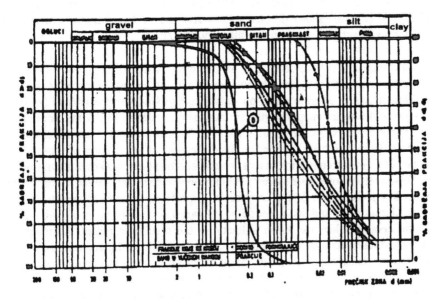

Figure 4. Granulometric characteristic of suspended solids in the Danube river

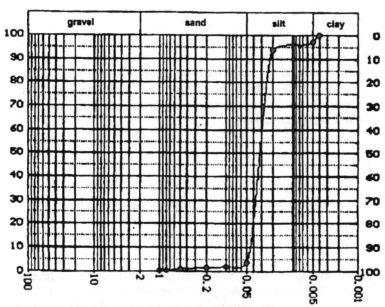

Figure 5. Granulometric characteristic of sediments on Tekija profile

As was stated above, large quantities of sediment are generated by sedimentation of undegraded organic matter, first of all bioseston. Formation of sediments with high organic matter content (COD-Cr in the range 50,000 -100,000 ppm) with high content of other substances creates serious problems for water use in this Danube section. The effects of such sedimentation are therefore numerous [3-7].

The Iron Gate reservoir bottom sediments have enormous significance, both short- and long-term for the whole river system. In addition to causing a sharp decrease of reservoir volume, sediment has an adverse effect on surface and ground water exchange (which is important for public water supply), with an increasing tendency [6]. Some preliminary laboratory experiments show different effects of bottom sediments on water quality [5].The lack of more thorough knowledge about the content and elution of harmful and toxic substances from bottom sediment as a "chemical time bomb" makes this system very uncertain. Results from some lake systems [6] indicate the need for a better understanding of water quality preservation of the Danube in Yugoslavia in order to prevent further degradation of this river and its surroundings. Reaeration constants in all subsections increase considerably when the mean velocity in the section increases. Although the values of the constants Kr and K2 lead to the conclusion that the biodegradation process and reaeration are balanced, and should bring about an increased content of dissolved oxygen at higher rates, the results of measurements in all hydrological conditions provide data on the constant increase in the deficit in the whole studied section, except for that of

oxygen deficit at the discharge of 11900 m^3. The balance of dissolved oxygen compensation does not include effects of wind, which in some periods of investigation considerably increases the reaeration effects. Similarly, the consumption of dissolved oxygen in the process of sediment mineralization may be quite important for oxygen balance. The organic matter content in the storage sediment is 15-120 mg O_2/g. of the dry mass of mud, depending on the place of sampling. In the main course, the organic matter content is minimum and increases in the parts of the reservoir with lower velocities. Mineralization in the deposit occurs in the transitory zone of Red-ox potential: oxidation processes take place in uncounsolidated mud layers while anaerobic ones occur in the homogenous deposits. The coefficient of mass transport on the liquid/soil interface is determined by the following expression [2].

$$K = D / Li$$

where : D - molecular diffusion of O_2 or BOD in the liquid and Li - thickness of diffusion layer determined by kinematics viscosity (μ) and dissipation energy (E) :

$$Li = (E / \mu^3)^{-1/4}$$

The increase of mean velocity in the reservoir heightens the effects of mass transfer on the liquid/solid interface and, thus, the total oxygen consumption too. This is certainly one of the reasons for the increased oxygen deficit in the conditions of high discharge and high reaeration.

CHANGES IN PLANKTON COMPOSITION ALONG THE SECTION

Throughout the years, plankton has had similar characteristics. The variety, and even more abundance, of phytoplankton and zooplankton decrease from the upstream to the downstream part of the studied section, (Figure 6). In phytoplankton, almost in all profiles, Bacilariophyta dominate both in quality and quantity. They varied most in the two upstream profiles and gradually less after that, while a small number of species (Asterionella formosa and Synedra) were observed all along the watercourse.

Chlorophyta are usually better developed in summer then in autumn. Most frequent were different species of Scendesmus and Pediastrum. Of the other algae groups, Heteroconta are present everywhere, while Euglenophyta and Zooplankton has been, as a rule, less numerous than phytoplankton. Rotatoria almost always dominate both in quality and quantity. The most frequently observed are various species of Brachionus and Keratella genera. Brachionus species dominate in the upstream and Keratella in the downstream part of the studied section. All the results, in accordance with the criteria for the evaluation of water quality, lead to the conclusion that the studied water is of water quality class II.

Production of organic matter in water systems is most frequently limited by some of the macro or micro nutrient [10,11]. The results of long term investigations of the water quality in the reservoir Djerdap I under all discharges conditions indicate that there is decreased plankton biomass production in the section under water retention affected by backwater, although the nutrient content provides conditions for the intensification of the process [5,6]

RELEVANT PROCESSES IN THE SEDIMENTS

Under the conditions of limited oxygen transfer in the bottom sediment polyvalent forms of nitrogen, sulfur and other elements becomes electron acceptors, by which red-ox potential of the media changes. Studies were done by means of a laboratory model, which made it possible well as conditions in the water-sediment contact layer. Change of the red-ox potential in the water column and the surface mud layer under conditions of limited oxygen transfer to the water phase and by this to the sediment are also shown in Fig. 7. Mineralization of organic matter in the mud is shown by an increase of inorganic carbon in the water column above the mud layer. Under the specific conditions of advanced organic matter degradation in the mud, which existed in the laboratory experiment, inputs of inorganic carbon from the deposit phase compensate to a great deal the carbon consumption in the trophic zone and it's availability for the organic production.

M. Perisic et al.

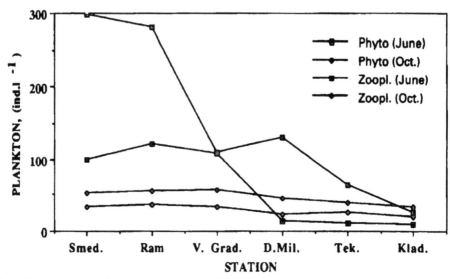

Figure 6. Changes in the plankton abundance along the section Smederevo-Kladovo

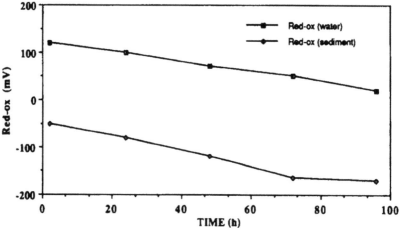

Figure 7. Redox potential changes in the surface mud layer

Changes of electron acceptors under the conditions of oxygen absence lead to the corresponding equilibrium between different valence forms of a number of metals (Fe, Mn and others) as well as to the changes of solubility as it was found with K (Fig. 8). The absence of higher valence nitrogen stages in the reservoir deposits is a reliable proof of the red-ox conditions during longer periods. A study of denitrification activity of these deposits taken from different location in the reservoir locations revealed the same properties as the typical sample shown in Fig. 9.

CONCLUSION

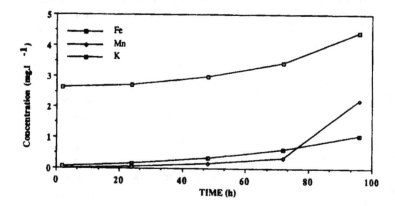

Figure 8. Increase of the efficiency of the metals elution with the development of negative redox processes

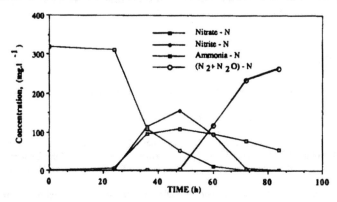

Figure 9. Typical denitrification activity of the mud

A feature of run-of-the-river reservoirs for electric energy production is the unstable flow regime, which creates specific conditions for biological and biochemical processes. The example of the reservoir Iron Gate 1 of the river Danube serves for the analysis of the characteristic changes relevant parameter for this type of reservoir. The unstable flow during the whole study period keeps the system out of the state of stratification. Major effects of sedimentation processes in the backwater-influenced conditions in an open reservoir result in a important reduction of the content of suspended solids and degradable organic matter. The released organic load is broken down in the reservoir with effects which are the better the longer the retention time and the lower the flow rate. Compensation for the oxygen dissolved in water and consumed in the process of biochemical degradation in the open reservoir is mostly made in the process of reaeration. However, this process does not produce a favorable oxygen balance, results in an increased deficit throughout the watercourse. The only exception to this are maximum water discharges, when the oxygen deficit in the downstream profile decreases. The sediment in the reservoir accounts for a major share in the consumption of dissolved oxygen, particularly in the periods of higher mean velocities. The reduction in primary production and decrease of abundance and plankton diversity, followed by the changes in hydrochemical characteristic of the water (absence of carbonate ion, lower dissolved oxygen content along the section) in the run-of-the- river reservoir is a consequence of hydrodynamic features of the waterbody. The lack of stratification has manifold negative impacts on the production. During a longer transport through water without light, the photosynthetic activity of the plankton is inhibited. Exposure of plankton material to the pressures equal to those in deeper zones of the reservoir destroyed part of the material.

The concentration over longer period of time of harmful and / or dangerous materials or materials that in certain processes generate harmful and / or dangerous substances with spontaneous sudden release from the sediment in the Iron Gate I have the characteristics of a chemical time bomb (CTB), a category of pollution with high risk, as this type of environmental pollution is classified in the current literature. Together with certain evaluations of characteristics of the sediment in the accumulation of the Iron Gate I conducted by hydrologists of countries along the Danube the analyses in question provide elements for planning the activities needed in this sector.

REFERENCES

1. Bruk. S., Varga S., Transport nanosa i prora~uni deformacije korita Dunava pod uticajem HE "Djerdap I", Vodoprivreda, 22, str. 135-157, 1991.
2. Levich,V.G. Fiziko-himi~eskaja gidrodinamika, FIZMATGIZ, Moskva,1959.
3. Peri{i} M., Tutund'i} V., ^uki} Z.,1987. Einge aspekte der qualitatsanderung des Donauwassers im speicherbecken "Djerdap I" wahrend der neidrigwasserperiode, 26. Arb. der IAD, Passau/ Deutschland.
4. Peri{i} M., M. Miloradov, V. Tutund'i}, Z. ^uki}. Changes in the quality of the Danube river water in the section Smederevo-Kladovo in the conditions of backwater effects. Wat. Sci. and Tech. Vol.22, No 5 , 181-188, 1990.
5. Peri{i} M.,V.Tutund'i} .Eigenarten der Produktionsveranderungin Durchflus- sakkumulationen . Limnologiishe Berichte der 28 Tagung der IAD , Ubersichtsreferate, 132-139. Sofia, 1990.
6. Peri{i} M., V. Tutund'i}, M. Miloradov. Selfpurification and joint effects in the Iron Gate reservoir. Verh. Internat. Verein Limnol. No.24, 1415-1420, Stuttgart, 1991.
7. Peri{i} M., A. Mihajlov, Pla{i} Lj., Sediments eutrophic lake and runoff the river reservoir with low level primary production, 6th International conference - Proceedings Environmental contamination, 456 - 460, Greece, 10-12 October 1994.
8. Stigliani, W. M., Andeberg, S. and Jaffe, P. R. 1993. Industrial metabolism and long-term risks from accumulated chemicals in the Rhine Basin. UNEP Industry and Environment July-September, pp.30-35, 1993.
9. Standard Methods for the Examination of Water and Wastewater .1980. APHA-AWWA-WPCF, pp.1193. Washington.
10. Varga P., Abraham M., Simor J., Water quality of the Hungarian Danube stretch and its major determining factors, Int.Conf. W.P.C. in the basin of the Danube, Novi Sad.1989.
11. Vollenweider, R. A., J. J. Kerekes . 1982. Eutrophication of waters, monitoring, assessment and control. OECD. Paris.
12. Water Resources Engineers,Inc.(1969) Mathematical models for preduction of termal energy changes in impoundments. US Env. Prot. Agency Water Pollut. Control Research Series 16130 EXT, US Gover. Print.Office ,Washin., DC, 157pp.
13. Wetzel, R. G., G. E.Likens. 1979. Primary productivity of phytoplankton ex. 14. 198-220., In Limnological analyses, W. B. Saunders company. London. measured

Proc. 30th Int'l. Geol. Congr., Vol. 23, pp. 53-58
Wang Sijing and P. Marinos (Eds)
© VSP 1997

Anchoring Engineering for Lianziya Dangerous Rockmass Controlling at the Three Gorges of the Yangtze River, China

YIN   YUEPING
China Exploration Institute of Hydrogeology & Engineering Geology, MGMR, 100081 ,China

Abstract

The controlling engineering for the Lianziya dangerous rockmass at the Three Gorges of the Yangtze River has been conducted. The paper introduces pre-stressed anchoring applied to the project. The anchoring is divided into eight zones and about   200 cables are fixed of 1000kN, 2000kN and 3000kN, in which the longest cable is over 61m.

Keywords: pre-stressed cable, anchoring, Three Gorges

INTRODUCTION

The Lianziya dangerous rockmass lies on the south bank of the Yangtze River in Zigui Country, Hubei Province. The rockmass is facing Xingtan landslide across the river that slided in 1985 and destroyed the Xingtan town, 27 km down to the dam site of the Three Gorges. It is an elongated cracked rockmass running from south to north, 700m long, 30-180m wide and over 400m high. It is narrow and low on the south, broad and high on the north. The potential unstable rockmass has been cut by more than 30 wide and huge fractures into three deformation sections, respectively, 80 thousand, 20 thousand and 2.5 million cube meters in volume from south to north. The fractures are 60~170m long, 0.5~5.0m wide and 50~110m deep.

The dangerous rockmass consists of Permian limestone intercalated with several carbonaceous shale and marl layers, underlain by weak and excavated coal layers. The strata strike N30~50E and dip NW at angles varying from 27 to 35. The potential unstable rockmass was formed under the influence of cliff gravity, working-out bottom (the hollow area covered 120000m^2 and mining height was 1.6~4.0m), water(surface and underground) processing, karstification and weathering with the help of favorable structure, well-developed fissures and small faults.

The severe hazards occurred since 500 years ago and the most severe one had stopped the navigation for 82 years. It would cause huge disaster if the dangerous rockmass falls to the Yangtze River. An steering group on the Lianziya   Huanglansi geohazard prevention was established in 1989, and the feasibility of the prevention was completed. Since 1992 , the design on the Lianziya dangerous rockmass controlling have been conducted. Special design for key parts of the rockmass will be carried out and modified during the working.

Font dangerous rockmass, also called "Wuwanfang", is most key part of the prevention. "Wuwanfang", 260 thousand cube meter in volume, over 100m high steep cliff, consists of huge-thick limestone intercalated with several soft layers. The pre-stressed anchoring engineering is applied to the controlling.

ANCHORING ZONING OF " WUWANFANG" DANGEROUS ROCKMASS

Parallel to the river, the "Wuwanfang" dangerous rockmass is cut by fracture system T11, T12 and soft layer R203 into triangle-shape mass, facing into the river. Vertical to the river, the dangerous rockmass is separated by the fracture system T11 and T13 with the steep cliff of 60~80m high into the thin plates. The anti-sliding and anti-toppling would be considered for the anchoring. A set of standard is applied as follow:

Designing standards for anti-sliding of anchoring

Design:
basic force :
gravity K>=1.2
special force: K>=1.1
gravity + earthquake(a=1.1)+hydraulic head (0.1H)
Calibration:
gravity+earthquake(a=1.48)+hydraulic head(0.3H)   K>=1.02
gravity+earthquake(a=1.10)+175m reservoir level   K>=1.02

Designing standards for anti-toppling of anchoring

Design:
basic force :
gravity K>=2.0
special force: K>=1.5
gravity+earthquake(a=1.1)+hydraulic head (0.1H)
Calibration:
gravity+earthquake(a=1.48)+hydraulic head(0.3H)   K>=1.2
gravity+earthquake(a=1.10)+175m reservoir level   K>=1.2

According to above designing standards, 310 thousand kN anchoring force is required. Anchoring grade is 1.1 million kN m, if the cable depth is 35m average. Now, the longest cable was fixed of 61.5m. The anchoring work is at the steep cliff, so the 3000kN, 2000kN and 1000kN cables are selected. 3000kN cables are mainly under the 140m elevation, anchoring force is 150000kN, one half of the total, about 50 cables. 2000kN cables are mainly between 140m and 160m elevation, anchoring force is 73000kN, one fifth, about 73 cables. So , the dangerous rockmass would be divided into three zones and eight sub-zones(Fig.1):

Zone-I(Yiwanfang west)

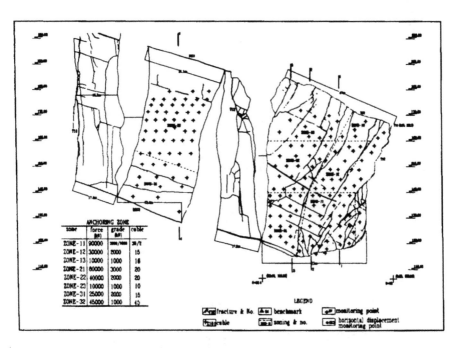

Fig.1 Anchor zoning of the Lianziya dangerous rockmass

Confined by T12, T16 and R203. Bottom width is 25m, upper, 15m, and 55m high. Total anchoring force is 130 thousand kN. Three sub-zones could be divided(Fig.2) :

Zone-11: The anchoring purpose is to prevent sliding of "Yiwanfang" west to north. It could be regarded as the "key block" of dangerous rockmass. total anchoring force is 90000kN. 30 cables of 3000kN are applied with 35m average and the max, 60m. Anchoring angle is 25 degree.

Zone-21: to preventing the toppling. 15 cables of 2000kN are applied with 35m, across the fracture T13 and T11. Total anchoring force is 30000kN. Anchoring angle is 15 degree.

Zone-13: to prevent the toppling and strength the T12 friction. 16 cables of 1000kN are applied with 35m long average, across the fracture T12. Anchoring angle is 15 degree.

Zone-II(Yiwanfang   East)
Confined by T14,T16 and R203. Bottom width is 15m, upper, 25m, 65m high. Total anchoring force is 110000kN. Three sub zone are divided:
R203 soft layer. Total anchoring force is 60000kN. Anchoring angle is 25 degree.

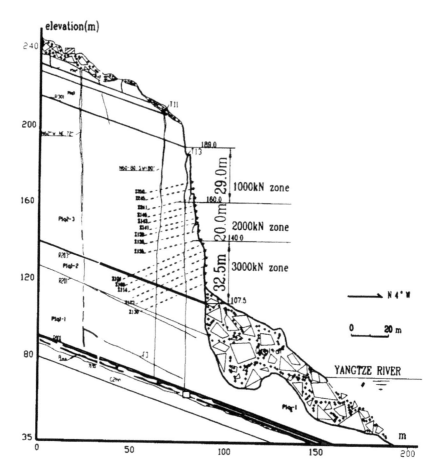

Fig.2   Section map of anchor zoning of the Lianziya dangerous rockmass

Zone-22: to prevent the toppling. 20 cables of 2000kN are applied with 35m long average, across the fracture T13 and T11. Total anchoring force is 40000kN. Anchoring angle is 15 degree.

Zone-23: to prevent the toppling. 10 cables of 1000kN are applied with 35m long average, across the fracture T13 and T11. Total anchoring force is 10000kN. Anchoring angle is 15 degree.

Zone-III(Wuwanfang East)
Toppling is generally failure pattern since the steeper cliff but the soft layer under bottom is thinner. Two sub zones could be divided:

Zone-31: 15 cables of 2000kN are applied with 35m long average. Total anchoring force is 25000kN, and angle, 20 degree.

Zone-32: 45 cables of 1000kN are applied with 35m long average. Total anchoring force is 45000kN, and angle,15 degree.

ANCHORING PARAMETERS OF WUWANFANG DANGEROUS ROCKMASS

Basic of anchoring parameters
Three types of cables are recommended as follows:

No.	grade(kN)	angle	depth(m)	numbers	note
1	1000	15	35	73	max depth < 45m
2	2000	15-20	35	50	max depth<50m
3	3000	25	35	50	max depth<65m

Selection of wires for cables
A low relax high strength steel wires is considered. The features are presented as follows:

ASTM A416-87a

N. diameter	15.24(0.6")
grade	270K
N.steel area(mm^2)	140.00
Min. breaking load(kN)	260.7
Weight per km(kg/km)	1102
Min. load at 1% extension(kN)	221.5
Min. Elongation(%)	3.5
Max relaxation after 1000hrs at 20	2.5

Suppose the safety index is 1.6, 19 wires is applied to 3000kN cables(failure cable per wire is 260.7kN), 12 wires to 2000kN, 7 wires to 1000kN.

Anchoring force lock
3000kN cables is recommended for the anti-sliding along R203. The safety index ratio between under basic force and special force 0.65 to 0.80. Since the rockmass is extremely broken along the R203 soft layer, the great deformation will be occurred if the 3000kN load is forced for per cable, the 20~40% anchoring force is recommended, i.e., the working anchoring force is about 600~1200kN. Same to the other cables, the 60% anchoring force is locked for 2000kN, i.e., the working anchoring force is 1200kN, the 70% anchoring force is locked for 1000kN, i.e., the working anchoring force is about 700kN. The anchoring force would reach at the design value as the rockmass further deforming and adjusting.

CONCLUSION

The pre-stressed anchoring is applied to the Lianziya dangerous rockmass controlling in the Three Gorges of the Yangtze River. For the key part of the rockmass, also called Wuwanfang dangerous rockmass, the anchoring is divided into eight zones and about 200 cables are fixed of 1000kN, 2000kN and 3000kN.

Borehole diameter and fixing tool
The borehole diameter and fixing tool are recommended as follow:

Fixing tool	grade(kN)	diameter	plate size
OVM15-7	1000	115	200×200×50
OVM15-12	2000	150	270×270×50
OVM15-7	1000	115	320×320×50

Selection of inner fixing length
Several methods are compared for determining the length of inner fixing part:
   a. Empirical formula
   b. case history comparison
   c. In-situs test

The length is recommended as follows:

grade(kN)	diameter	wires	safety index	fixing length
1000	115	7	2.0~4.0	3~4
2000	150	12	2.0~4.0	5~6
3000	175	19	2.0~4.0	7~8

Proc. 30th Int'l. Geol. Congr., Vol. 23, pp. 59-63
Wang Sijing and P. Marinos (Eds)
© VSP 1997

Reinforcement to the Foundation of the Plant Building of FUSHUN 1st Petroleum Plant

MA ZHAORONG, GAO GANGRONG

Research Institute of Mine Construction, Central Coal Mining Research Institute, Beijing, China

Abstract

The plant building of the Xiqiguo Workshop of FUSHUN 1st petroleum plant is located in the north side of the West Surfase Coal Mine, where the geological tectonics and the hydrogeological condition are complicated. The ground deformed suddenly. The plant biulding had suffered damage of different extent; which affected safety production. under these circumstances the grouting technique was adopted to reinforce the foundaton of the plant biulding. According to the geological tectonics and the hydrogeological condition, we selected the round curtain grouting metheod to close the foundation, which can't only improve the hydrogeological condition but also make the grouting size only diffuse into the foundation under the plant biulding , thus cut down material for grouting. The selection is both effective and economical. The construction sequence and the grouting technology were laid down reasonably. The material for grouting and the technical parameters of grouting were selected meticulously. The standard of quality was carried out strictly in construction. Finally the engineering quality was upto the standard. We had fully reached the expected ends, archieved good results that velocity of the surface subsidence reduced 40%. and the velocity of horizontal movement decreased 70%.

Keywords: Foundation Reinforcement, Grouting, Jet Grouting

INTRODUCTION

The Xiqiguo Workshop (a power workshop that serves the whole Plant) of 1st Petroleum Plant is located in the northern part of the West Surface Coal Mine. It is 120~130m apart from the mine boundary, and is rather near from the fault F_{1A}.

The south-east corner of the workshop foundation is convergencing unevenly, the connecting wall body in southern framework and northern framework occurs drawing breakage in eastwest directions. Surface of the workshop begins to displace from November 18, 1985 to August 17, 1991. Maximum surface convergence value of southern building wall nearby the mine is 588.5 mm, maximum horizontal deformation (movement to the south) is 591 mm, which is equal to 5.38 mm/m. In addition, according to determination observation hole, foundation rock body close to the workshop is moving to east-south direction, with displacement value of up to 26~65 mm. On these grounds, the Xiqiguo Workshop is placed on the first comprehensive control list for reinforcement and protection. Central Coal Mining Research Institute undertakes the overall design for the comprehensive control to northern slope of West Surface Coal Mine and surface deformation of No. 1 Petroleum Plant. A combination of grouting and concrete curtain is decided to take as the reinforced scheme.

ENGINEERING GEOLOGY AND GEOHYDROLOGY OF THE WORKSHOP

Geological conditions
The Xiqiguo Workshop is located on stage I bench alleviated by Hun River that belongs to primary geomorphic unit. The stratum of Quaternary system is composed of two parts   loose hand packing and original stacking. Its total thickness is 12~17 m. The first part is mainly of shale fragments and constructing wastes, the second part is mainly of mild clay, medium-sized sandstone and circular gravel. Most of the workshop is on the hanger of Fault F_{1A}. Basement rock is of Archaean mixed granite and mixed rocks. Lithological characters of strata between Fault F_{1A} and southern side of Fault F_1 belongs to fragmental sedimentary rock of Mesozoic group.

Geohydrological conditions
The Quaternary System underground water belongs to phreatic aquifer in pores. Burying depth of the water level is 6 m approximately. Water flow direction is from north to south. The average value of permeability coefficient K is 48 m/d. The K value of circular gravel layer at the bottom is up to 628~1443 m/d. The basement rock of Mesozoic Group belongs to weak pressure aquifer of crack-pore type. There are developed fissures and weathered rocks. Most of the fissures are argillaceously packed and the K value is small. Rock suction quantity at the shallow part of 16~20 m is 0.55 l/min.m.m .

CONSTRUCTION OF CONCRETE CURTAIN WALL AROUND THE WORKSHOP

Concrete curtain wall
To avoid leakage of medium and fine-sized fillings in gravel layer under the workshop basement and diffusion of grouting liquid to the outside of workshop, high pressure oriented jet grouting with triple pipes is adopted to construct the concrete curtain wall.

Constructing procedures and process

Constructing procedures First, construct the concrete curtain by oriented jet grouting at the southern side of workshop. Simultaneously, two machines grout to the east and to the west respectively. The grouting is carried out continuously according to hole numbers. After the southern side of concrete curtain wall is finished, go on with the concrete curtain walls in other directions of the workshop.

Constructing process The process flowsheet goes as the following:

(1) Dig furrow. Before constructing the concrete curtain, dig a furrow of 1 m × (1.5~2) m (width × depth) by hand in order to store liquid overflown from holes when jet grouting.

(2) Drill pilot holes. Drill pilot holes with cylinder alloy bit(108 mm) and two SGZ-

IIIA engineering drilling machines. Stop drilling as far as the bit goes into basement rock 1 m deep. Flush away the rock powder in holes. When met across with heavy leakage and hole collapse, protect the wall with slurry.

(3) Insert the triple pipe.   When the G-2A-50 high pressure jet grouting drill takes its place and is levelled , insert the triple pipe with special injection head to the bottom of pilot hole. Before inserting the triple pipe, make a test first to check if jet pressure and pipeline system are in normal operation.

(4) Jet grouting. When the triple pipe is inserted to the set depth and grouting parameters are adjusted to the normal, lifting and jetting can be carried out. When making the adjustment, make sure that the jetting direction of nozzles and the centre line of drilling holes form an intersection angle of 15^0 , so that the connecting quality of concrete curtain walls can be guaranteed. To those drilling holes with heavy leakage of flushing liquid at the lower part of gravel layer, first make quantitative and static grouting(general injection quantity 0.6~5.0 tons), and then lift normally and jet at the oriented direction. In the course of jetting, examine strictly if each parameter is up to the design requirement so that concrete curtain walls are up to the homogeneity.

(5) Refilling. Grouting liquid extracts water in the course of coagulation. Tops of concrete curtain walls form indentations because of contraction. Grouting liquid must be refilled repeatedly by special persons. This process is an important link in construction and must be treated conscientiously.

(6) Machine movement.   Both the engineering drills and high pressure jetting drills are installed on fixed tracks. When the drills move from the finished hole to the next one, hoist shall move the drills slowly.

GROUTING REINFORCEMENT FOR FOUNDATION AROUND THE FAULT ZONE AT THE SOUTHERN SIDE OF WORKSHOP

Reinforcement for the foundation started after the sealing ring of concrete curtains formed. Consolidation and induration of concrete curtain walls can prevent injection liquid from spreading to the outside of workshop buildings, and the injecting direction can then be induced.

Reinforcement is done with double-liquid   cement and water glass. In total,   30 grouting holes are drilled. 748.75 tons of cement and 138.12 tons of water glass are injected into the holes.

Process flowsheet
For process flowsheet of grouting reinforcement,   please see attached diagrammatic drawing.

Drilling of grouting holes

Grouting holes are drilled by SGZ-IIIA drill. The opening aperture is 150 mm. Finished aperture is 91 mm. Collar pipe is seamless steel pipe with   127 mm and 6.8~7.1 m long. When grouting to stratum of Quaternary System is done, insert the grouting pipe liner with   108 mm and 15~16 m long into basement up to 1 m.

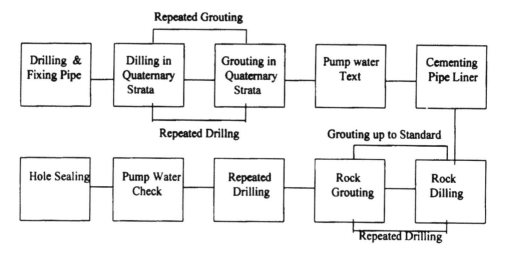

Fig.   Process Flowsheet of Grouting Technology

For specific heavily collapsed holes, hole walls are protected by slurry. Other holes are flushed with clean water.

Grouting machines & tools and grouting section

Grouting is made with YSB-120/250 hydraulic speed-regulating pump and 90/60 hydraulic grouting pump. Grouting transport pipelines are of high pressure rubber pipe and seam steel pipe.

Grouting sections of Quaternary system strata are divided according leakage quantity. When leakage is over 100 l/min., drilling should be stopped, and grouting should be done in 4~5 sections, and some up to 7 sections.

When grouting in basement rock section, it is generally divided into two sections for those holes with uneven fissures and lithological characters with big changes. The rest is grouted with one section.

Grouting materials and methods

Geological conditions have big difference. According to this characteristics, single cement grouting, cement-water-glass double grouting and single-double-single grouting are used alternatively.

For foundation reinforcement, cement with stable performance and long durability is taken as the grouting materials . The 425# Portland cement is used in the construction. The cement slurry composed of the Seagull cement has passed test on the spot.

In order to prevent slurry from leaking far away and from flowing into basement, quick-hardening liquid for cement-water-glass is used for some of the sections.

Standard for completion of grouting
Usually, there are three indexes for the completion of grouting , that is: final pressure, final grouting quantity and time of stableness.

When grouting for the Quaternary System strata, final pressure is 0.3~0.5 MPa, final grouting quantity is 38~50 l/min.. Time of stableness is no less than 20 minutes. Only when the rock suction quantity is up to 0.01 l/min.m.m approximately after pressurized water test, can the grouting be ended.

As to the grouting for basement rock, final pressure of prophase holes is 1.0~1.4 MPa, while that of anaphase holes is 0.5~0.7 MPa. Final grouting quantity is 35~45 l/min. approximately.

EVALUATION ON GROUTING QUANTITY

The gravel layer is the key point of reinforcing lower part of workshop foundation by grouting. Through multi-times of grouting, fissures are gradually filled, grouting pressure increases, and the standard for completion of grouting is fulfilled. By observing from the surface, slurry oozing points are all inside the concrete curtain walls. The horizontal distance between the farest slurry oozing point and the grouting holes is 30m approximately. By coring examination, cement slurry has cemented both the gravel and the loose layer , and consolidated one body together with the pipe liner. With strength similar to concrete, the grouted part has become retaining wall of the workshop foundation.

Oriented horizontal jetting through vertical drilling holes shall form a cement solid of 15~20 cm thick, of which the volume has exceeded design value. Opening check on the spot shows that the slurry liquid has been squeezed into cracks of 1 mm wide, good diffusion resulted. Finished walls are of good quality.

Two permeability tests and water-pressure tests show that rock suction quantity after reinforcement decreases by 90% compared with that before grouting. Permissibility coefficient of the concrete curtain wall is smaller than 1×10^{-5} cm/s. Through observation at Hole 2a , surface outside the reinforcement area deforms rather heavily. Speed of surface convergence within the reinforcement area decreases by 40% while the movement to the south decreases by 70%. The project has been up to the preset target.

Proc. 30th Int'l. Geol. Congr., Vol. 23, pp. 65-80
Wang Sijing and P. Marinos (Eds)
© VSP 1997

Offshore Geological Investigations for Port and Airport Development in Hong Kong

R.P. MARTIN[1], P.G.D. WHITESIDE[1], R. SHAW[1] & J.W.C. JAMES[2]

1 *Geotechnical Engineering Office, Civil Engineering Department, 101 Princess Margaret Road, Kowloon, Hong Kong*
2 *Coastal Geology Group, British Geological Survey, Nottingham, NG12 5GG, England, U.K.*

Abstract

Offshore geological investigations have played a vital role in Hong Kong's major new port and airport developments since the late 1980s. Knowledge of the Quaternary stratigraphy has been essential for identifying economically-dredgeable sand for use as reclamation fill and for designing offshore engineering structures. The initial exploration for sand was based on a simple geological model of Holocene marine muds overlying localised river sands in a variably-eroded surface of Pleistocene alluvial sediments. Investigations for the sand comprised extensive boomer seismic reflection profiling, correlated by means of continuously-sampled drillholes, vibrocores and cone penetration tests. Major economic sand deposits were found, where predicted, along Pleistocene drainage lines, with the volume of mud overburden generally thinnest where Holocene tidal currents were greatest. After detailed dredging assessments, c. 600Mm^3 of exploitable sand reserves were identified between 1987 and 1996, sufficient to meet Hong Kong's anticipated needs for marine fill up to the year 2015. The exploration programme has enabled the geological model to be refined and the stratigraphy expanded to include four Quaternary formations. Much of the geological information has been entered into computerised databases and compiled into maps and memoirs by the Hong Kong Geological Survey. This knowledge is being applied in the design and construction of reclamations, seawalls, breakwaters, submarine slopes and pipelines.

Keywords: offshore investigations, Quaternary sediments, natural resources, dredging, reclamation, Hong Kong

INTRODUCTION

Reclamation and offshore development in Hong Kong has increased dramatically in the past ten years, much of it related to the new airport at Chek Lap Kok, expanding port facilities and associated infrastructure. Offshore geological investigations have been a fundamental component of these developments. The two main objectives have been to find suitable sources of sand for reclamation fill and to provide the geological information needed for the design of reclamations and offshore structures. The investigations form part of a broader range of geotechnical work undertaken for port and airport development in Hong Kong [2].

RECLAMATION AND DEVELOPMENT IN HONG KONG

Because of the small area of the territory (about 1000 km²) and its mountainous terrain, Hong Kong has always been short of easily-developed flat land. Reclamation has been an integral part of Hong Kong's development for well over a century.

Fig. 1 shows a recent satellite photograph of Hong Kong. The most striking features are the large new reclamations which stand out as highly reflective tones. Clearly visible is the site of the new airport at Chek Lap Kok near Lantau Island, linked to the main urban area by strips of other newly-reclaimed land which carry the new road and railway routes. Other large new reclamations are in West Kowloon, for new container port development, and at Tseung Kwan O, for a new town residential and industrial purposes. In addition to these newly-reclaimed areas, much of the urban area of eastern Kowloon (including the present airport at Kai Tak) and the northern shores of Hong Kong Island are also on reclamation (Fig. 2). Further large new reclamations are planned over the next ten years, especially for additional container port facilities (Fig. 2).

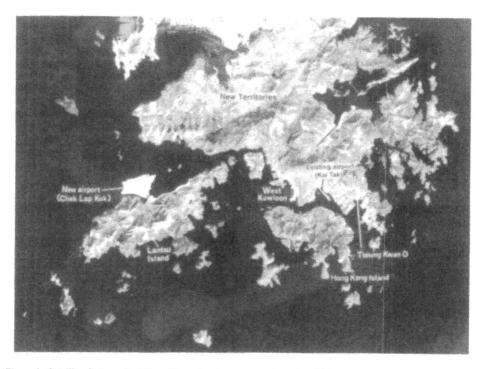

Figure 1. Satellite photograph of Hong Kong showing recent reclamations (light tones).

The total area of reclamation (1996) is about 6,000 hectares. This represents about 5% of Hong Kong's land area, but in economic terms the reclamation areas are much more important as they are the focus of most of the territory's major industrial and commercial facilities and house over 20% of Hong Kong's population.

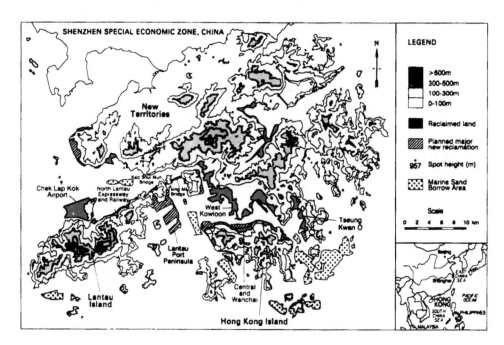

Figure 2. Topographic map of Hong Kong, showing reclamation, major coastal development projects and marine borrow areas identified for the Port and Airport Development Strategy.

Historically, the amount of reclamation has grown at an ever-increasing rate from the 1850s up to a peak in 1993 (Figs. 2 and 3). More than half has been formed in the last ten years, driven mainly by the Port and Airport Development Strategy (PADS) which formally commenced in 1989. The major PADS projects are shown in Fig. 2. Those currently under construction comprise the new airport, road and rail links to the urban areas, two major bridges between Lantau and Tsing Yi Islands, and the extensive new reclamations along West Kowloon and the north shore of Hong Kong Island. Civil engineering works for these projects were well advanced by 1996 and most will be completed by 1998. New port developments on Tsing Yi, North-east Lantau and near Tuen Mun will become the major PADS projects during the next ten years.

OFFSHORE QUATERNARY GEOLOGY AND INVESTIGATION TECHNIQUES

Systematic geological mapping of Hong Kong's offshore areas was started in 1982 by the Hong Kong Geological Survey (HKGS), part of the Geotechnical Engineering Office (GEO) of the Hong Kong Government. In response to Government's plans for new harbour reclamations and the feasibility studies for PADS projects, the GEO initiated a systematic exploration programme for offshore sand in 1987.

Initial Geological Model

As a starting point for the sand search, an initial geological model of the offshore Quaternary stratigraphy was conceived based on general knowledge of sea-level change since the late Pleistocene, experience from a number of dredging projects, and scattered published details of the local marine sediments [21, 4, 20]. The early model was schematic, comprising a sequence of Holocene fine-grained, predominantly marine sediments overlying mixed Pleistocene, predominantly alluvial sediments. The largest accessible sand deposits were expected to be on the late Pleistocene sub-aerial surface along old drainage channels in areas where Holocene scour channels had restricted or prevented the accumulation of Holocene marine mud. It was further expected that the Pleistocene drainage channels in the nearshore area would be closely related to the present catchment areas [1, 16].

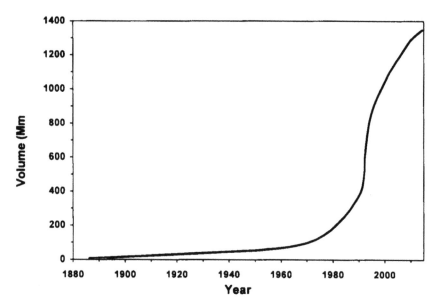

Figure 3. Cumulative volume of fill material used in reclamations in Hong Kong up to 1996 and predicted to be used up to 2015.

Investigation Techniques and Strategy

A phased programme of investigation was carried out, which provided appropriate information in the search for marine sand and allowed the geological model to be progressively refined. The main investigation techniques employed were shallow seismic reflection profiling using a high-resolution boomer source, drilling, vibrocoring and cone penetration testing (CPT)[18]. The boomer surveys achieved effective penetration up to 80 m below the seabed, sufficient to cover the full sequence of Quaternary sediments in most areas. Subsidiary techniques included gravity coring and side-scan sonar surveys.

Water depths around Hong Kong are generally less than 30m. Maximum depths are associated with the main Holocene tidal channels, particularly in the west towards the Pearl River estuary (Fig. 4). The seismic survey initially covered Hong Kong's western and central waters and was later extended into eastern and neighbouring Chinese waters to the south of

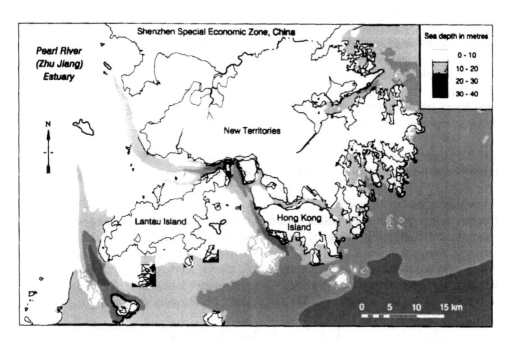

Figure 4. Bathymetry of Hong Kong and adjacent waters. The main Holocene tidal channels are clearly evident.

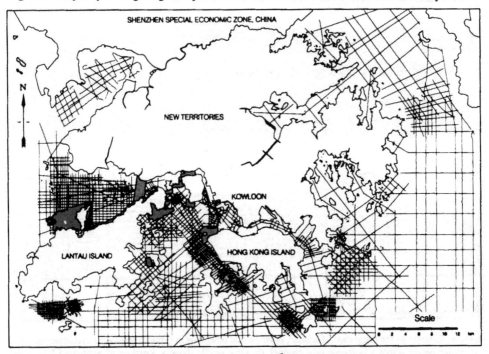

Figure 5. Location of seismic boomer profiles undertaken for offshore geological investigations in Hong Kong.

Hong Kong. The reconnaissance survey grid had an average line spacing of 3 km, with lines oriented broadly along and transverse to the main channels where sand was expected (Fig. 5). Once target marine sand bodies had been identified on the seismic records, more detailed second-stage investigations were carried out in selected potential borrow areas, often in two or three phases. Typically these investigations comprised seismic survey on a rectangular grid of spacing 250-500m, with drillholes, vibrocores and CPTs put down at grid intersections. The average density of the investigation holes or CPTs was about ten per 1 km². Ship board sampling and detailed logging was followed by laboratory testing of borehole samples, especially for particle size distribution.

Based on these investigations over the five-year period from 1987-1992, various marine sand borrow areas were identified and scheduled for detailed dredging assessments (Fig. 2).

Refinement of the Geological Model
Continuous integration of the results from the various investigation techniques was the key factor in confirming and refining the geological model [18]. An example of the integration process is shown in Fig. 6. Part (a) of Fig. 6 shows a typical example of a boomer profile. The high quality and detailed resolution of the seismic records allowed individual reflectors and groups of reflectors to be easily identified. With experience it became possible to recognise distinctive sedimentary sequences and to reliably infer the dominant sediment

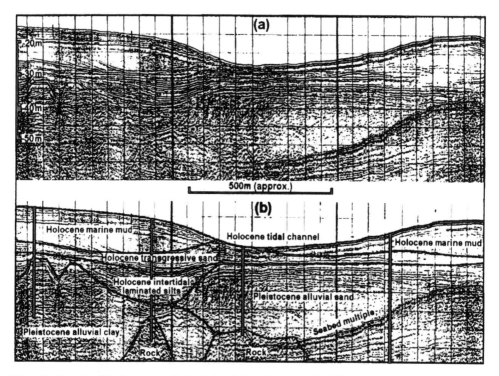

Figure 6. Example of the integration of seismic, borehole and other data in offshore geological investigations : (a) raw seismic profile across a Holocene tidal channel, (b) same profile showing locations of drillholes/CPTs, with interpretation of sedimentary units.

types. Seismic interpretations were then correlated with the drillhole logs, analysed samples and details of the insitu testing. Fig. 6b shows the same section with the major seismostratigraphic units interpreted and annotated.

Synthesis of all the offshore investigation data gathered over the last ten years has led to considerable development and refinement of the geological model. The original model has been expanded. Three formations have now been mapped, analysed and broadly dated, while a fourth has recently been mapped in south-eastern Hong Kong waters. The named formations, their ages, depositional environments and principal sediment types are summarised in Table 1. The basal Chek Lap Kok formation of Pleistocene age overlies weathered bedrock and is widespread in Hong Kong waters. The Sham Wat Formation is of limited extent and occurs mostly in western and southern waters, while the Waglan Formation is restricted to south-eastern waters. The Holocene marine muds of the Hang Hau Formation are ubiquitous except along coastal fringes and parts of the main tidal scour channels. Table 1 is a further development of the stratigraphy devised by the HKGS [7, 10, 8]. This model, which is based on a recent integration of all the seismic and other data, differs from an alternative proposed stratigraphy comprising ten subdivisions [22].

Table 1. Offshore Quaternary stratigraphy of Hong Kong.

Formation	Depositional Environments and Sediments	Age
HANG HAU	**Transgressive, marginal marine**: basal marine sand succeeded by silt and sand deposited in shallow, brackish conditions, overlain by soft clayey silt deposited in shallow, open marine conditions as at present	Holocene
WAGLAN	**Shallow, proximal marine**: transgressive shoreface sand deposited as a beach or as offshore barrier bars, overlain by silt and clay deposited in shallow conditions close to a coast	Late Pleistocene
SHAM WAT	**Marginal marine**: clay and silt deposited in an intertidal setting with alternating brackish and marine conditions causing the filling and overtopping of a complex channel network	Late Pleistocene
CHEK LAP KOK	**Fluvial to estuarine**: sand, silt and clay deposited in sedimentary environments fluctuating between fluvial channel and floodplain, with intertidal and possibly shallow marine deposits in places	Pleistocene

Five main types of sand deposits have been recognised within this stratigraphy [5, 15] viz : Pleistocene river channel sands, basal Holocene transgressive sands, localised marine channel sands, seabed residual sand sheets and nearshore coastal sands. The first type is present in large volumes and has been the most important source of sand, especially where present-day tidal channels coincide with the former Pleistocene drainage lines, resulting in a minimum cover of Holocene marine mud (e.g. Fig. 6).

The following two sections describe how this knowledge of the offshore sediments has been applied in two important areas of engineering development.

OFFSHORE FILL RESOURCES AND MUD DISPOSAL

Overall Fill and Mud Disposal Requirements

At the start of the 1990s the challenge for the PADS and related projects was to identify about 600 Mm^3 of bulk fill to meet requirements over the next ten years [3]. In 1990 it was anticipated that about half of this requirement would be provided from marine sources. Due to the declining suitability of land sources (mainly on environmental grounds), by 1994 the marine fill component required for the decade 1990-2000 had increased to about 400Mm^3. It is forecast that a further 200Mm^3 of marine fill will be required for the period 2000-2015.

From the outset it was foreseen that, to achieve efficient extraction at production rates suitable to meet the very tight PADS project programmes, the majority of the mud overburden and sand dredging would have to be carried out by large trailing suction hopper dredgers (Fig. 7). Depending on the geological conditions at the borrow area and the distance between the borrow area and the reclamation area, these dredgers can remove and place up to 250,000m^3 of material per week.

Figure 7. Trailing suction hopper dredger discharging sand into a reclamation in Hong Kong Harbour.

Over and above the marine fill required, in 1990 it was estimated that about 400Mm^3 of marine mud overburden would need to be dredged and disposed of elsewhere during the following 20 years. The main sources are reclamation sites (removal of mud prior to sand filling), marine borrow areas (removal of overburden prior to sand dredging) and shipping fairways (predominantly for deepening). Up to 1996 about 10Mm^3 of this mud has been classified as contaminated beyond levels acceptable for open sea disposal and has required special disposal in dredged seabed pits that are capped and isolated from the environment with clean mud [19]. The overall volumes of fill required and of mud to be disposed are shown in Fig. 8. Apart from marine fill, overall fill supply for the port and airport projects includes land-sourced bulk fill and special fill requirements [11].

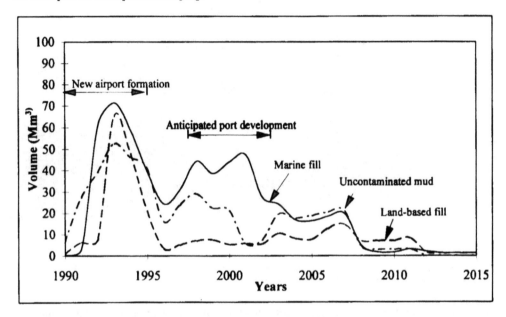

Figure 8. Fill demand and mud disposal requirements in Hong Kong, 1990-2015.

Dredging Assessments
To provide the technical basis for the huge offshore earthworks, detailed dredging assessments were required. These involved consideration of a number of factors including engineering, geological, environmental, marine traffic and submarine utilities. Ideal dredging conditions for large suction dredgers can be summarised as follows : where (i) water depths are from 10-40m (although exceptionally dredging up to 55m has been carried out), (ii) the stiffness/density of the dredged material is not greater than firm for mud and medium dense for sands, and (iii) borrow areas are close to the reclamation areas and designated mud disposal areas. In addition, for sand dredging, other desirable conditions are where the overburden ratio is less than one, there are no interbedded fine sediments, and the sand is not too coarse or angular. It is also beneficial if borrow areas are large enough for extraction to be phased in a series of individual pits, allowing the option of simultaneous sand borrowing and mud backfilling.

Many of the above dredging conditions are illustrated in Fig. 9, which shows a section through one of the marine borrow areas used to provide fill for the new airport. In this case the thicker central zone of marine mud interburden was used as a convenient line to subdivide the borrow area into two pits for separate excavations. Both pits still contain

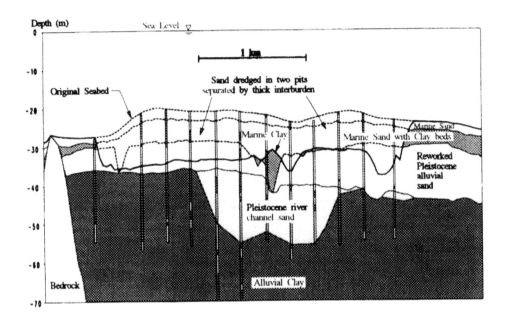

Figure 9. Cross-section through a marine sand borrow pit showing the location of drillholes/CPTs used for the dredging assessments and nature of the sedimentary units.

appreciable quantities of basal and reworked alluvial sands. On completion of borrowing, it is planned to reinstate the original seabed by using the worked-out pits as a disposal site for dredged uncontaminated mud.

Resource Management

In 1989 the Hong Kong Government established a separate policy committee, the Fill Management Committee (FMC), with the responsibility for management of the offshore sand resources and marine disposal of mud [17]. Creation of the FMC enabled a holistic approach to be adopted, as this one forum included representatives from both the project proponents and the various regulatory authorities responsible for environmental, navigational, planning and other matters. This enabled rapid decision-making on resource allocation and adherence to fast-track programmes.

A computerised GIS-based information management system was established to hold all the marine investigation data and allow efficient calculation of potential sand reserves [14]. Two-dimensional seismic survey data and one-dimensional drillhole and CPT data were digitised and integrated to provide a 3-D stratigraphic pattern of the different deposits.

Polygons were then constructed around each digitised fix point and volumes calculated by multiplying the thickness of the unit by the area of the polygon. Volumes of particular units within potential borrow areas are then easily computed by summing individual polygon values for each unit [17].

By mid-1992, five years after commencement of the offshore sand search, about 360 Mm^3 of readily-dredgeable sand reserves in nearshore waters had been fully investigated and detailed dredging assessments completed. From 1992 to 1996, similar investigations in waters further offshore to the south and east of Hong Kong have identified an additional c.250Mm^3 of dredgeable sand, bringing the total reserves above the amount required to meet the anticipated needs for marine fill in Hong Kong through to 2015. Environmental studies have been an important feature of all the borrow area investigations, in particular the assessment of potentially fragile marine species and their habitats [6].

As an indication of costs, the total amount spent on ground investigations, dredging assessments and environmental studies in the exploration phase was of the order of US$30 million. The net value of the nearshore sand reserves, based on average tendered rates for dredging and supplying fill to reclamation sites, is about US$1.5 billion at 1996 prices. Unit investigation costs were about US$6000/hectare in prospective borrow areas, or about US cents 4/m^3 of sand reserve identified.

DESIGN OF RECLAMATIONS AND OFFSHORE STRUCTURES

The huge scale of the PADS projects has led to an unprecedented expansion of marine civil engineering in Hong Kong in the 1990s. The main engineering applications of the offshore investigations are listed in Table 2 [12]. For each of these applications, characterisation of the sediments and classification into units of broadly similar engineering behaviour is necessary to ensure both efficient and robust designs and the selection of appropriate construction methods.

Reclamations

The majority of offshore reclamation sites in Hong Kong are underlain at the sea bed by soft, compressible Holocene marine muds. A crucial decision in reclamation design is whether to dredge some or all of these muds before placement of the fill. Traditional practice in Hong Kong up to the late 1980s was to reclaim over the marine muds without extensive dredging, in association with ground treatment to accelerate consolidation of the mud, usually by vertical drains. However, the tight construction programmes of some of the PADS projects have dictated dredging of the marine mud prior to sand filling, e.g. at the new airport site, where about 70Mm^3 of mud was dredged before placement of the reclamation fill and was disposed of at approved open-sea sites.

As an example of the approach used at the investigation and preliminary design stages of a major project, Figs. 10 and 11 summarise the offshore investigations used at the site of two proposed new container terminals as part of the Lantau Port Peninsula off north-east Lantau Island (Fig. 2). The area of reclamation, to be formed in two stages over a period of three to six years, is some

350 hectares. The seismic survey grid was run at a spacing of about 50m, while the spacing between drillholes was generally less than 200m. Information at this density was clearly necessary for assessing appropriate dredging levels for a dredged reclamation. In areas where seismic profiles were unobscured by gas blanking, it was determined that the thickness of the marine muds varies from <10m to >30m over the site, related to a pattern of late Pleistocene drainage lines whose courses were interpreted from a review of all the investigation data (Fig. 12)[9]. With a coarser grid, or less detailed interpretation of the palaeogeography of the site, the geometry of the marine muds would not have been established with sufficient precision for reliable reclamation design. In this case, to allow maximum flexibility and scope for detailed comparison of options, full designs for both dredged (removal of marine mud) and drained (retention of mud) reclamations have been prepared. Further drillholes and CPTs were carried out to resolve the uncertanties in the gas-blanked areas.

Table 2. Engineering applications of offshore geological investigations in Hong Kong.

Engineering Structure/Topic	Important Geological Features for Design and Construction
Reclamations	Disposition, nature and thicknesses of compressible sediments Dredgeability of compressible sediments Location of borrow areas and nature of reclamation fill Location of disposal areas for dredged mud
Seawalls and Breakwaters	Thickness and dredgeability of marine muds Density/consistency of alluvial sediments in foundation trench Location of borrow areas/quarries and nature of rock fill (pell mell, filter material, underlayer and armour rock) Location of disposal areas for dredged mud
Submarine Slope Stability	Nature and thicknesses of sediments above lowest excavation levels Location of disposal areas for excavated/dredged sediments Nature of dredged mud placed in spoil mounds
Ship Anchorages	Nature (especially consistency) of marine muds surrounding mooring sinkers or anchors
Submarine Utilities	Nature and dredgeability of marine muds along pipeline/cable trenches Location of disposal areas for dredged mud

Irrespective of the selected design option, knowledge of the geometry of the deeper sediments at the site has also been of great value in assessing the likely extent and magnitude of post-reclamation settlement that will be caused by consolidation of the lower silt and clay horizons within the Pleistocene sediments.

Seawalls and Breakwaters

In contrast to the main reclamation areas, traditional design practice for edge structures such as seawalls and breakwaters has been to remove all the soft marine muds to form a seabed trench, which is filled with granular material and rock to form the foundation of the structure. Fig. 13

shows a typical cross-section through a seawall in Hong Kong, together with a profile of the Quaternary sediments. As with reclamation design, it is not difficult to appreciate the

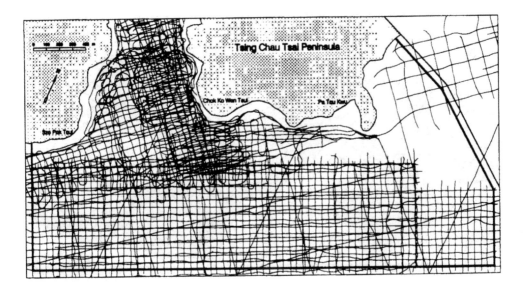

Figure 10. Seismic survey lines used in the investigation for Container Terminals 10 and 11, northeast Lantau.

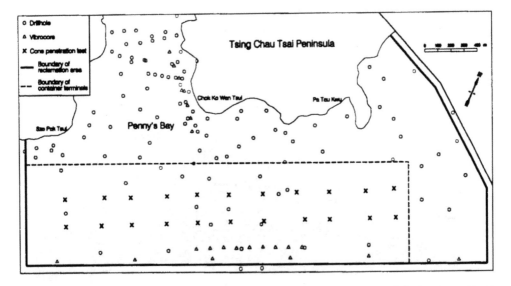

Figure 11. Locations of drillholes, vibrocores and CPTs used for interpretation of the seismic profiles and in the preliminary engineering design of Container Terminals 10 and 11.

importance of having adequate knowledge of the local geology in order to produce a safe and efficient design. Geological factors also have a bearing on the selection of various types of special rock fill required to construct seawalls and breakwaters (Table 2).

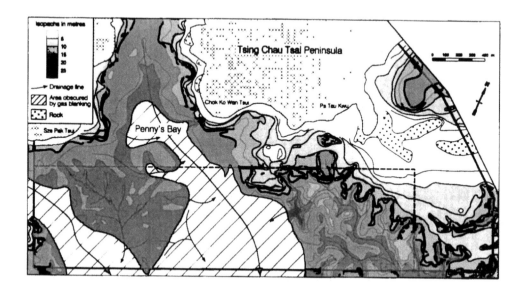

Figure 12. Map of the proposed Container Terminals 10 and 11 showing isopachs of Holocene marine mud derived from seismic and borehole interpretation. The courses of pre-Holocene drainage lines are clearly indicated. Areas of gas blanking coincide with organic-rich sediments along the main drainage lines.

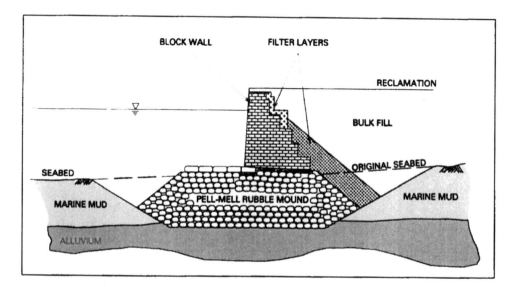

Figure 13. Cross-section showing the typical form of a seawall in Hong Kong.

Other Applications

Offshore geological investigations in Hong Kong are also important for the assessment of submarine slope stability [13], and for the design and construction of ship anchorages [12] and seabed utilities (Table 2). In addition, geochemical analyses of offshore sediments are being used to establish classification criteria for disposal of dredged material.

CONCLUSIONS

Offshore geological investigations have played a crucial role in the development of Hong Kong's new airport, port and other reclamation projects. The successful exploration for offshore sand resources, together with an improved understanding of the Quaternary stratigraphy, have been critical factors in engineering designs and in the timely and economic completion of the new infrastructure facilities.

Acknowledgements

This paper is published by permission of the Director of Civil Engineering of the Hong Kong Government. Mr James publishes with the permission of the Director of the British Geological Survey.

REFERENCES

1. R.S. Arthurton. Studies of Quaternary geology and the exploration of offshore sources of fill in Hong Kong. *Proceedings of the Symposium on the Role of Geology in Urban Development in Southeast Asia (Landplan III)*. P.G.D. Whiteside (Ed.). pp 229-238. Geological Society of Hong Kong, Bulletin 3 (1987).
2. E.W. Brand. Role of geotechnical engineering in Hong Kong's port and airport development. *Proceedings of the Twelfth Southeast Asian Geotechnical Conference*, Kuala Lumpur (1996). (In press).
3. E.W. Brand and P.G.D. Whiteside. Hong Kong's fill resources for the 1990s. *Proceedings of the Seminar 'The Hong Kong Quarrying Industry, 1990-2000: A Decade of Change'*, Hong Kong, 101-112 (1990).
4. C. Dutton. Geological investigations for new town planning at Tin Shui Wai, New Territories, Hong Kong. *Proceedings of the Conference on Geological Aspects of Site Investigation*. I. McFeat-Smith (Ed.). pp 189-202. Geological Society of Hong Kong, Bulletin 2 (1985).
5. C.D.R. Evans. Seismostratigraphy of early Holocene sand banks. In: *Marine Sand and Gravel Resources of Hong Kong*. P.G.D. Whiteside & N. Wragge-Morley (Eds.). pp 45-52. Geological Society of Hong Kong (1988).
6. N.C. Evans. Effects of dredging and dumping on the marine environment of Hong Kong. *Terra et Aqua*, 57, 15-25 (1994).
7. J.A. Fyfe. Towards a Quaternary stratigraphy for Hong Kong. *Geological Society of Hong Kong Newsletter*, 10:2, 5-10. (Discussion, 10:3, 19-23) (1992).
8. J.A. Fyfe and R. Shaw. *The Offshore Geology of Hong Kong*. Hong Kong Geological Survey, Geotechnical Engineering Office, Hong Kong (1997). (In press).
9. J.W.C. James. An interpretation of the marine geology of Penny's Bay and the site of Container Terminals 10 and 11, Hong Kong. *British Geological Survey Technical Report WB/94/35R*. Report to the Civil Engineering Department, Hong Kong (1994).
10. R.L. Langford, J.W.C. James, R. Shaw, S.D.G. Campbell, P.A. Kirk and R.J. Sewell. *Geology of the Lantau District*, Hong Kong Geological Survey Memoir No. 6, Geotechnical Engineering Office, Hong Kong, 173p (1995).

11. A.W. Malone and D.E. Oakervee. Providing aggregates for a major development : the Hong Kong container port and airport projects. *Quarry Management*, 20:11, 11-17 (1993).
12. J. Premchitt and R. Shaw. Marine geotechnical engineering for development projects in Hong Kong. *Proceedings of the International Workshop on Technology for Hong Kong's Infrastructure Development (Infrastructure '91)*, Hong Kong, 721-738 (1991).
13. J. Premchitt and N.C. Evans. Geotechnical aspects of the port developments in Hong Kong. *Proceedings of the International Conference on Port Development for the Next Millennium (Ports 2000)*, Hong Kong, 369-378 (1993).
14. J.R. Selwood and P.G.D. Whiteside. The use of GIS for resource management in Hong Kong. *Proceedings of the Eighth Conference on Computing in Civil Engineering*, American Society of Civil Engineers, 942-949 (1992).
15. R. Shaw. The nature and occurrence of sand deposits in Hong Kong waters. In: *Marine Sand and Gravel Resources of Hong Kong*. P.G.D. Whiteside & N. Wragge-Morley (Eds.). pp 33-43. Geological Society of Hong Kong (1988).
16. P.G.D. Whiteside. Preliminary indications from the Hong Kong Government's offshore sand search. In: *Marine Sand and Gravel Resources of Hong Kong*. P.G.D. Whiteside & N. Wragge-Morley (Eds.). pp 161-178. Geological Society of Hong Kong (1988).
17. P.G.D. Whiteside. Management of Hong Kong's marine fill resources. *Proceedings of the Seminar on Reclamation - Important Current Issues*, Hong Kong Institution of Engineers, 33-47 (1991).
18. P.G.D. Whiteside and J.B. Massey. Strategy for exploration of Hong Kong's offshore sand resources. *Proceedings of the International Conference on the Pearl River Estuary in the Surrounding Area of Macao*, Macau, 1, 273-281 (1992).
19. P.G.D. Whiteside, K.C.S. Ng and W.P. Lee. Management of contaminated mud in Hong Kong. *Terra et Aqua*, 65, 10-17 (1996).
20. N. Wragge-Morley. Dredging for Container Terminal 6 : a case history. In: *Marine Sand and Gravel Resources of Hong Kong*. P.G.D. Whiteside & N. Wragge-Morley (Eds). pp 121-129. Geological Society of Hong Kong (1988).
21. W.W.S. Yim. A sedimentological study of the sea-floor sediments exposed during excavation of the East Dam Site, High Island, Sai Kung. *Proceedings of the Meeting on Geology of Surficial Deposits in Hong Kong*. W.W.S. Yim (Ed.). pp 131-142. Geological Society of Hong Kong, Bulletin 1 (1984).
22. W.W.S. Yim. Offshore Quaternary sediments and their engineering significance in Hong Kong. *Engineering Geology*, 37, 31-50 (1994).

Proc. 30th Int'l. Geol. Congr., Vol. 23, pp. 81-87
Wang Sijing and P. Marinos (Eds)
© VSP 1997

The Deformation Monitoring of High Buildings in Xi'an City

JIN QIKUN, WANG HONGLONG, & NIE YANJUN
Xi'an College of Geology, Xi'an Institute of Mining, 710054 P.R.China

Abstract

The paper introduces the ground fissure and subsidence and their influence on the high buildings in Xi'an city. The authors have observed and researched on the deformation of high buildings in Xi'an for many years. They summarized the regularity of change of ground subsidence and analyzed the reasons of deformation of high buildings. And the technical principles and data processing methods for surveying the deformation of high buildings are presented. In addition, the protection and remedial measures for the buildings are also suggested.

Keywords: ground fissure, subsidence monitor, deformation surveying, accuracy evaluating

INTRODUCTION

Xi'an city is located in the geographic center in China. It is the northwestern China's political, economical and cultural center. In recent years, the high buildings with about a hundred meters are towering aloft in Xi'an. Geologically, Xi'an lies on second and third terraces of Weihe River, where loess has a quality of the subsidence with wet. And geological conditions are complicated such as the fall earth consists of varied materials, the layer shrinkage of earth is not equivalent and the base of the buildings is not very stable, etc. The earth fissures to damage the buildings in Xi'an have occurred since 1950's. The hazard is increased with time flying. We have been carrying on the research on the deformation monitoring of high buildings in Xi'an for many years. Now the results are given in the followings:

IMPORTANCE OF THE SUBSIDED MONITOR

Since the scientists found the first ground subsidence in Xi'an in 1959, the subsidence has been developing very fast. Up to 1971, the rate of subsidence was increased by a fact of three. Until 1993 the areas of subsidence have been 363 km², in which the region of 38 km² has the subsidence of 1000mm, while the southern part of Xi'an city has maximum subsidence about 1400-2100mm (see Table 1. & Table 2.). In 1993 there were 11 fissures in Xi'an city. They are located in the southern, eastern and north eastern areas of the city and the longest one is up to 10 km. The total length of the fissures is about 40 km.

Buildings were damaged by fissure and subsidence

It is more serious that the fissure and subsidence have destroyed some of the buildings and pipelines. For example, in a spot of the North-western University there are some fissures with a max. width of 8 cm. According to the incomplete statistics in 1991, the destroyed area of buildings is a number of 200,000 m^2. It made a directly economic loss about RMB 27 million Yuan. The destroying to the pipelines under the ground has made the economic loss more than a million Yuan. The constructions of high buildings are arising in Xi'an and the deformation monitoring for the buildings is very necessary, in order to clarify the deformation regularity and to analyze the affection of the deformation and to ensure safety of the constructions.

Table 1. The subsidence of ground in Xi'an city (1981~1989)

unit: mm / a

	Year	04. 1981 ~ 04.1982	04. 1982 ~ 04.1983	04. 1983 ~ 04.1984	04. 1984 ~ 04.1985	04. 1985 ~ 09.1988	09. 1988 ~ 09.1989
Location							
	Hujia Temple	100	100	99	113	110	90
old	Xiaozai Area	135	123	102	94	119	125
subsidence	Northwestern						
center	Polytech. University	96	98	93	97	86	63
	Dayan Pagoda	132	103	105	82	109	114
new subsi-	Western suburbs	37	45	54	57	82	108~182
dence area	Southern suburbs	48	58	55	76		167~187

Table 2. Accumulated subsidence(~1993)

unit: mm

Area	W. & N. part	Old city area	Bell Tower area	E. & SE. part	S. part
subsidence	50~200	200~800	639	1000~1900	1400~2100

THE METHODS OF MONITOR AND SAMPLES

The design of the datum network and establishing basic points
In order to monitor the subsidence of buildings we must establish 3~4 bench-marks as the datum of the leveling that located out of the subsided area (Fig. 1). The datum networks ought to be stable and able to be used consecutively. According to the information of investigations, the strata below the ground 6~10m are stable. So, we should bury the datum points deep about 10 m in the hard soil strata.

Establishing the observation points of subsidence
The general principles for establishing the observation points are as follows:
1. Setting up the points on the corners of the buildings or putting them on the basic pillars. (Fig. 2)
2. Setting up the points on the both sides of buildings which could connect the high building with shorter buildings, or the jointer of new and old buildings.
3. Setting up the points on the two sides of subsidence where there are the differences between buried base and natural base as well as artificial base of buildings.

4. Setting up the points on some parts of the buildings, where are often affected by the vibrating. Based on the above principles and considerations, and combined with the conditions of the buildings, we needn't set up too many points on the buildings so that we can reduce the amount of works.

Accuracy of monitoring
The accuracy of the deformation monitoring is depended on the purpose of the monitoring, the structures of the buildings and the basic types of the buildings. General principle adopted is the highest accuracy as we do as possible. For example, Minsheng Department Store Building was constructed in the first-stage, which is 41.10m in height, 36.68m in long and 24.28m in width. Base on the national first order leveling index, We estimate the accuracy of the observation points by the following formula:

$$m_H = \Delta * k / t \qquad\qquad (\text{ I.a })$$

where m_H is the root-mean-square error of height of a point; k is selected safety-factor; Δ is given allowable difference of subsidence; t is the rate of maximum error to mean-square error. The allowable base tilt for the Minsheng Building is 0.0015, then $\Delta = 0.0015 * 24.28 = 36$ mm and selecting k as 0.05 and t as 2, so

$$m_H = 36 * 0.05 / 2 = 0.89 \text{ mm}$$

That is to say, the monitoring of subsided points should be realized on the National Second Order Leveling Rule.

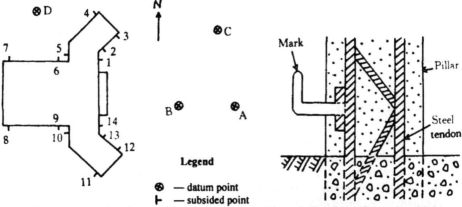

Figure 1. The datum network and subsided
points of the classroom building

Figure 2. The observation mark welded
on steel tendon

Instruments and methods of observations
In the surveying, we have used $S_{0.5}$ and S_1 types of geodetic level with two Invar staves. Every observation is in same conditions to satisfy the requirements of Code.

Accuracy evaluating of the lowest accurate point

The elevation difference of subsided points are measured on the basis of technical demand of Second Order Leveling. The root-mean-square error of a leveling station could be estimated as follows:

$$m_S = \sqrt{m_L^2 + m_A^2 + m_R^2} = \sqrt{(0.1\tau \cdot \frac{S}{\rho})^2 + (\frac{15}{v} \cdot \frac{S}{\rho})^2 + (0.1)^2} \qquad (\text{I.b})$$

where m_S is a root-mean-square error of a station,
τ is the value of level tube with 10 seconds,
S is sight length with 50 meters,
v is the magnification of the telescope with 40,
ρ is as 206265″, and
m_L is an error of leveling bubble,
m_A is an error of aiming,
m_R is an error of reading.
Then to calculate the m_S and to get the result:

$$m_S = \pm\ 0.277\ \text{mm} \qquad (\text{I.c})$$

In closed leveling-course the lowest accurate point lies on the center of the course. It was estimated that course of subsided leveling brings about 7 stations from the datum point to the lowest accurate point. So the root-mean-square error of the lowest accurate point is:

$$m_{low} = \pm\ m_S \ast \frac{\sqrt{n}}{\sqrt{2}} = \pm\ 0.277 \ast \frac{\sqrt{7}}{\sqrt{2}} = \pm\ 0.52\ \text{mm} \qquad (\text{I.d})$$

If we consider the bad condition, thus

$$m_{low} = \pm\ m_S \ast \sqrt{n} = \pm\ 0.277 \ast \sqrt{7} = \pm\ 0.73\ \text{mm} \qquad (\text{I.e})$$

where n is the number of leveling stations.
It is obviously that our selected instruments and surveying methods are satisfied to the accuracy of deformation monitoring, because the demand of the accuracy of monitoring is 0.89mm.

PRACTICAL PROJECTS

We have been monitoring several constructing buildings in Xi'an city, they are:
(1)Xi'an Minsheng Department Store Building;
(2)The third classroom building of Xi'an College of Geology;
(3)The Radio Communication Building;
(4)The stadium building of Shaanxi Opera House;
(5)The building of Shaanxi Publishing House;
(6)The test building of Ministry of Weapon-Industry.

The deformation monitoring of the classroom building of Xi'an College of Geology
The third classroom building with 40 meters' height was built in Xi'an College of Geology in 1987. The main part of the building is 11 floors and its both flanks are 2 floors, in which all constructions are steel-concrete structure. Uneven subsidence of the building has happened obviously since 1990, while some gaps on the wall of the connection parts between the building and its southern flank part have occurred. Authors have been

monitoring the deformation of this building since 1988. In the more stable place beside the building 3~4 datum points had established and 14 observation points had set up on the building(Fig. 1). The marks of observation points are as Figure 2.

Stability analysis of datum points
The observation of datum network had passed for four times at least, the results are showing as Table 3.

Determining the weights of leveling courses is as follows:

$$P_{ik} = C / n_{ik} \qquad \text{(II.a)}$$

where P_{ik} is the weight of leveling course from No. i point to No. k point, C is a constant integral number, and n_{ik} is the numbers of leveling stations.
The elevations of the points were obtained from the adjustment with rank-defects of a free network, and the changes of the elevations of datum points were resulted in Table 4.

Table 3. Observation data of the datum points(m)

Elevation difference	Time 1	Time 2	Time 3	Time 4	Layout of points
Δh_1	-1.1454	-1.1445	-1.1428	-1.1428	
Δh_2	+1.1522	+1.1585	+1.1632	+1.1645	
Δh_3	-0.0068	-0.0140	-0.0204	-0.0217	

(Fig.3)

Table 4. The variations of the elevations of points(mm)

v_i	Period 1	Period 2	Period 3	Period 4	Limited error	Average v_i
v_B	0	-2.7	-5.4	-5.8	± 1.3	-3.48
v_D	0	-1.8	-2.8	-3.2	± 1.7	-1.96
v_C	0	+4.5	+8.2	+9.1	± 1.4	+5.44

Where i is as the number of leveling points and v_i is as the change of elevation of paints. The values of v_i are decided on both factors of leveling error and rise-fall movement of datum points. If the absolute value of v_i is obviously bigger than leveling error, the v_i shows that the point rises or falls actually, i.e. it is not very stable point. In the opposition, we could consider that the v_i is only a surveying error. According to above principle, in datum network the point D and B are more stable, but point C is not a stable point.

Regression analysis of subsidence points

A linear relationship between the subsidence of points and the construction periods of the building has been found through the surveying of subsidence points on the building and drawing the related graphs, so the linear regression equation was adopted:

$$y = a + bx \qquad\qquad\qquad\qquad (II.b)$$

where y is as a subsidence value, x is as a related time, a and b are regression factors. This regression model was effectually verified after statistical tests of observation points.

Prediction and analysis to the subsidence of the building
The calculated results with regression model are as Table 5 and the predictions have been given based on the data. In the table 5, y_1 is the data in January, y_2 is in May and y_3 is in October of 1995. The unit of subsidence is mm.

Uneven subsidence tendency of the building
The subsided differences between the points No. 7 and No. 8 are 4.48, 4.72 and 5.02, and that between No.1 and No. 14 are 3.27, 3.40 and 3.55. So the building has a tendency of slope towards the south-western direction.

Subsidence tendency of southern side of the building
The subsided differences between the points No. 9 and No. 10 are 3.96, 4.10 and 4.28, and that between No. 14 and No. 13 are 0.11, 0.14 and 0.17. The increasing of subsided changes has been predicted.

Subsidence tendency of northern side of the building
Uneven subsidence between the building and its flank parts will be increasing obviously based on the data of the points No.1, No.2 and No.6 and No.5.

ANALYSIS AND SUGGESTION

Cause and protection
The subsidence and the movement of ground fissure are a complicated geological appearance, which is depended on many factors such as geological structure and environment conditions, etc. The falling of ground water is one of major effecting factors. So, the key task is to control ground water in good condition. Every effort ought to be made to save the water and control to fetch the ground water.

The way to deal with construction
In the construction projects can adopt the methods to fix the buildings, such as the method of "adding resist-cut roof beam."

Another method is soaking water to the parts of the site of buildings and to compensate the subsidence of buildings, etc.
The above methods can be used to correct the deformity of the building.

Table 5. The subsidence prediction and analysis (mm)

Points	y_1		y_2		y_3		Tendency
	Subsidence	Change	Subsidence	Change	Subsidence	Change	
1	+1.38		+1.43		+1.49		
		-0.33		-0.37		-0.43	Increase slowly
2	+1.71		+1.80		+1.92		
6	-0.95		-1.00		-1.08		
		-1.80		-1.88		-1.99	Increasing
5	+0.85		+0.88		+0.91		
7	+1.42		+1.49		+1.58		
		+4.48		+4.72		+5.02	Increase
8	-3.06		-3.23		-3.44		obviously
9	-5.23		-5.43		-5.69		
		-3.96		-4.10		-4.28	Increase
10	-1.27		-1.33		-1.41		obviously
14	-1.89		-1.97		-2.06		
		-0.11		-0.14		-0.17	Develop slowly
13	-1.78		-1.83		-1.89		

Acknowledgments

The authors thank Prof. Wang, Sijing for his good suggestions about this research paper, as well as Prof. Wang, Wenying, Dr. Li, Wenping, Assoc. Prof. Ma, Zhimin and Engineer Ms. Wang, Xianzhen. Their help and concern are appreciated.

REFERENCES

1. China general color metals corporation,*Engineering Survey-Code*, Beijing, China Planning Publishing House, 1993.
2. Chen Longfei, Jin Qikun, *Engineering Survey*, Shanghai,Tongji University Publishing House, 1990.
3. Jin Qikun et al, Settlement observation and analysis for high-rise buildings, *Journal of Northwestern Institute of Architectural Engineering*, No. 3, 1995.
4. Wang Honglong, The data processing and statistical analysis of land subsidence observation, *Northwest Water Power*, No. 3, 1996.

Proc. 30th Int'l. Geol. Congr., Vol. 23, pp. 89-94
Wang Sijing and P. Marinos (Eds)
© VSP 1997

Synthetic Site Investigation of Yangtze River Bridge of the Beijing Shanghai Express Railway

ZHU QUANBAO & XIA ZHIAI
Reconnaissance and Design Institute, Major Bridge Bureau, the Ministry of   Railway China

ABSTRACT

The Beijing_Shanghai express railway which will be built in the near future is an important railway.During the site investigation of Nanjing Yangtze River Bridge of this railway ,the Hydro-Acoustic Exploration method [HAEM] was mainly employed.

The HAEM which bases on the reflection principle of acoustic wave can explore the underwater geology and tectonics. The record profile directly reveals the undulation of the bed rock top.

The land shallow seismic reflection method which makes use of the CSP processing program can directly process the data on site. The profiles of the two methods compose of the complete bed rock cross section of the bridge site.
Based on the bridge site investigation example, the geophysical exploration , drilling method and their relationships are analyzed in this paper.

Keywords: Site Investigation   Yangtze River Bridge   Express Railway

INTRODUCTION

Beijing Shanghai express Railway (BSER), which is to be built in the near future, will became one of the important railway main lines and will also become the first express railway of our country (P.R.CHINA). In the first stage site investigation of BSER ,the Hydro-Acoustic Exploration method (HAEM) and the land shallow seismic reflected wave method (LSSRWM) were mainly employed and good engineering geology results were achieved .

HAEM is an advanced geophysical prospecting (GPP) method basing on the acoustic wave reflection principle to explore the underwater geology and tectonic. The bed rock top undulation can be directly displayed in the profile records of this method from the continuous navigating survey which is its advantage over any other prospecting method . LSSRWM has found wide use in recent years . CPS shallow seismic reflection (SSR) process software of our country is employed to process the SSR data on site. A rather complete bridge site bed rock cross section was acquired from the HAEM and LSSRWM data.

The synthetic site investigation of this bridge has riched the prospecting data , reduced the prospecting cost and shortened the prospecting cycle , and this is what we should pursue in the future.

THE BRIDGE SITE TOPOGRAPHY, LANDFORMS AND ENGINEERING GEOLOGY PROBLEMS .

The bridge site is situated in Nangjing of the lower Yangtze River. there are scheduled two bridge sites : the upper one and the lower one , the river width of the former is about 1.8km; the river width of latter approaches 2km. The distance between the two sites is more than 20 km. The water flows from SW to NE in this part of Yangtze River. There are plenty of shoals and the bank beaches are wide , the beat traffic is heavy . The topography of both banks is the middle and lower Yangtze River alluvial plain plus lower mountains and hills , the plain is the first-class Yangtze River terrace, zonally distributes along the river with even topography of an elevation of 5~7m, the lower mountains and hills have an elevation of 200~400m .

The bridge site tectonic belongs to the lower Yangtze tableland fold zone of Yangtze quasi platform. The crustal movements have frequented the site area since Mesozoic , caused this area fold and fault plasia . The bedrock is the red mud-sandrock of the group Pukou, the paleo Tertiary period. The folding and faulting leads to many tattered zone of faults , most of which are tension twist faults along the rock stratum strikes and the twist faults which slantly intersects the rock stratum strikes. The upper overburden compose of sand clay middle, super fine_ fine sand and the lower middle sand, large sand , gravel sand and gravel soil. According to the 'Nation Earthquake Intensity Districts Division Diagram '(1990 edition),the earthquake intensity scales class   . 'Railway bridge and culvert design regulations ' demands the bridge base should not be established on the unevenly soft and hard ground. The pier and abutment site must avoid the harmful foundations of faults, landslides, squeezed broken zones and etc.. Therefore , it's the main engineering geology problems to ascertain the undulation of the bed rock and the place and the distribution of the broken zone of fault in the site investigation .

THE SYNTHETIC SITE INVESTIGATION ORDER AND FIELD WORKING METHED .

Considering the aforesaid geology conditions and the engineering geology problems to be resolved , we employed the synthetic site investigation according to the following order: geology survey and geology drawing synthetic geophysical prospecting drilling exploring comprehensive geology analyses.

Geology Survey and Drawing(GSD)
In the engineering geology investigation the overall important fundamental work is GSD , whose exactancy and level will play a decisive role in the whole engineering geology work quality, especially the reasonable arrangement of the prospecting work. GSD

includes collecting the acquired data of district geology , engineering geology and earth quake geology, having a preliminary knowledge of the site geology and tectonic and advisable arranging various kinds of prospecting methods and their quantity . In this district , a great amount of site investigation work was performed in the late fifties and early sixties . So this has accumulated plenty of geology data, and has laid a good foundation for the arrangement and synthetic analyses of this bridge site investigation .

For example near the lower bridge site , it's known from the obtained riverbed bedrock geology diagram of the established bridge that there exist three fairly large broken zones of tension_ twist fault along the rock stratum strikes between pier 5 and pier 8 of the established bridge . It's impossible to prove the concrete extending region and width of the broken zones through drilling exploration only. Thus the comprehensive site investigation method of HAEM is mainly employed . The established bridge geology data serves as bases of the checking of the virtual bedrock reflection wave of the HAEM.

Combination of Synthetic GPP or various kinds of GPP .
Geophysical prospecting has the advantage of short prospecting period, low cost convenient equipment and large survey area. If there is the physical difference between strata of rocks or soils , a good geology result can be obtained . Geophysical prospecting is an un-direct survey method and its result must be checked by drilling exploration. Each kind of geophysical prospecting bases on physical properties different from each other and has advantage over each other , therefore , if different geophysical prospecting methods are practiced , they can help each other and check each other and finally we can get a exact knowledge of the rock stratum borders and the development of the faults.

HAEM
HAEM is an advanced GPP method basing on the underwater reflection of the acoustic wave to present the underwater geology and structures . Any rock or soil strata of obvious difference of wave resistance (DWR) can be differentiated in the recordings of HAEM, but rock or soil strata without obvious DWR is not easy to be distinguished . In the site investigation of the hydro_ engineering of bridges, tunnels, sharfts and dams, there exists an obvious DWR between the overburden and bedrock strata. The density of the strata of overburden also cause obvious DWR . All this sets up a stage for HAEM(Rock and soil strata without DWR can't form significant reflection and this reflects this respect of the bedrock). General Engineering Co-operation(GEC) of Ministry of Railway introduced a Meo-305 Serial Shallow Plotter of Telta company USA in 1992. And allocated it to our institute to undertake the prospecting work of the design institutes of GEC. This plotter was first used in the synthetic investigation of this bridge site in March , 93.

HAEM is a kind of continuos navigating survey . Discharge cable and hydrophone cable are drawn at the stern of the engineering ship the distance between the shot point and the detector (DSPD)is approximate 4m . The plane arrangement of survey profiles is like a network i.e. the distance of two profiles along or across the river is uniform . If the undulation of the bedrock top is great or the tectonic is abnormal, the uniform distance may be shortened . The survey positions are orientated by means of ahead-intersecting i.e. first datum lines are set on the both banks and then use two or three transits to measure intersectally , the intersection angles must be confined between 30~150 degrees to ascertain the orientation exactancy of survey points .

Before formal survey , adjust the survey instruments, carry out experiments of survey method and select reasonable working parameters to make the instruments work perfectly ; and adapt the survey method to the local geology conditions. Parameter-selecting mainly includes scanning time , trigger periods , band-pass filters and etc.. During formal survey the engineering ship must maintain a speed of 2m/s or so , always watch the survey results of the destination strata and regulate the parameters of the instruments .

LSSRWM

LSSRWM originates from its applications in the sources explorations of petroleum, natural gas coal and etc.. Recent year's improvement and wide use of computer technology has laid a foundation for its application in the engineering geology prospecting .LSSRWM can be compared with the shallow refracted seismic wave method (SRSWM) in the following aspects: (1) SRSWM demands that rock stratum wave speed increase as the depth increase. i.e. the speed of the lower stratum must be greater than that of the upper stratum , but physical property parameter of LSSRWM is the difference of wave resistance(DWR) between two rock strata , even if the DWR of the lower stratum is greater or less than that of the upper stratum , there always exists a significant reflection wave , thus, LSSRWM can be applied more widely . (2)SRSWM can only receive the refracted seismic wave outside the blind area , so demands a large working space, but LSSRWM has no such confinement. (3) Because of the blind area ,the detector of refracted wave must be placed quite far from the shotpoint, so a great enough amount of energy must be excited by a seismic focus of a large charge. But this is forbidden in the city proper , factory , mining area or near embankments . But the DSPD of LSSRWM is far less than that of SRSWM . Its necessary exciting energy is a fraction of that of SRSWM. This advantage over the SRSWM sees the convenience of its application . (4) LSSRWM can use more information than the initial arrival time of elastic waves , which is the only valuable information in SRSWM , to resolve more complex geology problems . Thus the application of LSSRWM is more urgent than that of SRSWM in our institute's bridge site investigation in which the seismic focus of a charge isn't allowed and the working space is narrow. Since the end of 1991, we have purchased a series of LSSRWM equipment, such as seismograph interface, portable 286 computer, CPS LSSRWM processing software, light covering cable,200000 JOULE seismic focus of electric sparkle, 100HZ vertical detector, floating cable and R24 seismograph of EG&G Co. Limit. for our LSSRWM .

The survey order of LSSRWM is: by expanding spread method carry out noise analyses to decide a perfect reception window, stipulate the eccentricity distance (ED)and the group interval of two detectors (GITD) , regulate the frequency range of band-passfilter and other parameters . Perform many times CDP covering survey on the regulated parameters and finally use CSP software to process the LSSRWM data on site .

In this bridge site investigation , the survey line was the central line of the bridge , ED 40m ,GITD 20m, band-passfilter frequency range 100-500Hz , the max energy of electric sparkle seismic focus 18000J .

Land Direct Current Depth-gauging Method (LDCDGM)

Often hired to resolve simple geology problems is LDCDGM ,in which the physical property of electric difference between rock and soil strata is used to capture such information as geoelectric stratum delimitation (GESD) , crosswise resistively change and etc., , and to provide qualitatively and quantitatively the data of the hidden depth of bedrock , change of rocks , abnormality of tectonic and etc.

From the depth-gauging data near the drilled holes , the resistively of the strata of the bridge sites is , from top to bottom, clay soil 20   m , sand soil 25   m , gravel sand and gravel soil 40   m and sand mudrock 20   m . The little difference of the above figures makes it not easy to quantitatively explain the thickness of the overburden . (e.g. the hidden depth of gravel sand and gravel soil is more than forty meters , but the thickness is only several meters which can only be called a thin stratum .) But we can employ the change of the curves of LDCDGM to deduce the change of rocks .

In the synthetic site investigation of this bridge , D.C.depth -gauging points were set along the bridge central line , four electric poles were arranged symmetrically , the max pole distance AB/2 was 420m.

Through synthetic geophysical prospecting , proven is the undulation of the bedrock top of the bridge site . Especially , the bedrock reflection waves , captured by HAEM and LSSRWM has subjectively outlined the top undulation and tectonic along the bridge central line. In waters , the data from HAEM revealed a wide tension-twist-fault along the bedrock strike in the river course of the lower site , and a crushed broken zone of fault in the river course of the upper site .All results above are not easy to be gotten by other geology prospecting methods .

Combination of Drilling Prospecting and Geophysical Prospecting

Engineering geology site investigation depends mainly on drilling prospecting. During prospecting data can ascertain the reason of abnormality of GPP and verify the conclusions of GPP data. The hidden depth of bedrock, determined by drilling, provides the interpretation of geophysical prospecting data with an exact time-depth transform factor . The geophysical prospecting results of the abnormality , rock borders , broken zones and etc. , direct a reasonable arrangement of drilling holes , curbing the possible blindness .   A synthetic analysis and compression of the two kinds of data above can enlarge the application range of drilling data to reduce the working amount of drilling .

E.g.1   near the lower bridge site waters , there lays the international communication cable across the river. Archoring is forbidden , an underwater drilling hole was placed 400m away from the site central line down the river.   In the survey course of HAEM , we purposely watched, from the point of the drilling hole to the correspondent point of the central line , the bedrock reflected waves to make up the defect because of no correspondent drilling hole on the central line.

E.g.2   Through LDCDGM near the drilling holes of the both bridge sites , the GESD data of every stratum was obtained . A careful analysis of all LDCDGM curves and the crosswise resistively change explained change of rocks , determined the rock borders and enlarged the application range of drilling data to verify the tectonically broken zone inferred by the LSSRWM . Through because of the no Archoring   near the lower bridge site , two drilling holes were placed 200m and 400m away from the broken zone respectively . The rock core of the two holes displayed the tectonic rub traces on the faces of the cracks which were filled with gypsum . All this verified the existence of the technical effects . The next stage survey will further the verification .

CONLUSIONS AND COMPREHENSIVE ANALYSES OF THE RESULTS

The comprehensive analyses of the synthetic site investigation include the following contents . Base on the data of geologizing and geology drawing to determine the geology periods ,strikes , tectonic and its distributions of the rock strata and the change of rocks. Continuously analyses the GPP data and compare them with the relative drilling data to combine the physical delimitation with geological delimitation to convince the geophysical prospecting results . E.g. LSSRWM recorded four groups of virtual reflected waves, we analyzed them and compared them with drilling data; analyzed the reflection characteristics such as the strength of the reflected energy,   continuity of reflected waves and the density of the reflected waves; analyzed the different characteristics which displayed the deposition environments, analyzed the reflection abnormality such as inversion diffraction and etc. of bedrock ; and exactly delimited the overburden, determined the hidden depth of bedrock and the concrete position and distribution of broken zones of faults . Above all , the synthetic analyses have helped us come to fairly integrated and convincible conclusions as follows.

The data of drilling holes and the reflected waves of HAEM and LSSRWM have subjectively presented the undulation of the bedrock top of the bridge site central lines .
The positions and distributions of the faults and broken zones have been resolved by the HAEM .Combining the drilling with the LSSRWM on the upper bridge site, the strata of overburden of the Quaternary Period has been exactly de limited. The crosswise resistively of LDCDGM has qualitative explained the change of rocks and extended the application range of drilling data.

The site investigation of BSER, according to the site geology conditions and the design demands , the synthetic site investigation method was chosen. And acquired has been the complete and convincible engineering geology data , especially, the concealed faults and broken zones and their distributions near the waters of the two sites. Were it not for the synthetic site investigation method , drilling exploration couldn't have accomplished the results in such a short time. So synthetic site investigation method is to be what we search for in our future prospecting.

Proc. 30th Int'l. Geol. Congr., Vol. 23, pp. 95-102
Wang Sijing and P. Marinos (Eds)
© VSP 1997

Repair and Strengthening Stone Masonry Buildings in Seismically Active Areas of Pakistan

M.ARSHAD KHAN
Institute of Geology University of Azad Jammu and Kashmir Muzaffarabad, Pakistan

Abstract

This paper presents methods to repair and strengthen the stone masonry building in seismically active areas of Pakistan..

Keywords: stone masonry building, dynamic test, epoxy material, steel jacket

INTRODUCTION

The concept of repair and strengthening of existing buildings in Pakistan was nonexistent till 1993. This concept was launched in 1994 after a long discussion with the Professors of the Institute of earthquake engineering and seismology Republic of Macedonia. The concept of strengthening of existing buildings connected with interventions prior to earthquakes aimed at increasing the original strength of the structure and eliminating damage sources. The second priority is to repair of the damaged systems and future planning. The problem of mitigation of the seismic hazard by means of prior strengthening has received relatively little attention in spite of its practical and social relevancy. There is no A Seismic Design and Construction code in Pakistan for previous existing buildings or for new constructions. Five hundred [500] old damaged buildings were selected in Quetta and Sindh region from a practical point of view. Prior strengthening and repair show very similar features. Similar techniques can be used to prevent damage and to restore the building to its undamaged configuration and to increase the original strength. An orthogonal wall separation is event occurred in a stone masonry building during a strong quake. After local repair has been performed horizontal steel tendons are frequently used to improve the overall structural organization of the building. The influence of various strengthening procedures on ductility and ultimate strength has not yet been made clear nor a design code has been made to cope with all the aspects connected with strengthening and the resistant mechanism yet available. The building types dealt with here, are the oldest types in human settlements, the whole subject still seems at a rough and ready stage of development and at this point perhaps the experimental approach is the most appropriate.

In this paper results of static tests carried out on models scaled 1:2, of stone masonry buildings loaded with a set of forces simulating earthquake actions is discussed. 10 models were tested. Of these 8 were all similar in principle. Six of them were strengthened using different devices and two were tested in their original configuration in order to provide a reference pattern for the failure mechanisms. After ultimate conditions attained the models were repaired and again tested. The description of the complete set of models represented.

TESTS

In Fig.1 the model with no plaster applied in its original configuration, represents schematically strengthening techniques adopted. The original configuration of the models intended to represent typical rural houses without rigid slabs and without particular connections of orthogonal walls. In grouted models an average of 6 holes were drilled per square meter in the walls. The water at 3-4 kg/cm^2 was injected. Next a cement plaster of 1.5 cm was applied and a 1:1 cement water mixture was grouted into the walls at about 2 kg/cm^2. In order to access the upper limit of the effectiveness of the grouting the 4th model was built with no mortar at all [dry masonry] in order not to prevent the mixture spreading. To provide bond between orthogonal walls a r.c. slab [10 cm thick] was applied to model 5 and diagonal tie rods to model 6 (Table 1,2). The slab was locked to each supporting wall by means of two wedges approximately, 10x10 cm through the wall thickness. Vertical tendons were placed, inside plastic sheets in drilled holes and grouted outside to fill the drilled cores. As a result tendons could slide with respect to the wall. The cross-section area of tendons was 0.87 t for pier tendons. These forces were chosen to give an additional vertical stress of 1 kg/ cm^2 on the normal horizontal sections of the walls. Horizontal tendons were placed on both sides of the walls above and below opening. Each couple of horizontal external tendons was given a precompression on 1.5 t. All the tendons referred above were of the strand type with a cross-section area of 0.87 cm^2 [Fig. 1-10].

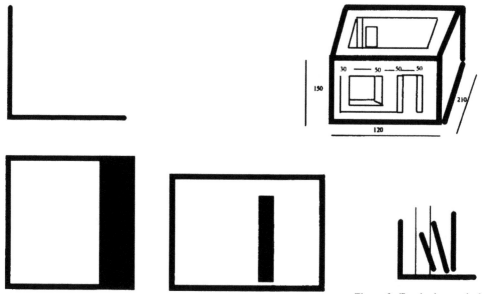

Figure 1. Grouting and wall connection by steel drilled bars

Figure 2. Tensioning vertical tendons in central piers

Table 1. Description of the parts of damaged buildings strengthened using modifications in different elements

MODEL	DESCRIPTION
1.	Unstrengrthened
2.	-
3.	Partial cement grouting
4.	Fully cement grouting
5.	Insertion of rigid slabs
6.	Tie bar connection of ortogonals walls
7.	Vertical tendons in drilled cores at corners.
8.	Vertical tendona in drilled cores in central piers.
9.	Horizontal tendons external to walls on both sides
10.	Horizontal tendons plus vertical at corners.
11.	Horizontal tendons plus vert. in central piers
12.	Vertical tendons in central piers and at corners.
13. steel	Model 1 repaired by grouting connections of orthogonal walls by inclined rods at cornors drilled cores space 0.5 m
14. cm reinf.	Model 2 rep. by insertion \of r.c columns and band columns dim: 10x10
15. external to vertical	Model 3 repaired by local grouting of cracks and horizontal tendons walls on both sides. Horizontal precompression 1kg/cm2on the nominal crossection.

Table 2. Ultimate lateral load, Lateral load of the first important crack in central pier, Lateral load of first separation of transverse wall, ratio of ultimate load and ultimate lateral load in the actual model, ultimate load of the repaired model and original model.

Model	1	2	3	4	5	6	7	8	9	10	11	12	13
ou	35	4	45	100	25	47	42	55	75	60	80	42	45
afc	30	35	35	-	22	15	35	50	65	50	75	32	42
as	15	15	15	-	22	15	15	15	-	-	30	-	-
f	3	2	4	-	3	2	2	2	3	2	3	3	-
p	-	1	-	3	1	6	1	1	1	1	2	-	-
r	-	-	-	-	-	-	-	-	-	-	-	-	-

ou	ultimate lateral load
afc	lateral load of the first important diagonal crack in the central pier
as	lateral load of first separation of transverse wall
f	displacement at point
p	ultimate lateral load
r	ultimate load in the repaired model

Both original and repaired structures were loaded by a set of static forces simulating earthquake forces. The load system represented on storey plus roof above the model under test. Vertical forces V1 were kept constant as a horizontal forces F1 and F2 increased, while vertical forces V2 varied with horizontal forces in order to represent the effect of the overturning movement. The horizontal force F2 was accounting for the construction to total horizontal load of the transverse wall to infer about the effect of the out of plane collapse on the total serviceability of the structure. In what follows horizontal forces will be expressed in terms of their ratio 'a' to the total vertical load [self weight p of the model plus V1 and V2] multiplied by 100. $a = F1+F2/v1+v2 +p \times 100 -1$. According to the law of similitude, this quantity is related to the homologues quantity a' for modeled structure, i.e., the real building , in the following way. The similitude is defined by the conditions. $I =0.5I'$, $a = a'$, s $=Bs' -2$ where I, and S are respectively length, stress,and density measures in the model , and I, S' and S' are the homologues quantities pertaining to the model structure. Eq. 2 depends upon quality of mortar chosen, which was intended to simulate the real conditions of the actual mortar in rural buildings. Eq. 2 on the other hand determine further degree of freedom is available through the choice of the stone material from Eq. 1,2,3, it can easily be shown that the relation between a and the value a' likely to be measured in the real building is

$$a = 2/B \, a' \quad 3$$

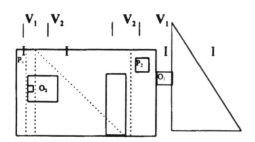

Figure 3. Loading Scheme

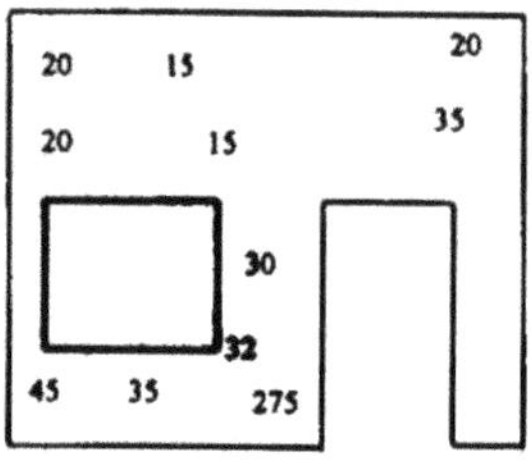

Figure 4. Cracked pattern of model

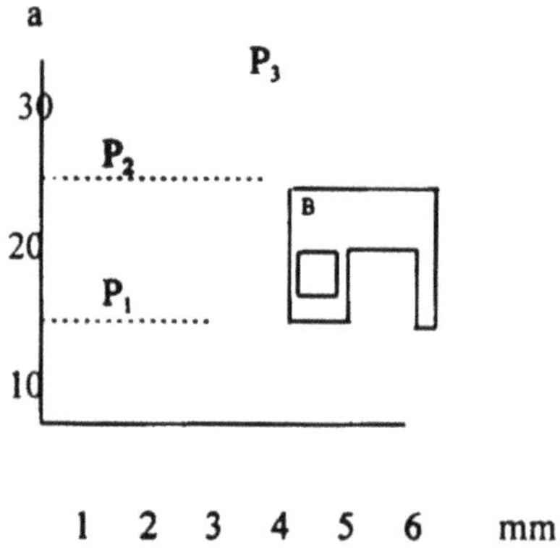

Figure 5. Load displacement curves

Figure 6. Corner tie-bars model

The model masonry density was $S = 2.8$ t/m3, which is rather greater than met with in practice. Loading of the unstrengthened systems showed the typical failure modes recorded in stone masonry buildings during strong earthquakes. This is depicted in Fig.4 which refers to the cracked portion of the first failed wall in model 1. Each crack is marked by a number denoting the value of a of its first appearance. Two basic items are pointed out by Fig.4a the shear failure of the central pier occurring at 86% of the overall ultimate load and b] the separation of the orthogonal wall Fig.3 which beings to appear for a =15. Two mechanisms begin to show the out of plane collapse of the orthogonal wall and the central pier failure. The wall separation prevents the left hand pier of Fig.4 from giving any further contribution to the overall resistance, hence a new distribution of horizontal forces arises leading to an increase of the load acting on the central pier. For slightly higher value of a neither the orthogonal wall nor the central pier are able to contribute to further horizontal resistance. At this stage real seismic behavior can no longer be represented by a static test, it is dependent in particular on the duration. The quantitative overall behavior can be described by a displacement load diagram drawn in terms of horizontal displacements of a point belonging to the spandrel beam, versus the corresponding value of a .This curve can be approximated quardilinear curve [Fig. 5] delimiting 4 different stages of behavior. The first is the linear stage ending at point p . In this phase a slight reduction of the initial stiffness [20%] takes place due to partial settlement of the material. Displacement in this stage are of order of 10 mm with residual displacement of 9.5 mm. In the second phase the first visible cracks appear preceded by a considerable stiffness degradation probably due to microcracks in the mortar.

After the first diagonal crack in the central pier there is further reduction of stiffness , which singles out the point p2, where the second phase ends. At this load level the failure mechanism begins to appear , i.e., initial collapse of the pier and marked walls separation . The failure mechanism develops through the third phase during which the ultimate load is attained [point 3] . The overall stiffness in this phase is continuously reduced starting from a value of about 7% of the very initial stiffness at point p2, up to zero.

The failure mechanism recorded in the models where basically of the two types described a] collapse of the central pier and b] transverse walls separation. Point p2 refers in all cases to the full development of the failure mechanism in all cases p2 occurred at a load level not much differ from the ultimate load at p3. A possible ductility measures is the ratio of the horizontal displacement at p3 to the displacement at p2. This quantity is reported with the other significant test results in table.2.

DESIGN SUGGESTIONS

In considering the results of table 2 two points must consider 1] the strengthening system of model 6 [Fig.6] was to assure bond between orthogonal walls. From this point of view the method proved to be quite satisfactory : relative displacements between orthogonal walls were negligible at the increase of the lateral load. Nevertheless the model performance was rather unsatisfactory if compared with the reference results. This may be due to damage suffered by the model during transportation.2] Model 4 showed very good performance. The resistance exceeded the capacities of the equipment. As a matter of fact at the increase of load [a>80MPa] no crack was recorded and the structure tended to rotate rigidly with a significant separation of the right transverse walls from the foundation slabs. In the case in which a high number of diffusion paths exist for the cement mixture the system tends to become almost a monolithic assembly with high strength. Obviously the results of this test have to be considered as a theoretical upper bound for grouting effects. The value of R represents the efficiency of the considered strengthening methods with respect to the ultimate lateral resistance . Apart from the stated exceptions results are rather good ranging from an increase in strength from 10% to more than 100% . The application of the rigid slabs did not produce increase in the ultimate strength . This confirmed that was expected due to the symmetry of the structure and of the loading . However, a better connection between the orthogonal walls was achieved. The repair techniques investigated in tests 13-14 ,15 gave good results. In this case u seems to be independent of the type of strengthening , this fact is however, influenced by the other factors. The displacement recorded at p3 is quite different . For the reference model U3 was of the order of 4.2 mm. The use of horizontal tendons allowed displacements 3-4 times greater while drilled vertical tendons gave rise to horizontal displacement at ultimate twice as great as the reference ones. This is due to the following reasons: one of the major components of failure mechanism is wall separation [collapse of transverse wall], horizontal tendons tie to gather the structure and pavement, this mechanism allowing for greater displacements and the full resistance of vertical piers to be developed. In the absence of devices to improve the connections between longitudinal and transverse walls, separation occurs rather early. The presence of slabs delay it little and makes it appear in the bottom part of the wall connection . The failure of the central pires was of two types 1] shear and 2] flexural [models 7,14] in which steel was inserted in the central pires. in the form of drilled tendons and r.c columns . In all other cases the shear mode occurred with the first diagonal crack in the central piers. The shear stress in piers at the first important diagonal crack can be assessed on the basis of the following formula, tfc = tk\ 1+o0+ox/1.5tk +o0ox/225tk2   tk =on/1.5 -4, tfc is the limit shear stress averaged in the pier section; o0 = the vertical normal vertical stress on the section ; ox is the horizontal precompression stress, is the ultimate tensile stress of the mortar. The value of tk for stone masonry range tk= 0.3 -0.8 kg/cm2. For full grouted masonry ranges k=0.8+1.2. In our case model 13 an average value related to full grouted was assumed. The assumed value of tk in this case is rather high due to good mortar quality. Determination of the vertical normal stress is made by considering the effect of the over-turning movement, in addition to the construction of vertical permanent loads. Basing on test loading system the effect of the overturning movement was represented by considering the distribution of forces v2 on piers. The construction of the slender piers like the one aside the door in the nodal need be neglected . This assumption is justified by the fact that it was soon sheared off at its top. The construction of the lateral piers , aside the window , seems evident in this pier infact diagonal cracks appeared at the end of the loading history in both cases. The results derived with the above procedures are summarized in Table 1. and are compared with the experimental ultimate load of the first failed wall. The proposed procedure fits the experimental data reasonably well provided the real vertical normal stress and the normal stresses due to tendons are accounted for. Steel in the central pier causes a change of the failure mode from shear to flexural. This is clearly shown by Fig 8,9, which represent the cracked patterns of mode 13 and 14 and the relevant stiffness degradation curves. In these the ratio of the actual to the initial stiffness is plotted against the load parameters a. In both cases curves are compared with the   analogous stiffness degradation of the reference unstren-gthened models which showed a shear type failure. As can be seen in the final stage the two strengthened models exhibits a slower decrease of stiffness with respect to previous stages and to the original models thus denoting that the vertical steel has come fully into operation in the resisting mechanism. The mechanisms arising in the two cases differ from each other. Insertion of r.c column [model14] allows the following to be made in order to explain experimental evidence. The interpretation of the test results performed by various authors on masonry piers with vertical steel showed that the ultimate lateral strength can be predicted by the use of simplified r.c formulas. Moreover, pier failure is considered to take place at yielding of tension bars. Inspection of Fig. 9 shows that lateral piers are practically uncracked while the central one exhibits the typical bending cracks in tension zone. particularly are the base while at the left hand corner, compression zone of the base crushing crack exist. The lack of a rigid slab allows the rotation of the upper base with respect to the bottom one. Basing on this all, following can be derived: 1] the central piers works at full height and takes the full lateral load of the wall. 2] ultimate conditions for the wall are attained in the cantilever type behavior of the pier determined by yielding of the steel. The failure load can be determined basing on the Fig. 9, Turnesk, et al (1970). The following formula can be obtained: $Fu = E\, Aidi/H\, Xfy$ -6, where H is the total height of the pier and di the distance of the ith bar of area Ai from the compression toe and fy the weight [$3500kg/cm^2$. For model 7 however the use of the above procedure was not able to predict the experimental ultimate load. In the model behavior two stages can be identified. before and after diagonal cracking of the central pier. The use of eq. 6 in which o0 accounts for the additional normal stress explains the increase of afc with respect to the reference system. Up to this point the mechanisms are rather similar. When the diagonal crack has developed the two parts of the pier tend to move apart and a hold down action exerted by the tendons. From this point on the resisting mechanism seems to be of frictional type. the normal stress acting on crack being dependent also on the tendons action on this phase. The collapse of the model was due to wall separation. When this was prevented by adding horizontal restraints a considerable increase of strength could be obtained. However, the nature of phenomena involve in this resisting mechanism needs further investigation, Khan, (1994).

CONCLUSIONS

The epoxy material and steel jackets are the better material for strengthening in the area. Assess is needed to estimate what extent the results obtained in the static field to describe the effect of the various devices for strengthening can be applied to predict their behavior during a real seismic action.

The dynamic tests are needed for repair and strengthening stone masonry buildings.

Parts of the damage observed after Friuli 1976 and Norcia 1979 earthquakes in Italy can be compared with the damages in Quetta region. The laboratory tests and real damages are little contradicted .

The setting up of reliable method for strengthening masonry systems needs further research. This is also important with respect to the strategy of intervening in existing buildings and particularly with respect to priorities. It must be decided which buildings are to be substituted with new ones and which are weaker and to be strengthened first.

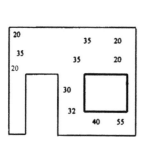

Figure 7. Cracked pattern of the model with rigid slab

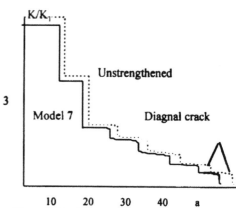

Figure 8. Cracked pattern and stiffness degradation of model 7

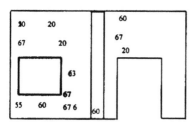

Figure 9. Compression Toe

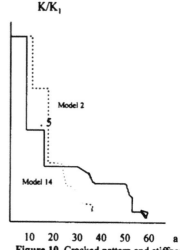

Figure 10. Cracked pattern and stiffness degradation od model 14

REFERENCES

Khan, M. A. Aseismic Design and construction. Unpublished Research report [PGD] Republic Macedonia, (1994).

Tumsek, F. Cacovic, V. Some experimental results on the strength of brick masonry walls. Proc. II Int. Brick Masonry Conf. Stake on Trent. P.20-30, (1970).

Proc. 30th Int'l. Geol. Congr., Vol. 23, pp. 103-110
Wang Sijing and P. Marinos (Eds)
© VSP 1997

Research of Railway Rock Slope in China

XIE QIANG & JIANG JUEGUANG
Southwest Jiaotong Univ. , Chengdu, China, 610031

Abstract

In this paper, the change of the stress in various rock slope sections and the relation between slope stress and slope deformation were analyzed. The quantitative analysis methods determining the slope angle of railway rock was presented, based on the rockmass quality. As an example, the stability of a river bank slope with a height of 250m under the bridge load was analysed and the appropriate position of the bridge foundation was given.

Keywords: railway, slope, stability

INTRODUCTION

In China, there are vast mountain areas with complex geological conditions. In construction of mountain railway, an important problem, which is often faced with, is how to correctly and reasonably determine the rock slope height and to evaluate its stability. Railway rock slope includes cut slope, slope of the entrance or existence to tunnels, natural slope, the natural river bank slope and so on. Artificial excavation slope height usually is less than 30m or 40m, but these slopes are numbers and wide-ranging. On the other hand, the natural slope on railway bed or river bank slope under bridge foundation probably reach hundred metres or several hundred metres in height and are very steep, especially in high mountains or deep gorges. The stability of this railway rock slope is not only closely related to the safety of railway transportation, but also determine the economic rationality of slope engineering. Therefore, how to correctly and reasonably decide the rock slope height and to evaluate its stability have important practical significance.

This paper, according to authors research of railway rock slope, presents several achievements on the deformation and failure of slope rockmass, the quantitative method of railway rock slope angles determined by the rock quality, and the stability analysis of railway rock slope in different conditions.

RESEARCH OF THE MECHANIC CHARACTER OF RAILWAY SLOPE

The Stress Character of Excavation Slope

The Stress Character of Polygonal Slope In the construction of section the railway cut slope, the natural slope is partly excavated, that is, the whole slope section occurs as polygonal shape, which is steep in the toe and gentle in the top, in most case. The stress feature of those slopes, which had certain slope angles and height by excavating the natural slope, is analysed with the FEM(Finite Element Method) by author. Analysis shows that the maximum tension stress in the change point of slope angle may be result in the tensile failure on slope, and the maximum shear stress in the toe of slope may be result in shear failure on slope[1].

At the same slope height, with the increase of slope angle excavated, the increasing stress in the toe of slope is shown in figure 1. And at the same slope angles, with the increase of slope height, the changing of stress in the toe of slope is shown in figure 2. Contrasting between figure 1 and figure 2, we can understand that the stress variation caused by slope angle with change of 1° is equal to that caused by slope height with change of 4m. Thus, for the stress of slope, the effect caused by the slope angle variation is more sensitive than that caused by the slope height variation.

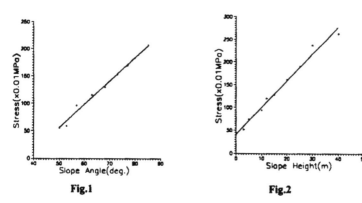

Fig.1   Fig.2

According to the above analysis, the key to slope design is to think over the tensile strength in the top of new excavation slope and the shear strength in the toe of slope, for the homogeneous or statistical homogeneous rockmass. In particular, the mechanical properties of those joints, which suits to the stress direction, have to be carefully researched. To the two parameters: slope height and slope angle, the variation in slope angle has more effect on the stress than the that in slope height does. So, in the process of slope design, the first thing is to consider how to choose slope angle, especially with a gentle slope.

The Stress Character of Bench Slopes For bench slope, the stress is concentrated respectfully on the up-toe and down-toe of bench. Bench height and width affect the stress level in slope. Figure 3 shows that the stresses of up-toe and down-toe of bench slope vary with the bench width. From the figure we can know that the stress of down-toe gradually decreases with the increase of bench width, and the stress of up-toe rapidly increases at first, then tends to a fix stress level with the increasing of bench width. This process is begun at one time width-height ratio. By calculating, enlargement of the bench width can improve the stress concentration of the down-toe, which is useful for the stability of slope. However, at the same time, the increasing of bench width can result in the increase of the stress of up-toe. Certainly, considering the two factors synthetically, we can find an optimistic bench width, which is near the point of intersection of two curves.

Figure 4 shows the relation of the up-toe and down-toe stresses of slope with the bench height. Contrasting to figure 3, the stress of down-toe increases and the stress of up-toe obviously decreases with the increase of bench height. It can be seen from figure 4 that, when bench height is about less than one third of the original height, the stresses of toe of slope is more than three fifths of original stress. By analysis, it can prove that several low benches are the better scheme to high slope.

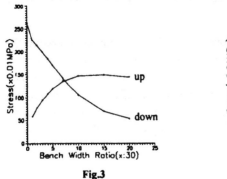

Fig.3

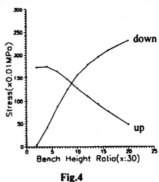

Fig.4

Synthetically considering various factors, we recommend that bench height be about less than one third of original slope height, bench width-height ratio be not less than 0.2–0.3, that is, a slope may be designed to several bench slopes, each with a low height.

The Character of Slope Deformation and Failure

Progressive failure analysis[2] and DEM(Discrete Element Method) analysis[3] show that excavating slope easily take place the tensile failure caused by tensile stress. If crack has formed in top of slope, then the range of slope failure maybe increase with cutting the slope, and the stability of slope also will decrease.

Progressive failure analysis and model tests[4] show that the unstability of slope would increase with the increase of dip angle of interline or the included angle of two joints in a wedge. For the slope with close-jointed bedded rockmass, if the dip angle of bedding opposing slope dip is over 60°, rock fall will be easy to take place. If slope consist of flag rockmass or plate rockmass, desquamation failure easy to occur.

THE DETERMINATION OF STABILITY ANGLE IN RAILWAY ROCK SLOPE

Today, the slope angle is determined by experience in railway rock slope in China. All over the country 183 representative rock slopes from 12 main railway lines were carefully investigated by author and so on. By the statistical analysis for these slopes and the correlation analysis for factors influencing slope stability, a quantitative relation between the quality index of slope rockmass and slope angle and the analysis methods for slope stability were established. This technique has been successfully applied to a serial of engineering practices.

Slope Stability Classification and Influence factor analysis
By investigation and research, according to slope stability, slopes were divided into three types: stable, sub-stable and unstable. Statistical analyses show that many slope built are too safe, thus it can be considered to increase slope angle; on the other hand, a few of slopes designed are too steep, so that they are unstable. In addition, the bedding slide failure takes place in almost all the slopes in which the layer occurrence inclined to line and the slope angle is over the dip of strata, whether slope angle is steep or not. It is showed that it is no possibility that only decreasing slope angle improve the slope stability for these slopes in many cases.

Factors influencing the slope stability are as follows: the rockmass rebound number R, the rockmass intact coefficient I_c, slope height H, apparent block size D, extended length L, joint roughness coefficient JRC, joint mean dip angle β_j, mean dip difference between slope and joint $\Delta\omega_j$, mean dip angle of joint faces intersection line β_w, mean dip difference of joint faces intersection line $\Delta\omega_w$, and slope angle α. With the grey-relation analysis, the interrelation between those parameters and the slope angle is given in table 1:

Tab.1: relation degree between slope angle and main factors[5]

factors	I_c	R	JRC	H	$\Delta\omega_w$	β_f	D	β_w
relation deg.	0.9829	0.9796	0.9794	0.9630	0.9487	0.9443	0.9215	0.9200

To determine rock slope angle, the influent degree of factors must be considered in sequence according to relation degree number. We defined that lithological character, weathering level and rock strength feature are synthetically indicated by rebound number R, and rockmass structure feature by rockmass intact coefficient I_c. For water, its influence will be converted into a reduction coefficient for rockmass quality is used for the calculation of the slope angle.

The Determination of Railway Rock Slope Angle
Based on the above research, the quantitative relation used to determine the slope angle α in the core of rockmass quality RQ is presented as follows:

$$\alpha = -40 + 38\log\left(\gamma_w \cdot RQ\right), \quad (H < 20m)$$

$$\alpha = 21\log_e\left(\gamma_w \cdot RQ\right) - 78, \quad (H \geq 20m)$$

$$RQ = R \cdot I_c \tag{1}$$

$$I_c = \sum_{i=1}^{n} a_i g(x_i)$$

where, H is slope height, a_i is weight function, x_i is a quantitative index on joint set number n, joint extended length L, joint density f, coherence degree of joint face C, layer thickness t, and the rock block size D. These quantitative indexes are easily got in field. $g(x_i)$, a contribution value, is calculated as follows:

$$g(n) = 100e^{-0.38n}$$
$$g(f) = 100e^{-0.3f}$$
$$g(l) = 100e^{-0.18l} \qquad (2)$$
$$g(D) = 100(1 - e^{-0.15D})$$
$$g(t) = \begin{cases} 100(1 - e^{-0.4t}), & \text{bedded rockmass} \\ 0, & \text{un} - \text{bedded rockmass} \end{cases}$$

$g(c)$ is shown in table 2:

Tab.2: structural surfaces contribution value

connection	very good	good	common	loose	very loose
g(c)	100	70	50	30	0

The weight number of factors, bedded rockmass and un-bedded rockmass, is as follows:

Tab.3: weight number of the bedded rockmass and un-bedded rockmass

factors	n	f	l	D	C	t
un-bedded	0.15	0.25	0.25	0.3	0.05	0
bedded	0.1	0.2	0.2	0.2	0.05	0.25

γ_w is the reduction coefficient for rockmass quality by influence of water on rock slope, $\gamma_w = (C_1 + C_2 + C_3)/3$. C is shown in table 4.

Tab.4: Rockmass quality reduction coefficient based on water

	I	II	III	IV
Weathering Level	un-weather	slight-weather	weather	strong-weather
C_1	1.0	0.85	0.70	0.55
Water	dry	moist	dropping	linear-flow
C_2	1.0	0.9	0.75	0.6
Filling	un-filling	slight-filling	filling	all-filling
C_3	1.0	0.85	0.70	0.55

It is found that the slope angle of 70 percent of 183 slopes investigated, which were re-designed with this relation, can increase about 8 degree in general. It corresponds to the result of field investigation.

Stability Analysis of Railway Rock Slope and Field Verification

Stability analysis must be made for the slope angle designed on *RQ*. With field investigation, classification matrix B is formed according to the stability of slope, and fuzzy matrix R is based on the rebound number R, rockmass intact efficient I_c, mean stability coefficient K, mean apparent-persistence P, and unstable block type number N.

The weight function matrix **A** is calculated with the parameters. **B** is calculated as follows:

$$B = A \cdot R \tag{3}$$
$$B = (b_1, b_2, b_3)$$

Where, b_i is percentage number of slope under stable, sub-stable, and unstable state, respectively. According to the sorting percentage number, we may judge the stability of slope. The detailing calculation sees in Ref.6

This system of slope angle determination and stability analysis is applied to Tumen–Hunchun, Benxi–Tianshifu, Guangtong–Dali, Shuicheng–Xiaoyunshang, Nanning–Kunming Railway and some highway slopes to verify. The result shows that the system has not only good effect on optimising design and reducing the engineering investment, but also on scientifically reasonably evaluating slope stability and ensuring slope safety.

THE RESEARCH OF RIVER BANK SLOPE STABILITY AND BRIDGE FOUNDATION POSITION

To river bank slope set up bridge pier, its stability is not only affected by geological feature and slope height as cut slope, but related to the position, type, and load level of bridge pier. Therefore, researching on bank slope stability and the position set up bridge foundation under load is different from researching on common slope stability analysis.

Analysis of Stress in River Bank Slope under Bridge Foundation
To the rockmass loading bridge pier in river bank slope, its stress and deformation is analysed with FEM. The results show that the rockmass stress field and deformation field varied with the difference distance of foundation to the bank face under the loading of pier. When the distance increase to certain value, the rockmass stress and deformation will occur as drop, the stress and deformation have been improved.

The form and buried depth of bridge foundation affect bank slope stress. Deepen or changing foundation would better improve slope stability.

Synthetic Analysis Method of Bank Slope Stability and Bridge Foundation Position
When bridge is built on gorge slope, we must not only ensure the stability of bridge foundation and bank slope, but also conform to the engineering reasonableness. Therefore, bridge foundation stability analysis differs from common slope stability analysis. It is a synthetic analysis work of geology and engineering. By the stability analysis for high-steep bank in Bei-pan River Bridge, Qing-shui River Bridge, Xia-lao-xi Bridge, and Feng-jie Long River Bridge under bridge load, a synthetic analysis method for slope stability and bridge foundation position have been established and have excellently been applied to engineering. The methods include following several contents:

(1) On the basis of site investigation and engineering geological analysis, bank rockmass geological model is abstracted.

(2) Using qualitative or quantitative analysis method and technique such as numerical analysis, physical model test, theory calculation etc., the stress, deformation and failure feature of rockmass geological model are analysed and researched under the bridge foundation load. The slope angle and bank stability are calculated and researched to determine.

(3) According to the result of above analysis and the form of bridge foundation, position set bridge foundation and its safety coefficients are evaluated and calculated.

(4) Recommend the engineering measure if necessary.

Engineering Example
Outline of Bei-pang River Bridge Bei-pang River railway Bridge in Shuicheng-Xiaoyunshang Railway is situated in Guizhou Province in China, where there are the heavily deep-cut gorge with width of only 130m and vertical bank wall with height of up to 250m. Bridge is designed to an arch bridge with single span of 180m and clear height of 305m.

The stratum of upper Permian(P_2), composed of medium-thick bedded limestone, is exposed at bridge area. Occurrence of strata slightly incline to outside of slope(apparent dip angle is about 6°). Three relaxation cracks were at the top of bank wall, which is 2–10m to the edge of bank face, whose virtual depths reach 40m. A regional small rockfoll occurred on the top of wall. On the wall there exist long and large joints, which spread within distance of about 30m and cut the bank. The dead load of bridge pier is 5500t in horizon(direct to inside of slope), and 12000t in vertical.

*Rockmass Stress Feature and Failure Model in Bank Slope:*Field investigation shows that rockmass of bank slope mainly composed of is huge block and hard rock. The movement of block cut by bedding-plane and joint is main form of the bank failure. When the bridge pier was set up to the 30m and 40m from bank face, rockmass deformation was analysed by DEM and was tested by Base Friction Test. The result shows[3] that after the bridge pier being rested, the lateral pressure will directly cause the block in the top move quickly. When the pier is set up to 30m from the wall face, the rockmass deformation is strong, plastic zone is greater. When the pier is set up to 40m from the wall face, the rockmass deformation and plastic zone were observably improved. The failure range of top of slope is about 12m–20m.

Calculation for Stability of Bank Slope Rockmass. By site investigation, numerical analysis and model test, it shows slope stability can be analysed by topple and bend failure in plate structure rockmass. The calculation show, whether there is the bridge load or not, bank slope all is stable.

According to Eq.1 in this paper, the stable angle of the bank slope in Bei-pang River Bridge was calculated. By field survey, the parameters used for calculation are: γ_w=0.98,

I_c=52.5, R=40. Inserting these parameters into the slope angle quantitative relation, the stable slope angle is 82.2^0. With the profile section of the bank, the maximum failure width in top of bank slope is 21.2m. When the position is 30m from pier to the wall face, the safety coefficient of the bank slope only is 1.42. But when position is 40m, the coefficient can reach 1.86.

CONCLUSIONS

From aforementioned research, the conclusion can be given as follows:

(1) In cut slope, the tensile stress in top of slope and the shear stress in toe of slope are key to slope design and construction, we must carefully analyse the structural plane feature and rockmass mechanical feature suiting to the stress direction for these points in order to ensure slope stability. The measure of designing bench can be taken to high slope. We recommend that bench height should not more than one third of the whole slope height, and the bench width would no over two tenths or three tenths of the single bench height.

(2) The angles determine of railway rock slope based on the RQ and the stability analysis method with the fuzzy synthetic judgement have achieved good results in practice. The methods afford a new way to the design of railway rock slope, from experience to quantitative evaluation.

(3) Synthetic analysis method for bank slope stability and bridge foundation position not only effectively analyse the inter-action between bridge load and bank rockmass and the stability of bank slope, but also offer feasible evaluation method for the bridge pier safe distance to slope face and buried depth on the river bank slope.

REFERENCES

1.   J. G. JIANG & Q. XIE: Finite Element Analysis for Stress in Slope, *Proceeding of 6th Int. Conf. Numerical Method in Geomechanics*, INNSBRUCK, Austria, 1988
2.   J. G. JIANG: Rock Slope Wedge Failure Analysis, *Journal of Hebei Institute Geology*, Vol. 12, No. 2
3.   Q. XIE et al.: Model and Analogue for Rockmass Failure Feature of Bank Slope in Bei-pang River Bridge, *Proceeding of '95 Geotechnical Mechanics*, 1995, Chengdu Univ. of Science and Technology Press
4.   H.G. QIAN: Model Test on Incline Failure, *Proceeding of 35th Birthday of Engineering Geology Department of Southwest Jiaotong Univ.*, 1993, Southwest Jiaotong Univ. Press
5.   J. G. JIANG et al.: Quantitative Analysis for Railway Rock Slope Stability Angle, *Application of Numeric Method in Geotechnical Engineering*, 1990, Tongji Univ. Press
6.   Q. XIE: Fuzzy Synthetic Judgement for Railway Rock Slope Stability, *Journal of Southwest Jiaotong Univ.*, 1991, No. 2

Proc. 30th Int'l. Geol. Congr., Vol. 23, pp. 111-121
Wang Sijing and P. Marinos (Eds)
© VSP 1997

Integration of Discrete Element Method and Time Series Analysis Technique to Predict Deformation in Blocky Rock Slopes

PAN SHIBING

Institute of Hydrogeology & Engineering Geology Techniques, MGMR, 071051, Baoding City, Hebei Province, China

Abstract

This paper presents a computational modeling method, which incorporates discrete element method (DEM) and time series analysis technique, to predict the deformation in blocky rock slopes. In combination with time series analysis of data obtained from the field measurements, DEM is improved and it can be applied to displacement and time prediction. A transformation function, which can be determined by the displacement fitting approach, is defined to link the two modeling methods. Based on both observation data and rock physical and mechanical parameters, the coupled computational modeling method would be expected to be more accurate. Model application to the analysis and prediction of the deformation in Lianziya slope located in Three Gorges of Changjiang River of China is also discussed in detail in the paper.

Key words: discrete element method, time series analysis technique, coupled modeling method, deformation, Lianziya slope.

INTRODUCTION

People have been paying a growing attention to the study of the modeling methods for the purpose of the prediction and control of geological hazards. Many analytical and computational methods have been developed for predicting the deformation in slopes during recent decades. They may be generally classified into two categories, (i) determined modeling methods based on rock physics and mechanics, such as limiting equilibrium analysis, finite element method (FEM), boundary element method (BEM), discrete element method (DEM) etc., and (ii) undetermined modeling methods based on observation data obtained from the field measurement, including curve fitting approach, Pearl's growth curve model, time series analysis technique and so on.

Discrete element method firstly developed by P. A. Cundall [1971] is a numerical simulation method for discontinuum, and it is considered as one of the most effective and powerful method to simulate the progressive deformation in blocky rock slopes. This method requires physical and mechanical parameters of rock mass and a well understanding of the structure of the slopes. However, due to the difficulties to obtain the precise parameters of rock mass and the uncertainties in geological structure, the displacements simulated by DEM usually do not agree with the observations obtained from the field measurement.

The series of observation data reveals the general rules of the development of the deformation in slopes associated with all kinds of factors. By time series analysis technique, the observation data which are subject to measurement errors can be treated through smoothing and filtering process, and models can be developed for the purpose of prediction. However, it is often difficult to give a reasonable interpretation of the prediction results in geology, especially when predicting beyond the range of data. This is because the models are not constrained by any physical conditions.

In this paper, a computational modeling method, integrating the discrete element model with time series analysis technique, is proposed for the purpose of prediction of the deformation in blocky rock slopes. Its application to the analysis and prediction of the deformation of Lianziya slope which is located in Three Gorges of Changjiang River of China, is also discussed in detail in the paper.

GENERAL DESCRIPTION OF METHODS

Discrete Element Method

Discrete element method is based on the second Newton's law of motion. It considers the rock mass as a number of discrete blocky elements, and any blocky element as a separated part will cause the deformation and motion by the action of forces applied by the adjacent elements. The interaction of blocks depends on the characteristic of the discontinuity and fissures in the rock mass. DEM uses the dynamic relaxation iteration to solve the motion equations. The central differential scheme of the iteration formulas for the movement of elements can be written as

$$v_{i(t+\Delta t/2)} = [v_{i(t-\Delta t/2)} (1-\alpha\Delta t /2) + (F_{i(t)} /m +g_i)\Delta t]/ (1+\alpha\Delta t /2)$$

$$x_{i(t+\Delta t/2)} = x_{i(t)}+v_{i(t+\Delta t/2)}\Delta t , \quad i = 1,2. \ldots\ldots,n \tag{1}$$

where x_i, v_i denote the displacement and velocity of the i-th element respectively, the damp coefficient, F the net force acting upon the each element, g the gravity acceleration, m the mass of each element, Δt the time step used for iterations, n the number of discrete elements. The displacement and velocity of each element at different time steps can be determined by solving the above system of equations.

The main advantage of discrete element method is that it can be used to simulate the deformation in blocky rock mass under various environmental factors simultaneously, to analyze and evaluate the stability of the rock mass in static or kinetic state. It should be noted that the damp coefficients are introduced into the equations in order to obtain a stable solution, the time steps used for the iteration do not represent the physical time. Therefore, discrete element method cannot directly be applied to the prediction of the displacement and time.

Time Series Analysis Technique

The purpose of time series analysis is to smooth the observation data of displacement of the slopes which are used for the prediction of the deformation. Adequate modeling methods are successive moving smoothing models, such as the second moving-average method . Let $x(t)$, ($t = 1, 2, ..., n$) be the original observation series, the first moving-average formulas can be written as

$$M^{(1)}(t) = [\, x(t) + x(t-1) + ... \, x(t-p+1)\,]\, /\, p \tag{2}$$

where p is the number of observations used for smoothing for each interval. The smoothing results will be retarded as compared with the original data series. The second moving-average is carried to form a linear equation to calibrate the retardation, if data series possesses a linear trend. The aim of the second moving-average process is to find the smoothing coefficients which are used to establish the mathematical model, it can be expressed in an iteration form

$$M^{(2)}(t) = M^{(2)}(t-1) + [M^{(1)}(t) - M^{(1)}(t-p)\,]\, /\, p$$

The mathematical model for the displacement prediction is developed as following

$$Y(t+T) = a(t) + b(t)\, T \tag{3}$$

in which $Y(t+T)$ is the prediction at the period of $(t+T)$, $a(t)$ and $b(t)$ are smoothing coefficients given by

$$a(t) = 2\, M^{(1)}(t) - M^{(2)}(t), \quad b(t) = 2\, [M^{(1)}(t) - M^{(2)}(t)\,]\, /\, (p-1)$$

We can make a reasonable understanding of the rule of development of deformation in slopes in future. Higher order moving-average method or cubic spline function approach can be employed if the observation data demonstrate a distribution with the curvature. It should be pointed out that it is hard to determine the deformation when predicting beyond the range of data.

Coupled Modeling Method
The coupled modeling method incorporates the discrete element model and time series analysis technique by using a transformation function. The transformation function $w(t)$ is defined as

$$w(t) = Y^{\wedge}(t)\, /\, X^{\wedge}(Lt) \tag{4}$$

in which $Y^{\wedge}(t)$ is the observation series after smoothed and filtered with the second moving- average method mentioned above, and $X^{\wedge}(t)$ is the initial simulation series with DEM , one simulation value every L iteration times. The value of L can be found by the model trials. The transformation function can be determined by the best orthogonal polynomial fitting approach, it functions as a link of the two modeling methods. The prediction equation of displacement $Xp(t)$ is established as following

$$Xp(t) = w(t) X^{\wedge}(Lt) \tag{5}$$

The procedure for the coupled modeling method is firstly to determine the objective of prediction and to recognize the geological condition of rock slopes, and then carry out the following steps, (i) To select the data series and deal with the original observation data by using the second moving-average method, we obtain the data series $Y^{\wedge}(t)$ which has been smoothed; (ii) To develop the geological concept model and find $X^{\wedge}(t)$ by using DEM; (iii) Model calibration, to find the transformation function $w(t)$ by using equation (4); (iv) Model verification, to run the coupled model and make a comparison between the calculations and original observations. If the calculations fit the original observations well, the model can be applied to prediction. If they do not, return to step (iii). The data series used for model verification should be different from the data for model calibration; and (v) Model prediction, using equation (5).

MODEL APPLICATION

Background
Lianziya slope is located in the south bank of Three Gorges of Changjiang River, Zigui country of Hubei province of China. The slope is only 26 kilometers upstream from the site of Three Gorges Dam, of which the construction is going on at present (see Fig.1).

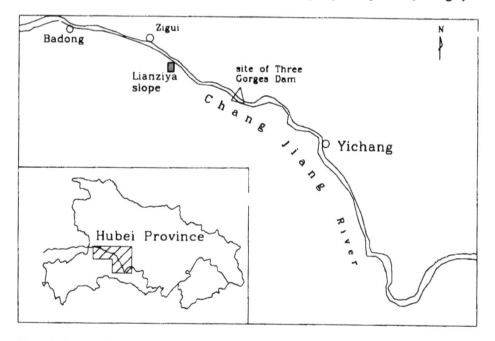

Figure 1. Location of Lianziya slope in Three Gorges of Changjiang River

Running from the south towards the north, the slope is 700m long and 30-180m wide, with an elevation over 100-250m, being narrow and low in the north, wide and high in the south. The most common rock type in the area is limestone of permo-carboniferous period. The strata strike N 30   -35   E and dip NW at an angle varying from 27   to 35   . The rock mass is characterized by being composed of the blocky rock masses separated by dozens of deep openings and several soft layers (see Fig. 2). The openings which cut nearly vertically the slope into sections are well developed with 60-170 m long, 0.5-5m wide and 50-105m deep, dominantly stretching from the west to the east. There exist excavated coal seams underlying the bottom of the rock mass, and the mined-out area reaches about 120000 m^2 with a height of 1.6-4m. The slope has been gradually impaired by the long period of geological processes such as the action of cliff gravity, water pressure, karstification and weathering, and the slope, especially its north part, becomes a potential danger to the shipping of Changjiang River.

The project of controlling of the slope has been undertaking since 1994. The project involves a unified engineering system mainly consists of drainage, pre-stressed anchoring, bonding walls of reinforced concrete for load-bearing and anti-sliding. The prediction of the deformation in rock mass is an essential aspect to guarantee the safety during the construction, to guide the feedback design of engineering and to evaluate the effectiveness of the project when being fulfilled.

Previous investigation and research
Many investigation and research work had been finished during the last decades, some of recent work are summarized in Tab.1. A perfect monitoring system mainly using geodetic measuring technique had been established since early 1970s. Long term regular displacement observation data obtained provide the fundamental data for the analysis of the deformation mechanism. It had been concluded that the slope is stable under natural condition but it is likely to cause sliding under dynamic condition regarding earthquake and water storage related to Three Gorges Dam.

Coupled modeling of Lianziya slope
It is adequate to use DEM to simulate the deformation in Lianziya slope. The discrete element model previously developed by Yin et al. [1994] has been adopted for further discussion in the paper. The model considered both the action of gravity of the rock and the hydraulic pressure in openings as the main factors to govern the movement of the rock mass. The distribution of sections of discrete element model is shown in Fig. 3.

The rock physical and mechanical parameters of the rock mass see Tab. 2. The values of normal and shearing rigidity (Kn, Ks) are given by [Y.P. Yin et al , 1994]

Kn = E / [(1-r)(1+r)h] ,   Ks = 2E / [(1+r)h]
where E is elastic modulus, r is poison's ratio and h is the thickness of element.

The annual horizontal and vertical displacement data of several observation points are analyzed for formation of models using the second moving-average method. The distribution of observation points is also illustrated in Fig. 3. Only the calculation results for Gs and Sc observation points are given here. The results show that the values of prediction are generally larger than those of observations (see Fig. 4).

Observation data from year 1979 to 1990 are used to model calibration to find the transformation function. The results are shown in Tab. 3. Data from year 1991 to 1993 are used for model verification. The results show that the displacements calculated by the coupled model fit the original observations very well (see also Fig. 4). So, the model can be used to predict the deformation in the slope.

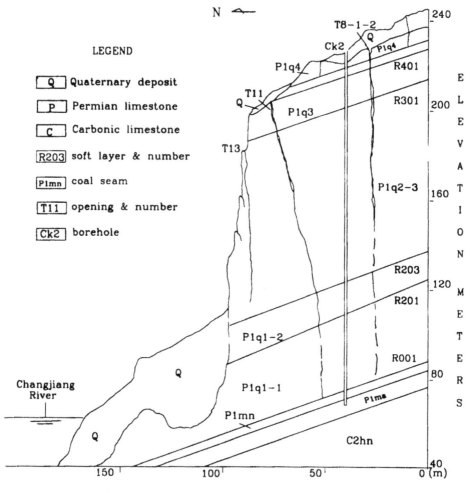

Figure 2. Profile showing the characteristics of the structure of Lianziya slope

The prediction results from year 1994 to 2000 (see Fig. 5) indicate the deformation in the rock mass is gently slow during the period under natural condition. According to the feather of the variation of displacement with time , the period from 1979 to 2000 can be divided into two stages, relatively fast from 1979 to 1989, and slow from 1990 to 2000.

This suggests that Lianziya slope may be in a uniformly sliding state, which is so-called the second stage of creep slippage with respect to M. Satio's creep theory curve.

DISCUSSION AND CONCLUSIONS

Based on the discussions of the strength and weaknesses of both the discrete element method and time series analysis technique, a coupled modeling method has been put forward, and it can be applied to the displacement and time prediction of the deformation in blocky rock slopes. A transformation function is introduced to connect the two modeling methods. The coupled modeling method would be expected to be more accurate since the method is based on both the observation data from the field measurement and the rock physical and mechanical parameters of the rock mass. Consequently, the coupled modeling method increases the requirements for data including historical measurement data, rock parameters and the quantity parameters for the descriptions of the weaknesses in rock slopes.

Table 1. Main research work accomplished recently 1990s in Lianziya slope

No.	model testing	numerical simulation	research contents	author(s) or organization	finished time
1	plane model, three-dimensional model	FEM (two- and three-dimensional)	deformation mechanism in static and dynamic states	Institute of Geology, CAS	1992
2		DEM (two-dimensional)	stability analysis	F. Ren et al	1990
3		rigidity limiting equilibrium	stability analysis	C.M.Xu et al	1990
4		plane limiting equilibrium, spatial blocky body analysis, FEM, DEM	study on the feasibility of controlling project	Committee of Water Conservancy of Changjiang River, MWC	1992
5	blocky stock-pile with inclined plate		slope failure mechanism	Chengdu Institute of Technology	
6		DEM (two-dimensional)	analysis of deformation and optimal anchoring design	Y.P.Yin et al	1994
7	three-dimensional gravity similarity model	FEM using orthogonal anisotropy representative continuum	stability analysis of blocky rock mass	Q.L. Ha and R.X. Zhang	1993

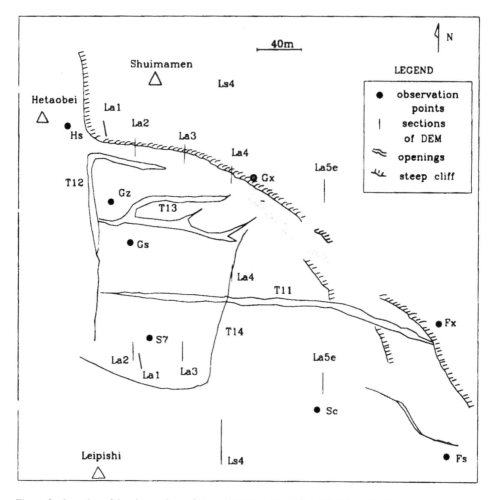

Figure 3. Location of the observation points and DEM sections (after Y.P.Yin et al, 1994)

Table 2. Rock physical and mechanical parameters used by discrete element model

rock type	unit weight (kn/m³)	cohesion (MPa)	friction coefficient	poison's ratio	elastic modulus(GPa)
knotty limestone	25	0.50	1.10	0.25	35
lumpy limestone	27	0.55	1.15	0.20	45
tumulus limestone	26	0.50	1.10	0.25	40
marl and shale					0.19
thick-bedded limestone	27	0.60	1.20	0.20	50
coal seam	23			0.37	

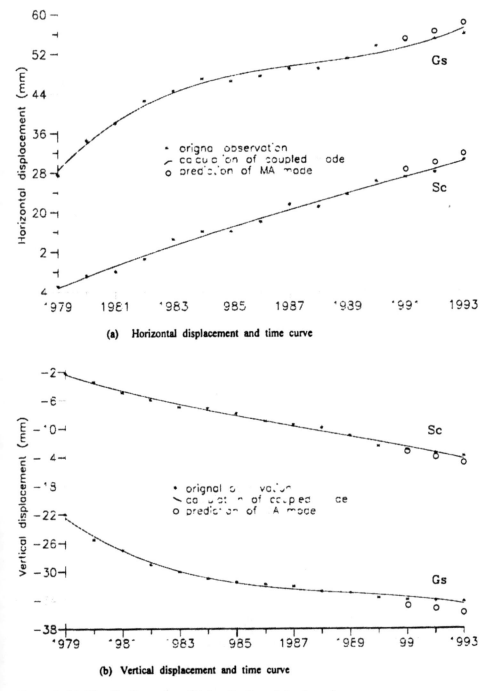

(a) Horizontal displacement and time curve

(b) Vertical displacement and time curve

Figure 4. Model calibration and verification for Gs and Sc obs. points

Pan Shibing

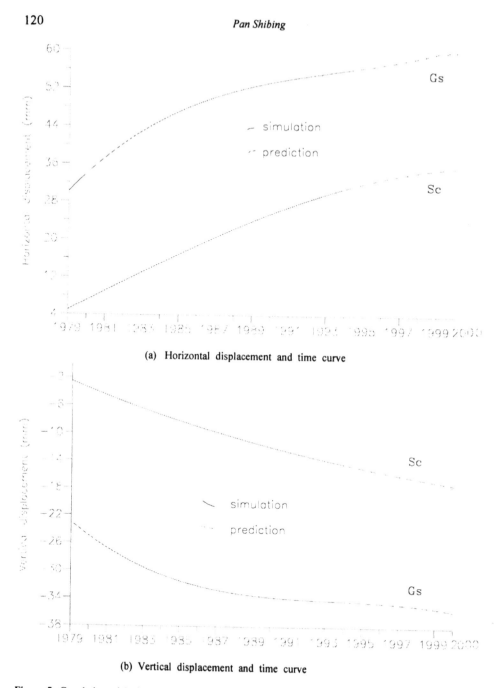

(a) Horizontal displacement and time curve

(b) Vertical displacement and time curve

Figure 5. Coupled model simulation and prediction for Gs and Sc obs. points

Table 3. Model calibration for Gs and Sc observation points, Lianziya slope

No of obs. points		w(t)	MA model
Gs	horizontal	$w(t) = 13.2825 + 3.9763\ t$ $- 0.3538\ t^2 + 0.01166\ t^3$	$Y(t) = 1.7156\ t + 32.6678$
	vertical	$w(t) = -24.1613 - 3.5887\ t$ $+ 0.3389\ t^2 - 0.01271\ t^3$	$Y(t) = -0.7757\ t - 24.6914$
Sc	horizontal	$w(t) = 7.3215 + 2.1741\ t$ $- 0.2262\ t^2 + 0.009347\ t^3$	$Y(t) = 1.7946\ t + 3.8769$
	vertical	$w(t) = -31.2662 - 4.2154\ t$ $+ 0.4326\ t^2 - 0.02562\ t^3$	$Y(t) = -0.8114\ t - 2.2452$

Notes: Time step is given in year, displacement in mm. Value of L in equation (4) is 1000.

Model application to Lianziya rock slope has been successful, this prove that the coupled modeling method may be a useful computational technique to predict the deformation in blocky rock slopes. The prediction results will be very meaningful to the analysis of the stability of the slope. The changes in environmental conditions, such as water storage related to the Project of Three Gorges Dam and controlling engineer-ing, should be taken into considerations in the modeling for further research.

Acknowledgments

The author wishes to acknowledge Dr. Yueping Yin, Prof. D.R. Tang and Mr. R.L. Zhu for the detailed directions and suggestions. Special thanks are due to Dr. Yin for providing me with computer programs and data. Thanks also are due to my colleagues for helping me generously in the preparation of this paper.

REFERENCES

1 Brown E. T.(ed.) 1987, Analytical and computational methods in engineering rock mechanics. London: Allen & Unwin Ltd.
2 Cundall P. A. 1971, A computer model for simulating progressive large scale movement in blocky rock system. In rock fracture, Proc. Int. Symp.
3 Ha Q.L. et al, 1995, Rock slope engineering--Study on the stability of Lianziya dangerous rock mass in Three Gorge of Changjiang River, China. Publishing House Of Chongqing University.
4 Huang Y.Q. et al 1991, Numerical simulation and imitation in engineering geoscience, Hydrogelogy and Engineering Geology, *Vol.18 No.5*
5 Lemos, J. V., et al, 1985, A generalized distinct element program for modeling jointed rock mass. In fundamentals of rock joints. O. Stephansson(ed.),
6 The Committee of the Water Conservancy of Changjiang River, MWC., 1993, Report of the study on the feasibility of the project of protection and control of Lianziya dangerous rock mass in Three Gorge of Changjiang River, China
7 Yin Y.P. et al , 1994, Report of the preliminary design of the project of protection and control of Lianziya slope in Three Gorges of Changjiang River.
8 Yin Y.P. et al, 1995, Geo-engineering support system and Lianziya dangerous rock mass anchoring, Geologic Publishing House, China.

Proc. 30th Int'l. Geol. Congr., Vol. 23, pp. 123-139
Wang Sijing and P. Marinos (Eds)
© VSP 1997

A New Method Of Estimation Of Slope Stability Composed Of Steeply Dipping Strata

A. N.RYUMIN
VNIMI, St. Petersburg, Russia

Abstract

The conventional engineering methods for prediction of slope stability are directed to evaluation of conditions of limit equilibrium of rock mass immediately preceding ultimate failure and practically do not consider a possibility of partial losing of partial losing of strength of rock mass far before failure. Such an approach is valid in case when a slope is composed of homogeneous unlayered rocks. In other case if a slope is composed of layered rocks the losing of slope stability may be connected not with the failure of rocks themselves, but with the failure of week contacts between layers, what provoked bending deformations due to lack of bending rigidity in foliated rock mass. There is a characteristic case of such a process-an undrained slope composed of subvertical dipping strata-under consideration in this report. The main condition of limit equilibrium moments of bending forces and strength(friction and cohesion), resisting to sliding layers on each other, is the following:

$$\frac{\gamma_0 \cdot h_{cr}^3}{6} = \sum_{i=1}^{n} \left[\int_0^{h_i} \left(\sigma'_{c_i} \cdot \tan \varphi_{c_i} + c_{c_i} \right) dz \right] \cdot m_i \quad (1)$$

Here γ_0 is the density of water, h_{cr} is the critical height of the slope, h_i is the height of the i-th layer, m_i is the thickness of the i-th layer, φ_{c_i}, c_{c_i}, σ'_{c_i} are the angle of friction, cohesion and the effective normal stress on the contact between the i-th and the i+1-th layer, correspondingly, n is the number of layers, composing the slope, z is the vertical distance form the base of the slope.

In the important for the practical use case of supposingly homogeneous structure of the slope, composed of layers with equal thickness and strength

$$h_{cr} = \frac{3 \cdot c_c}{\gamma_0 \left(\tan \alpha - \tan \varphi_c \right)} \quad (2)$$

where α is the angle of slope inclination, φ_c, c_c are the strength caracteristics of the contacts between the layers.

Calculation of the horizontal displacement v of the slope points may be fulfilled using the following formulae:

$$v = A(3Z_n^2 - 4Z_n^3 + Z_n^4)$$
$$v = A(6Z_n^2 - 4Z_n^3 + Z_n^4) \quad (3,a,b)$$

Correspondingly, for the cases when the layers are dipping outside(3a) and inside (3b) the rock mass.

Here $A = \dfrac{\gamma_0 \cdot h^4 \left(\tan \alpha - \tan \varphi_c \right)}{6Em^2}$, h is the height of the slope, m is the thickness of a layer, $z_n = z/h$, z is the height of the layer.

Keywords: Slope stability, Dipping strata

INTRODUCTION

The conventional engineering methods for prediction of rock mass behavior at open-pit coal mining are based on the assumption that the rock mass is a loose medium being stiff as the elastic strains develop until failure, which results in total disturbance of the rock mass in the slope or in movement of the entire slope body into the mined-out space along a closed sliding surface cutting out the slope area from adjacent rock mass, on which the condition of limit equilibrium is satisfied /Fisenko, 1965/.

This approach ignores bonding deformations of rock masses which may be very large and may change conditions of stability especially in case of slopes composed of stratified rocks. Rock mass of steeply dipping strata subjected to substantional displacements gradually loses its stiffness because of disturbing the planes of weakness between the layers. The failure processes of this type are especially characteristic of openpit mining of coal deposits where there are steep layers (i.e. the vertical layers or the layers deviating from the vertical by 10-20 degrees) in the slope and in the footwall*). It is not creeping, because appearance and amplitude of displacements are closely connected with mining activity and, usually, all the sliding processes will stop as soon as the mining operations breaking natural equilibrium cease. Besides, these deformations are typical to rather hard rocks with uniaxial crushing strength of about 10-100 MPa.

For the first time, the deformations of the footwall were observed in the Urals openpit coal mining in the seventies and later in Kuzbass and Ekibastus. The sedimentary rocks within the opencut slope area were separated into individual sheets or packets with a thickness of 5 to 20m and deflected towards the mined-out space, with the opencut benches laterally moving without failure(at a slope height of about 100m the displacements reach 20-30m).

All observations carried out in situ at numerous deformed opencut slopes showed, that there is a new type of slope deformation during which the slope body does not move translatory as a rigid body cut off from the rest of rock mass but deflects under bending loads.

The detailed analysis of the situation indicated the existence of relationship between the rock mass deformations and the active lateral earth pressure as a result of stress relief in horizontal direction due to extracting rocks and the coal. It revealed also a very considerable role of subsurface water pressure that contributed to increasing external loads and decreasing friction between the layers. All the observed deformations in the stratified rock mass were clearly bending deformations and therefore, in order to formulate a mathematical approach to evaluate the strata displacements and deformations, the theory of plates and shells should be used in its simplest version.

As regards possible objections to the simplicity of the mathematical approach applied in

*) Footwall of the mine is, here, the zone of underlying strata adjacent to the opencut slope.

this paper one could say the following. As a rule, the information about the rock mass strength and deformation behaviour to be required for prediction and calculation procedures in practical slope stability estimation is highly insufficient and it is often advisable to use back calculations, i. e. to obtain the desired parameters from the in situ investigations conducted at an initial stage of slide movement of slopes. Under these conditions, the theory of applied type may not be useful if it is overburdened with details of little significance.

STATEMENT OF THE PROBLEM

The present investigation deals with the estimation of equilibrium conditions and the calculation of displacements of opencut slopes in steep-dipping strata under ground water pressure and the earth pressure.

In order to evaluate the bending deformations in the slope area, it should be reasonable to consider what is known as prism of resistance or a slope body(Fig.1) which is cut off from the rock mass both by one horizontal plane passing through the pit bottom and three vertical planes, one of which separates the slope body from the adjacent rock mass (or footwall)and is directed along the strike whereas two other planes are directed across the strike and pass at the left and the right ends of sliding slope area. The rock prism in the slope area is considered further as a composite plate that is pinched along its edges and from below (Fig.1a).The deformation behaviour of the plate can be treated as a simultaneous bending of two beams one of which is a frontal beam regarded as infinite on the strike and rigidly fixed at the open-pit bottom level (Fig.1b) while the other is a longitudinal beam that is restricted from below and pinched on the strike (Fig.1c).Each of two imaginary beams is subjected to a bending load (q_x, q_z)which is a component of the total plate load (q_1). In order to obtain a satisfactory solution, it is important to determine the role of either beam in forming the total bending strain. Having assumed at first, that the beams are bending independently under the loads q and q, we obtain the formulas to find the bending strain components $dy_1(z)$and $dy_2(x)$ which are caused by bending moments applied to the frontal beam and longitudinal beam, respectively. Then, using the equation of static equilibrium and the principle of solution multiplication, we may obtain two relationships between q_z and q_x to be found, on the one hand, and the known resultant plate load q_1, on the other hand.

Of great importance is knowledge a limit size of either beam, i. e. critical height h_{cr} and critical length l_{cr} when the composite beam loses its initial high bending rigidity as a solid unbroken body and begins to behave as a group of thin simple beams with little bending rigidity. At that moment, rock mass within the slope area deflects towards the mined-out space as a group of layers sliding on each other. The critical equilibrium state of the slope is such a special state when the limit equilibrium is reached only on the planes of contacts between layers, since the deformations have appeared but the rock mass strength on the planes of fixing is still high enough to prevent the slope body from its total failure.

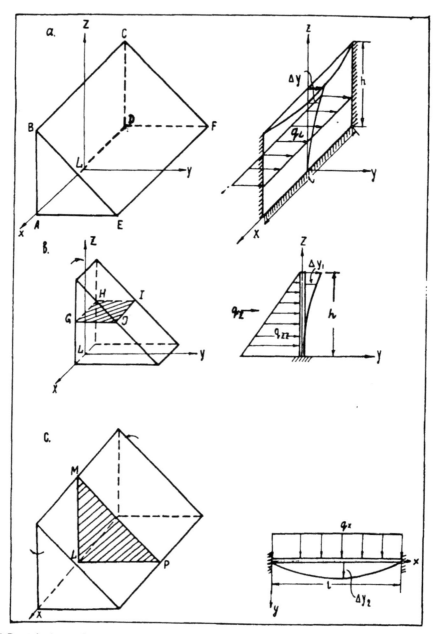

Fig.1. General scheme of open-cut slope bending calculation.

a) Slope-body prism cut off from adjacent rock mass is considered as a plate, restrained at its longitudinal edges and from below, and subjected to a lateral load with intensity of q   (z) per unit area (or q per unit length)on its rear side.

b) Vertical strip of unit length, cut out of the slope-body prism, as a frontal cantilever beam fixed from below and subjected to a load with intensity of q   (z) per unit area (resultant force q)

c) Slope-body prism undercut from below at the level of the open-pit bottom as a longitudinal beam fixed at both ends and subjected to uniformly distributed load with intensity of q per unit length.

   The arrow indicates the direction of bending and shaded sections show the cross sections of beams used in calculation of moments of inertia I and I.

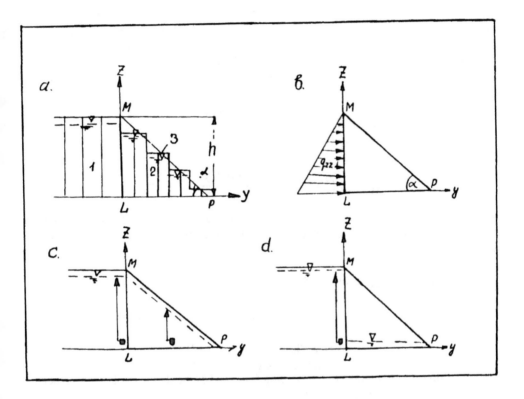

Fig.2. Cross sections of the slope-body prism.

a)General view of layered slope. 1-footwall area, 2-slope area, 3-surface of a bench, -ground water level.

b)Distribution of external load with intensity of q(z) per unit area subjected to vertical edge of slope-body prism.

c) Totally watered lope area. High ground water level is attributed to a poor drainage effect due to low permeability of rocks across the strike of the stratified structure.

d) Totally drained slope area. (Usually due to horizontal drainage wells bored from the bench at the level of pit bottom)

Shaded point with arrow shows position of water pressure gauge.

REDUCTION OF SOLUTION ALGORITHM

In order to obtain a physically clear pattern of calculation of the plate deformation without complex calculations of the theory of plates and shells, we can combine the required solution of two-dimensional problem from two solutions of one-dimensional problems, using the principle of solution multiplication that is valid for the functions which satisfy the Poisson equation. In accordance with this principle (Carslaw, 1945), the two-dimensional problem solution for dy (z,x), might be presented as the product of two one-dimensional problem solutions that have been derived under properly transformed boundary conditions. For the plate considered (Fig 1)we obtain

$$dy(z,x_0) = dy_1(z) \cdot \frac{dy_2(x_0)}{dy_{2\max}(x)} \qquad (1)$$

or

$$dy(z,x_0) = dy_2(x) \cdot \frac{dy_1(z_0)}{dy_{1\max}(z)} \qquad (2)$$

where $dy_1(z)$ and $dy_2(x)$ are the horizontal displacements of the frontal and the longitudinal beams, respectively, at the points z and x under the bending loads in the one-dimensional problem.

Since $dy(z_0,x_0)=dy(x_0,z_0)$, it follows that

$$dy_{1\max}=dy_{2\max} \qquad (3)$$

and since the bending strains of the beams and the acting loads are linearly related, we may write using (3):

$$dy_{1\max}=q_z f_1(z)_{\max} \qquad (4)$$
$$dy_{2\max}=q_x f_2(x)_{\max} \qquad (5)$$

where $f_1(z)\max$ and $f_2(x)\max$ are the bending functions of the deflection line of the beams at the points of the maximum bending strains.

Taking into account the equation of the statics.

$$q_1=q_z+q_x \qquad (6)$$

Where q_1 is the load acting on the unit length of the plate, we obtain

$$q_z=q_1/(1+\lambda) \qquad (7)$$
$$q_x=\lambda q_1/(1+\lambda)$$

where

$$\lambda=f_1(z)_{\max}/f_2(x)_{\max} \qquad (8)$$

FRONTAL BENDING DISPLACEMENTS

The pit slope is considered as an infinite beam[*] bending in frontal direction. The bending deformation of the frontal beam can be described by comparing the moments of loads and resisting forces acting on the planes of weakness in rock mass /Ryumin, 1979/. The frontal beam of the unit length (Fig.3) is loaded by external lateral line load q   (evenly

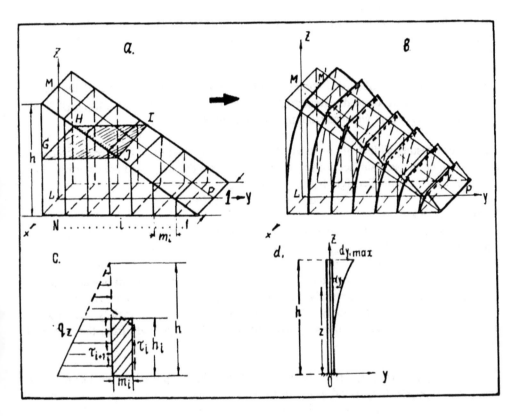

Fig.3. Bending of the frontal beam.
General view of the slope strip before (a) and after (b)bending; I-number of layer; GHIJ is the section at which the bending rigidity   is determined.
c) Diagram of external load q   subjected to the i-th layer; shearing resistance stresses   create a moment preventing the i-th layer from sliding relative to i+1-th and i-1-th layers in vertical direction.
d) Deflection curve of the beam before and after bending.

distributed along the plate length (Fig.1a)) with triangle intensivity q along the beam height (Fig.1b),such as lateral earth pressure q_{zz1}

$$q_{zz1}=\xi\gamma(h-z) \qquad (9)$$

and hydrostatic underground water pressure q_{zz2} (Fig.2)

$$q_{zz2}=\gamma_0(h-z) \qquad (10)$$

acting on the plane at the boundary between the slope area and the footwall.

Here ξ is the lateral earth-pressure coefficient expressed as $\xi=\nu/(1-\nu)$, ν is the Poisson's ratio, γ is the unit weight of rock, h is the height of opencut slope, z is the vertical distance measured from the plane of pinching the beam, that is, from the open-pit bottom, and γ_0 is the unit weight of water.

Out of the two types of lateral forces which are horizontal components of the active resultant force, the force of the greatest magnitude, i.e. the force $q_{zz}=max(q_{zz1},q_{zz2})$**) has to be taken into account when calculating the moment of loads.

The forces resisting sliding of one layer of the composite beam upon another are cohesion and friction on the side surfaces of strata within the beam discussed and are determined by means of the Coulomb-Terzaghi type equation.

$$\tau=\sigma_{eff}\tan\varphi_1+c_1 \qquad (11)$$

*)That is a case $\lambda=0$, $q_z=q_1$

**)In accordance with the Terzaghi-Bishop principle of effective stress, $\sigma_{total}=\sigma_{eff}+p$, where σ_{total} is the total stress, σ_{eff} is the effective stress, and p is the fluid pressure. In order to obtain a magnitude of the horizontal (total)force acting on the slope body, remind that since σ_{eff} must not be negative, then, if $\sigma_{eff}=0$, σ_{total} will be equal to the water pressure, $\sigma_{total}=q_{zz2}=p$; in case if σ_{eff} is strictly more than 0, σ_{total} will be equal to the reaction force, i.e. $\sigma_{total}=q_{zz1}$. And since usually $\gamma_0>\xi\gamma$, here below we consider mainly the case of hydrostatic water pressure as the only load, acting on the slope body from the foot-wall mass.

Where, $\sigma_{eff}=q_{zz}(z)-p(i,z)$, $q_{zz}(z)$ is the external lateral load, applied to the beam at the height z, and p(i,z)is the ground water pressure at the contact planes between the i-th and I+1-th. layers at the height z above the pit bottom, c_1 is the cohesion and φ_1 is the angle of friction on the contact surfaces of layers (Fig.3)

Furthermore, a reactive force and its moment are induced at the pinched section of the beam and hence prevent the layer rotation and displacement within the pinched portion.

We now wish to dwell. In more detail, on the general theoretical aspect. Instead of usual analysis of limit equilibrium conditions for a solid body after elastic deformations have ceased, it is rather more important to know whether the stratified rock mass has sufficient strength along the planes of weakness to resist elastic bending deformations, affected by horizontal forces.

The solution of this problem may be general one, but in order to simplify calculation procedure, consider a special case of the strongly vertical layers forming the slope body.

In this case, gravitational forces do not act on the vertical planes and the beam is subjected to the horizontal load and prevents the layers from sliding relative to each other only owing to the strength along the contact planes of layers.

The deformation process (in the sense mentioned above)in the frontal beam is initiated as the moment of shearing forces M_1 and that of resisting forces[*] M_r are equal . It is simple to show, that

$$M_1 = q_{zz}(h-z)^2/6 \qquad (12)$$

and

$$M_r = \sum_{i=1}^{N}\left[\left\{\int_0^{h_i}\left(\sigma_{eff,i}\cdot\tan\varphi_{1,i}+c_{1,i}\right)dz\right\}mi\right] \qquad (13)$$

[*]Forces resisting layers sliding, i.e. friction and cohesion, are internal forces occurring on the contact planes of layers under external line load q_z per unit length applied to the interface of area of the slope face and area near the slope face ABCD(Fig 1). Homogeneous medium , with $\varphi_{1,i}$ and $c_{1,i}$ and m_i constant .

where m_i and h_i are the thickness and the height, respectively, of the i-th layer out of N layers within the beam in question.

As can be seen from (13), the expression for the moment of resisting forces is obtained by addition of moments of frictional and cohesive forces acting between the layers. This expression, in general, can not be simplified to the algebraic formula in the case of varying values of $\varphi_{1,i}$ and $c_{1,i}$ and m_i for different layers.

Since, as a rule, individual properties of layers and their contact planes are not known, it is important to derive a practically applicable expression for an idealized

Assuming that $m_i = m_0$ =const and taking into account that $i=(h-z)/(m_0\tan\alpha)$, one may express the critical height of the open pit slope as a function of the strength parameters on the contact surfaces as follows:

$$h_{cr}=3c_i/\gamma' \qquad (14)$$

where γ' is the conventionally taken value of shear forces on contact surfaces, as defined depending on the lateral force and ground water conditions (Fig.2) within the slope area (Table 1).

Table 1

Ratio q_{zz1}/q_{zz2}[*]	Groundwater conditions within the slope area	
	totally drained slope[**]	toally watered slope
more than 1	$\xi\gamma(\tan\alpha-2\tan\varphi_1)$	$\zeta\gamma(\tan\alpha-2\tan\varphi_1)+\gamma_0\tan\varphi_1$
less than 1	$\gamma_0(\tan\alpha-2\tan\varphi_1)$	$\gamma_0(\tan\alpha-\tan\varphi_1)$

As the limit equilibrium of moments of loads and resisting forces is reached the separation of contact planes between the layers will occur and as a result of lack of both coherence and bending rigidity the bending deformation will take place.

*)If $q_{zz1}>q_{zz2}$, then $q_{zz}=q_{zz1}$, otherwise $q_{zz}=q_{zz2}$.

**)The case of totally drained opencut slope takes place in particular while draining through horizontal wells when in the area above the horizontal well level the condition $H(z)=z$, $p=0$ is fulfilled (H is the ground water head).

The horizontal displacement dy within the deflection line of strata can be defined by means of the differential equation

$$\frac{d^2(dy_1)}{dz^2} = \frac{dM}{EI_1} \qquad (15)$$

at

$$dM=M_1-M_r' \qquad (16)$$

EI_1 is the bending rigidity of the beam and is defined by

$$EI_1=N(z)Em_0/12 \qquad (17)$$

$N(z)$ is the number of layers within the beam section, M_r' is the moment of resisting forces bue to the friction between the layers which move relative to each other on their contact surfaces while bending within the previously intact beam/plate; it is assumed that in consequence of disturbing the contact planes the cohesive forces along them will not exist and hence M_r' has to be evaluated by (13)at $c_{1,i}=0$

By carrying out the summation of $dm_i/EI_{1,i}$ over all the layers in the beam we obtain

$$\frac{d^2(dy_1)}{dz^2} = \frac{2\gamma'(h-z_n^2)}{Em_0^2} \qquad (18)$$

from where, at $dy_1=0$(at $z=0$) and $d(dy_1)/dz=0$(at $z=0$), the horizontal displacement dy1 may be obtained by

$$dy_1 = \frac{\gamma' h^4 Z_n^2 (Z_n^2 - 4Z_n + 6)}{6Em_0^2} \qquad (19)$$

where z_n is the normalized value, $z_n=z/h$.

The maximum of the horizontal displacement at $z_n=1$ is expressed by

$$dy_{1max}=\gamma'h^4/(2Em_0^2) \qquad (20)$$

and used for calculation of λ.

LONGITUDIONAL BENDING

In order to analyze the deformation behaviour of the plate under study, one should dwell on the question as to how important is the effect of finite dimensions of this plate Ryumin, 1990; 1992a;1992b/. Consider for this purpose the plate as a longitudinal beam being cut off from the rock massif on the pinching plane of the frontal beam (see above) and restrained along the strike(i. e the case λ is equal to infinity, see(7),$q_x=q_1$(Fig.4).

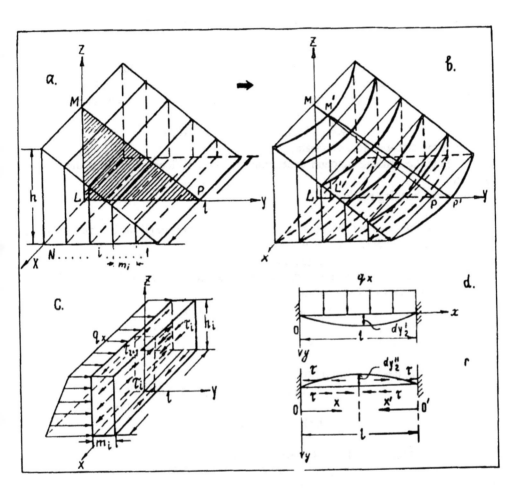

Fig.4. Bending of the longitudinal beam.

General view of the slope undercut from below before (a) and after (b) bending deformation; LMP is the section at which the bending rigidity is determined.

c) Longitudinal bending of i-th layer subjected to external uniformly distributed load with intensity qx per unit length; cohesional and frictional strength   create a moment preventing the i-th layer from sliding along adjacent layers in longitudinal direction.

d) Deflection line of the beam due to external load with intensity of   per unit length only.

e) Imaginary deflection line of the beam subjected to resistance frictional forces only.

The analysis of the conditions that contribute to initiation of deformations of the longitudinal beam is similar to that of the frontal beam above/see Ryumin, Lasarev, 1990/.

The critical length of the open-pit slope at which the bending rigidity is lost as a result of rock mass loosening is expressed by

$$l_{cr}=4c_{ih}/\gamma' \qquad (21)$$

if $h=h_{cr}$ and if $c_{ih}=c_i$ and $\varphi_{ih}=\varphi_i$ are not dependent on the movement direction of the layers. The horizontal displacement of the beam subjected to the excessive load q_x uniformly distributed along the beam's length l, may be described by the equation/Kinasoshvili, 1975/.

$$dy_2=q_x l^4 f(x/l)/(EI_2) \qquad (22)$$

Where x is the distance from the beam,

f(x/l) is the bending function depending on the conditions of pinching the ends of the beam,

EI_2 is the bending rigidity.

If the central axis moment of inertia I_2 is defined for nonlayered media (that is before reaching the critical state)by means of equation

$$I_2=h^4/(36\tan^3\alpha) \qquad (23)$$

it will be defined for the layered (that is after reaching the critical state) rock mass composed of N layers, at the layer thickness m_0,by

$$I_2=h_2 m_0^2/(24\tan\alpha) \qquad (24)$$

that is, the bending rigidity will decrease by at least $2/3N^2$ times, N being the number of layers which form the opencut slope.

The active load q_x has to be determined as

$$q_x=\xi\gamma h^2/2 \qquad \text{if } \xi\gamma>\gamma_0$$

or

$$q_x=\gamma_0 h^2/2 \qquad \text{if } \xi\gamma<=\gamma_0 \qquad (25)$$

In order to find dy_{2max} and $f_2(x)_{max}$, the approximated relation

$$dy_{2max}=\gamma'' l^4/(6Em_0^2) \qquad (26)$$

may be used, where γ'' is to be adopted from Table 2.

Table 2

Ratio q_{x1}/q_{x2}	Groundwater conditions within the slope area	
	totally drained slope	toally watered slope
more than 1	$\xi\gamma(\tan\alpha-2(h/l)\tan\varphi_{1h})$	$\zeta\gamma(\tan\alpha-2(h/l)\tan\varphi_{1h})+\gamma_0\tan\varphi_{1h}+\gamma_0(h/l)\tan\varphi_{1h}$
less than 1	$\gamma_0(\tan\alpha-2(h/l)\tan\varphi_{1h})$	$\gamma_0(\tan\alpha-(h/l)\tan\varphi_{1h})$

EVALUATION OF λ

With substitution of dy_{1max} (20)and dy_{2max} (26) into the equations (4),(5) and then, taking into consideration, that here $q_x=q_z$ we obtain from (8)
$$\lambda=3\gamma'h^4/(\gamma''l^4) \qquad (27)$$
By the time the greater deformations have occurred at $h=h_{cr}$ and $l=l_{cr}$, the value of λ is about 1 and hence the active load is distributed equally between the frontal and longitudinal beams. With increasing the opencut slope length, λ rapidly decreases whereas at $l/h_{cr}=3$, $\lambda=0.03$ only.

COMPARISON TO ACTUAL DATA

In April 1972, the sliding-down of the upper bench at the Vakhrushev open pit of the Kuzbass coal basin occurred for the first time under similar conditions, accompanied by significant surface ground movement in the vicinity of the open pit (Ryumin, 1977). The results of open-pit surface survey and of in situ measurement of bench slope deformations indicated (Fig.5) that the slide movement was not restricted to the overlying loams, which was supposed from the visual observation, but involved also opencut benches composed of the basement rock. At openpit district No 2 (western part)where the slope the slope slide took place, the average slope height is 100m, the slope angle is 30 , the deformed slope length is 300m, coal-bearing strata are sandstones, claystones and siltstones and the Permian coal seams are inclined at 70to 75degrees into the rock massif The upper bench containing loams of about 15-m thickness was entirely destroyed Cracks, grooves and steps as a consequence of sliding were encountered300m along the strike of the slope face and 150m deep in the slope body, The water tables in the blasting holes on the pit benches before the slope failure stood about 3to 5m below the berm level and therefore one may use the calculation pattern applied for the to tally watered opencut slope. The condition of limit equilibrium is satisfied at the following values of cohesion and friction angle on the contact planes of layers for $h_{cr}=100m$ (see eq(14),at $\gamma'=\gamma_0(tan\varphi_1-tan\alpha)$, Table 3)

Table 3

c1(kpa)	197	150	100	50	20	0
φ_1^0	0	8	16	23	27	30

From these parameters, the values of $c_1=100kpa$ and $\varphi_1=16^0$ are supposed to be a most probable pair of parameters for the Kuzbass coal field and correspond to the least values of contact plane strength characteristics obtained in the tests on the one-plane shear test device.

Knowing the maximum. Value of horizontal displacement, $dy_{max}=10m$, the average friction angle, $\varphi_1=16^0$ and the layer thickness, $m_o=3.7m$, one could calculate from (20)the effective Young modulus, $E=10^6kpa$.

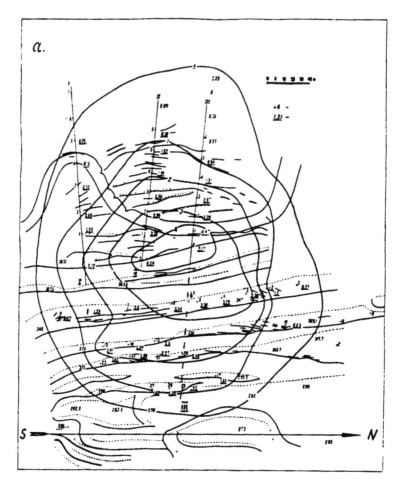

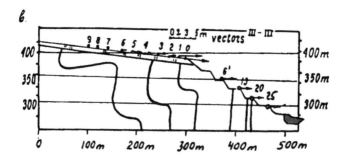

Fig.5. Mining plan and profile of the landslide area in extraction district No2 Vakhrushev openpit.
a) Mining plan with isolines of displacements in meters(summer 1972); 04 is the number of the fixed poinnt;1.51 is the displacement(m) from 29.4.72 to 17.10 72; dotted lines correspond to cracks.
b) Profile 111 with displacement vectors (from 29.4. to 17.10.72); " " " " strikened are flat-lying loams. In the coal-bearing series composed of intermittent, fine-grained sandstones, siltstones and coal beds, only the coal seams are shown in the Figure.

The measurements of the slope movement were carried out with different degree of accuracy, first, because the great movements that occurred in April 1972 were found to be quite unexpected for the open-pit surveyors and , secondly because the marked points that were fixed on the pit benches in the spring of 1972 became almost inaccessible to observe after the bench failure in the winter of 1972-1973

Table 4contains the results of the measurements of horizontal displacements carried out at different times.

Table 4

Elevation of pit benches(m)	May,1972		Oct. 1972		Apr. 1973	
	Height above pit bottom(m)	Horiz. Displ.(m)	Height above pit bottom(m)	Horiz. Displ.(m)	Height above pit bottom(m)	Horiz. Displ.(m)
385	100	10	110	-	110	16
365	80	9	90	-	90	-
345	60	4	70	3.3	70	7.3
325	40	2	50	2.5	50	4.5
305	20	cracks	30	2.5	30	2.5
285	0	-	10	0.8	10	0.8

The values obtained in the mine survey on the 15th of May1972were determined with a low accuracy of about ±1 to 2m which corresponded to the accuracy requirements of the general opencut mining plan before the slope failure. The instrumentally measured displacements of the marked points from April to October 1972(Fig.5) were determined with an accuracy of 0.01m. The magnitudes quoted as of April 1973 were obtained as follows the horizontal displacement magnitudes on the pit benches for the period of 1972/1973were summed up while the maximum displacement of the upper bench edge was then measured using a newly established line of marked points on the surface of the central part of District No 2 above The results obtained during the mine survey on May 15,19772were compared with the values calculated from eq. (19) (see Table5).

Table 5

Elevation of pit benches(m)	height above pit floor(m)	Horizontal displacements(m)	
		measured	calculated
385	100	more than 10	10.0
365	80	10	7.7
345	60	4	4.8
325	40	2	2.5
305	20	Cracks	0.7

Taking into consideration , that there are the layers of the thickness ranging from 0.5m to 10m and even up to 15m within the open-pit slope area and that the different lithological

features of thin -lamin ated claystones, on the one hand and of the massive coals and
sandstones, on the other hand , predetermine a rather significant variabilityof Young s
modulus of different layers, one may consider that the measured and calculated horizontal
displacements evaluated without distinguishing individual properties of layers agree
reasonable well.

Fig 5 shows the results of ground movement measurements that have been carried out
from April to October 1972 by using the marked points established on the berms of pit
benches. The measured values are compared to the calculated ones . By means of the
formulae(1)(2)(19)(26)the horizontal displacements at the points x and z have been
calculated. Using Eq.(26) the nondimensional relationships $\beta_1(z)=dy_1/dy_{1max}$ and
$\beta_2(x)=dy_2/dy_{2max}$ are round and then $dy=dy_{1max}\beta_1(z)\beta_2(x)$ is obtained ,.assuming $\lambda=0$ (Fig.
6)

As seen from Fig 6, the general trend to decreasing horizontal displacements downwards
and away from the center of the center of the deformed area is kept quite strictly, The
actual displacements are not symmetric with respect to the center line and are greater than
the calculated ones, The fact that the actual deformations fail to follow the symmetrical
pattern can be explained by that rock in the southern end of the pit is pinched more rigidly
than in the northern end , because the brachysincline in the South is a closed fold and
forms the boundary of mining activity(and so the profile pinching), whereas the northern
boundary of the slope slide area is given by the extensive zone of gradual decrease of the
height and the slope angle of the pit wall. The discrepancy of the actual and calculated
amplitudes of displacements, significant in particular at the lower portion of the pit slope,
arises form the fact that the deformations in the summer of 1972 were intensified with
increasing depth of coal mining followed by the expansion of the slide area along the
strike.

Thus, the difference of actual and calculated deformations principally results from the
adopted calculations pattern based on the assumptions which are realized at the working
opencut slope only partially.

An additional check-up of the reliability of the calculation method took place after the
open pit was deepened down 112m deep at the same value of the pit slope angle in 1972-
1973. The greatest displacements of the upper bench observed in the spring of 1973
reached 16m and had a reasonably agreement with the predicted greatest displacements
calculated by

$$(dy_{ma}x)''=(dy_{max})'(h''/h')^4 \qquad (28)$$

where $(dy_{max})''$ and h'' are the displacement the height of the openpit after deepening
whereas $(dy_{max})'$ and h' are the same parameters before deepening.

CONCLUSIONS

The above approach to the calculation of opencut slope deformations that does not
pretend to the mathematical strictness and accuracy because of rather rough

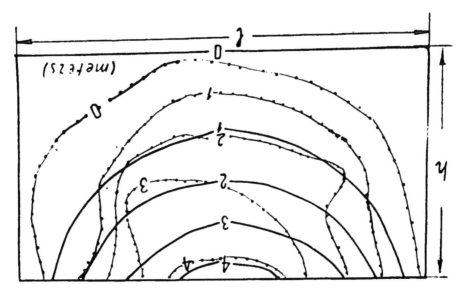

Fig.6. Comparison of calculated (solid lines) and measured (dotted lines) values of displacements at district No 2 (in meters, from 29.4. to 17.10.72).

approximation as far as the bending deformation analysis of the plate is concerned enables nevertheless a reliable qualitative assessment of the effect of bending deformations on the rock mass rigidity factors. Algorithm of using the suggested relationships is intended for obtaining the necessary quantitative strength and deformation characteristics of rocks as far back as at initial stages of mining operations. The strength and deformation behaviour of the rock mass within the slope area is analyzed using the results of rock movement observations on the opencut slope as a whole under the state of stress approaching the limiting state of stress within the slope part subjected to deformations. Based on these strength and deformation data one may design stable non-operative open-pit banks under boundary conditions with taking into account the permissible surface displacements and ,if it is necessary, one may plan a drain system or introduce any changes into excavations order with variations of the permissible working front length along the strike.

REFERENCES

1. Carslaw H. S.(1945). Introduction to the mathematical theory of conduction of heat in solids.2nd ed. New-York, Dover Publication.
2. Fisenko G. L.(1965). Stability of slopes and dumps in open-pit mines. Nedra, Moscow (In Russian.).
3. Kinasoshvili R. S.(1975). Strength of materials. Moscow (In Russian.).
4. Ryumin A. N.(1977). Research Report 02-15-02. Fondy VNIMI, S.-Ptb., 27-105. (In Russian.)
5. Ryumin A. N.(1979). The role of ground water in the deformation process of the open-pit slope containing steeply dipping layers. Trudy VNIMI, Sb. 112,45-51 (In Russian.).
6. Ryumin A. N., Lasarev I. A. (1990). Investigation of the role of ground water heads in development of the landslide at the Angren open-pit mine. Trudy VNIMI, Sb. 136,126-129(In Russian.).
7. Ryumin A. N. (1992a). Analysis of opencut slope stability. Inzhenernaya Geologia, No 1. (In Russian).
8. Ryumin A. N.(1992b). On the role of loosening of clays in the development of opencut slope slides. Ugol' No 4.(In Russian.).

Proc. 30th Int'l. Geol. Congr., Vol. 23, pp. 141-150
Wang Sijing and P. Marinos (Eds)
© VSP 1997

A Study on the Safety Thickness of Top Plate above Shallow Underground Cavities

HUANG RUNQIU   CHEN S. Q.
Chengdu Institute of Technology, Chengdu, Sichuan 610059, China

Abstract

In this paper, the factors affecting the safety thickness of top plate above shallow underground caves in the mountain-city are analyzed under the reference of previous study. The failure mechanism of mutual process between the foundation and underground caves, the factors affecting on the safety thickness of top plate and the related variability laws are studied by means of FEM numerical simulation. The safety thickness of top plate forecasting model is obtained by means of stepwise regression analysis. The paper is also studied how the fractures in top plate affecting the safety thickness of top plate.

Key words: Shallow underground cavities, Foundation, Numerical simulation, Safety thickness of top plate failure mechanism, Forecasting model, Chongqing City

INTRODUCTION

As a result of historic factors, abundant underground caves were built in the famous mountain-city, Chongqing, China' biggest city in southwestern. Along with the development of the city in resent years, many underground constructions have been or will be built including underground railroads, traffic tunnels and storehouse. The caves have common characteristics: shallow and was built in bed rocks. Since the short supply of urban land resource of the city, it is unavoidable for buildings to be constructed above the caves. In order to keep stability of surface and underground constructions, the safety thickness of top plate of the caves have to be studied. The problem has been paid close attention to along with higher buildings have been constructed since 1980's.

As early as the end of 70's, Protodynakonov's theory had been used to determine the safety thickness of top plate in Chongqing. Later, somebody studied this problem by boundary numerical simulation. Obviously, Protodynakonov's theory is only suitable for loose rocks not for intact bedrocks. In the second theory, it is not only conservative to use the critical depth as the safety thickness but also difficult to confirm the critical depth . The critical depth changes depend on above load, rock structures and the shape of caves. In this paper, a method of determining safety thickness of top plate, based on FEM numerical simulation, is put forward.

ENGINEERING GEOLOGICAL CONDITIONS AND THE FACTORS AFFECTING THE SAFETY THICKNESS OF TOP PLATE

The research area is located at the southern plunging end of the Long Wang Dong anticline, eastern Sichuan fold belt, Chongqing city. The area has poorly developed fault structure. Joints are mainly the original torsion or tension ones vertical to rock strata. Basic seismic intensity is less than VI.

The rock strata, which is almost horizontal, in the research area are mainly sandstone and mudstone of Jurassic Upper Shaximiao formation (J2s) with complete structure and the thickness of about 200 meters. There are three layers of sandstone, which are mainly gray or purplish red medium fine to medium coarse grain feldspar-quartz sandstone with calcareous or siliceous cements. The layers have similar physical characteristics: Elastic modulus (E) is about 6000MPa, Passion's ration (μ) is between 0.15 and 0.25, C is between 4 and 6MPa, tgφ is about 1.0 and tensile strength (St) is about 1.2MPa.   There are also three layers of mudstone, which is mostly thick purplish red and silty, with different physical characteristics: E is about 4000MPa, μ is between 0.2 and 0.35, C is between 2.5 and 3.5MPa, tgφ is about 0.9 and St is about 0.2 MPa.

According to pre-research achievement [1] , the safety thickness of the top plate is influenced by the followings factors:

1) Lithological characters and the strength, especially the tensile strength;
2) Foundation pressure and its location, width and depth of foundations;
3) Shapes of caves, including span and height-span ration.

NUMERICAL SIMULATION OF FAILURE MECHANISM AND THE SAFETY THICKNESS

Establishment of computation model and selection of parameters.
According to lithological characters of the shallow caves in the research area, rocks above top of the caves are mostly sandstone while mudstone at the two sides. Fig. 1 is a basic model of the caves.

Based on above model, the factors and their combinations are considered   separately during the computing.   Main factors are: foundation pressure (P), span (L), width of the foundation (B), load eccentric rate (e, $e = \dfrac{e_0}{L}$), height and span rate (H , $H = \dfrac{H_0}{L}$) and top plate tensile strength (St). The computation model (Fig. 2) was established according to above geological model. It should be pointed out that the foundation can be perpendicular, oblique or parallel to the direction of the caves. The most dangerous situation, the parallel situation, is considered in our study. Table 1 is the computation parameters used in the model.

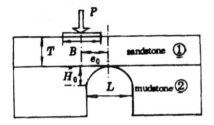

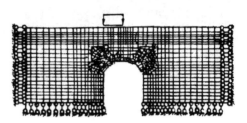

Fig. 1. Basic Model   **Fig. 2.** Computation Model

Table 1   The table of computation parameter

	E(MPa)	μ	C(MPa)	φ	C_r(MPa)	$φ_r$	γ(kN/m³)	S_t (MPa)
Sandstone	6000	0.2	6.5	50	1	30	24.65	1.0
Mudstone	4000	0.4	3.5	40	0.5	20	25.32	0.1

It is indicate that St is a most important parameter.   The values of St in table 1 is defined according to experience and in-situ tests.

Mechanism analysis of cave and foundation interaction system
In order to understand the failure mechanism of cave-foundation interaction system, a common situation was selected to be a computation model: L=10m, B=2m, T=3m, H=1/3 and e=0. The foundation pressure (P) keeps gradually increasing until the cave completely damage, thus the deformation and failure distribution of the top plate is studied during this course.

Fig. 3 shows the deformation and failure situation in the case when the foundation pressure is 1MN, it was stable all around the cave and the top plate. Tensile stress was accumulating at the top of the cave, but was not large enough to damage the cave.   The largest displacement downward was only 2 mm.

While the foundation pressure is inceased to 5MN, tensile stress is still accumulating at the cave top, which corresponds a 0.5m tensile failure area as shown in Fig. 4.   The deformation is getting larger and the displacement reaches to 5 mm, but the system of foundation-cave is still at the stable condition.   There is a certain degree of damage at the cave top if we consider the cave alone as a system.

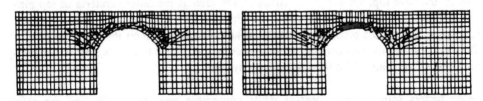

Fig. 3 The figure of failure area while P=1MN   Fig. 4 The figure of failure area while P=5MN

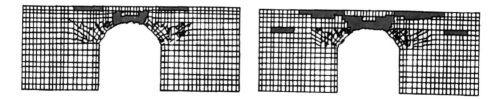

Fig. 5. The figure of failure area while P=10MN   Fig. 6. The figure of failure area while P=16MN

When the foundation pressure is 10MN (Fig.5), tensile stress is made further accumulation at the cave top. Furtherly, It has little accumulation on the surface near the foundation and results in some small scale damage which makes the tensile failure area near the cave top develop upward. The deformation is getting larger with displacement of 10 mm. The cave has been seriously damaged.

When the foundation pressure increases up to 16MN (Fig. 6), the displacement near the foundation and the cave top has reached up to 16 mm, tensile stress is highly accumulated between the foundation and the cave top. The surface tensile failure area and the tensile failure near the top of the cave has been joined up together. The foundation will start shearing slip downward along the tensile surface. The foundation-cave system has been totally damaged.

According to above description, the whole failure process can be sum up as: Foundation pressure action-Tensile stress accumulating on the cave top-Development of the tensile failure area and joined up to the surface -Foundation shearing slip-Totally damaged. The failure mechanism of foundation-cave system can be summarized as "tensile-shearing failure".

Safety thickness analysis of top plate
In accordance with above failure mechanism analysis, under the action of   upper load of foundation,   failure of foundation-cave system start from the top of the cave, then develops upward, and finally join up to the surface and shearing slip happen.   Before the failure joined up to the surface, the whole foundation is stable, but the top of the cave has been damaged.   This is not allowed for traffic tunnel and underground storehouses. The key problem is to find out the critical thickness of the top plate to make the foundation-cave system at stable condition and to ensure both foundation and cave will not lose their functions of usage.   Since it is not necessary to ensure no tensile failure area on the cave top, we use 0.5m tensile failure area as the failure creteria of top plate. The computation method is: adjusting the thickness of top plate and make numerical simulation during this process until 0.5m failure area occurred on the cave top (Fig. 7). The thickness at this moment can be considered as the critical thickness, that is, the safety thickness of top plate of the foundation-cave system.

Other factors should keep constant while computing one factor affecting the safety thickness of top plate. Constant values are common values for the factors. Table 2 is constant and changing values of the factors.

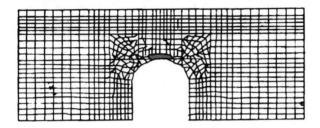

Fig. 7. The figure of typical failure zone

DISCUSSION OF RESULTS

In accordance with above idea, we have got numerical simulation results under 30 different factor compositions (Table 2). The relations between each factor and the safety thickness of top plate are showed from Fig. 8 to Fig. 13.

Fig. 8 is the relation curve of L and T. It is clear that when L getting smaller, the safety thickness of top plate (T) getting smaller. Otherwise, it will getting larger. The best relation between L and T can be described by a exponential function as listed in table 3.

The relation curve of P and T (Fig. 9) indicates that T will increase along with the increasing of P. The curve slopes more gently while P is getting bigger. The best relation equation of P and T is listed in table 3. From the mathematical equation provided in the table 3, when p approaches to infinity, T will not convergence. Actually, the foundation pressure P can not be infinity. According to the data in the research area, the largest bearing capacity of sandstone foundation is not larger than 15 Mpa (about 30MN foundation pressure if the foundation wideth is 2m). Thus, it can be defined safety thickness of top plate (T_0) while P=30MN as the safety depth of the cave. In this case, it can be considered that the cave has not affection to foundations. Of course the safety depth is related to foundation pressure, shape of the cave and tensile strength of top plate. The estimated safety depth is about 15 m when L=10m, B=2m, H=1/3, e=0 and St=1.0Mpa.

Fig. 10 and Fig. 11 show that T will reduce rapidly if B of e increase. It can be considered that the cave has no affection to above construction when B or e increases to a certain value at which T equals to 0.

Fig. 12 indicates when H is smaller than 1/3, T will increase rapidly if H decreases. The safety depth is at the largest position when H=0. T will keep almost stable when H = 1/3.

Table 2. The composite of affecting factors and the outcome of computation

	No.	L (m)	P (MN)	B (M)	e	H	St (MPa)	T (m)		No.	L (m)	P (MN)	B (M)	e	H	St (MPa)	T (m)
	1	4	10	2	0	1/3	1	3.6		16	10	10	2	0	1/3	1	5.2
	2	7	10	2	0	1/3	1	4.3		17	10	10	2	2/5	1/3	1	4.4
L	3	10	10	2	0	1/3	1	5.2	e	18	10	10	2	3/5	1/3	1	4.1
	4	15	10	2	0	1/3	1	6.5		19	10	10	2	4/5	1/3	1	3
	5	20	10	2	0	1/3	1	8.8		20	10	10	2	1	1/3	1	2.4
	6	10	6	2	0	1/3	1	3.4		21	10	10	2	0	1/2	1	4.4
	7	10	10	2	0	1/3	1	5.2		22	10	10	2	0	2/5	1	4.5
P	8	10	16	2	0	1/3	1	8.7	H	23	10	10	2	0	1/3	1	4.7
	9	10	20	2	0	1/3	1	10.5		24	10	10	2	0	1/4	1	5.3
	10	10	24	2	0	1/3	1	11.8		25	10	10	2	0	0	1	7.2
	11	10	10	7	0	1/3	1	2.1		26	10	10	2	0	1/3	1	5.2
	12	10	10	5	0	1/3	1	3.3		27	10	10	2	0	1/3	1.2	5.7
B	13	10	10	3	0	1/3	1	4.9	St	28	10	10	2	0	1/3	1.3	3.2
	14	10	10	1	0	1/3	1	5.5		29	10	10	2	0	1/3	1.4	2.4
	15	10	10	0	0	1/3	1	5.9		30	10	10	2	0	1/3	1.5	2.1

Table 3. Equations between safety thickness of top plate and factors

		The best relation equation	Linear relation equation
L	T	$T = EXP(0.0548 \times L) \times 2.924$,　$r = 0.995$	$T = 0.3189 \times L + 2.108$,　　$r = 0.980$
P	T	$T = P^{0.8932} \times 0.7190$,　　　　$r = 0.996$	$T = 0.6439 \times P + 1.0286$,　　$r = 0.994$
B	T	/	$T = -0.55 \times B + 6.1$,　　$r = 0.987$
e	T	/	$T = -0.6216 \times e + 5.208$,　　$r = 0.993$
H	T	/	$T = -5.9303 \times H + 6.9793$,　　$r = 0.968$
St	T	$T = EXP(-1.86 \times St) \times 33.91$,　$r = 0.999$	$T = -6.338 \times St + 112.4324$,　　$r = 0.980$

The relationship between St and T is exponential function (Fig.13) as described in table 3.

PREDICTIVE MODEL OF SAFETY THICKNESS OF TOP PLATE

Above discussion shows how each single factor affects the safety thickness of top plate. Actually, the safety thickness of top plate is affected by muti-factors. Thus we establish a multi-factors equation to describe it based on above analysis.

Firstly, we assume:

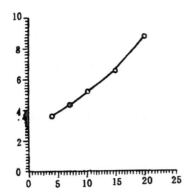

Fig. 8. The relation curve of L - T

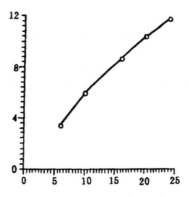

Fig. 9. The relation curve of P - T

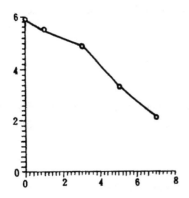

Fig. 10. The relation curve of B - T

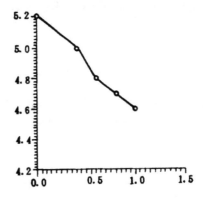

Fig. 11. The relation curve of e - T

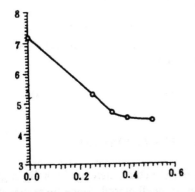

Fig. 12. The relation curve of H - T

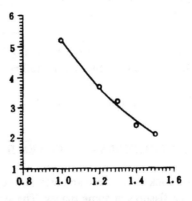

Fig. 13. The relation curve of St - T

$$T_0 = A_0 + A_1 \times L + A_2 \times P + A_3 \times B + A_4 \times e + A_5 \times H \qquad (1)$$

In this equation, A_0, A_1, ... A_5 are constants. According to table 4, we can find the constants A_0, A_1, ... A_5 by using stepwise regression analysis.

The following equation can be obtained by using data in table 2 and stepwise regression analysis:

$$T_0 = 0.0019 + 0.32 \times L - 0.5 \times B - 2.53 \times e - 5.7 \times H \qquad rr = 0.992 \qquad (2)$$

Here, rr is complex relation coefficient. Stepwise regression precision analysis indicate that the absolute error of stepwise regression value Tc and the computation value of FEM T_f, is less than 0.2m.

Then, a strength revision coefficient (Ks) is defined according to the top plate tensile strength (St) . The equation (1) is been revised by Ks and the revised value of Ts is the safety thickness of top plate:

$$T_s = Ks \times T_0 \qquad (3)$$

Ks can be calculated by the following method: Take the safety thickness of top plate Ts_0 as a standard while St equals 0. Ks equals to the safety thickness of top plate (T0) divided by Ts_0:

$$Ks = \frac{Ts}{Ts_0}$$

The values Ks show in the table 4. Ks has its distinct meaning which indicates the reducing times of the safety thickness of top plate if the tensile strength (St) increased by one time.

Table 4. Computation of Ks

St	T	Ks
1	5.2	1
1.2	3.7	0.71
1.3	3.2	0.651
1.4	2.4	0.46
1.5	2.1	0.404

The relation between top plate tensile strength (St) in table 1 and the strength revision coefficient (Ks) can be described by the following equation:

$$K_s = 6.52 \times EXP(-1.86 \times S_t) \qquad r = 0.991 \qquad (4)$$

THE SAFETY THICKNESS AFFECTED BY TOP PLATE FISSURES

Although most rocks in the research area have complete structure, there are high dip angle (>75°) shearing fissures at some places. The fissures are well closed. Since there was no reliable and unified evaluation method before, the thickness of top plate had been estimated conservatively by experiences.

Experiences indicate that fissures affecting the safety thickness are closely related the position in a cave. In order to describe the position of a fissure, the position coefficient S_f is defined as:

$$S_f = \frac{D_f}{0.5L}$$

D_f is the distance between the fissure and the center axis of the cave.

It is clear that $S_f=0$ means the fissure is at the top of the cave while $S_f=1$ means the fissure is right at the top of the side walls of the cave.

In order to study the affection of different position fissures to the safety thickness of top plate, let's take a group of conjugated high dip angle fissures as examples. The safety thickness with fissures ($S_f=0$, 0.5 or 1, Fig. 14) has been compared with the thickness have no fissures.

We have calculated three different safety thickness with the fissures at position , and . The shape of the cave and its loading are: L=10m, H=0.333, B=2m, e=0, P=10MN. The result is listed in the table 5.

We have calculated three different safety thickness with the fissures at position , and . The shape of the cave and its loading are: L=10m, H=0.333, B=2m, e=0, P=10MN. The result is listed in the table 5.

It is clear that fissures near the top of a cave have distinct affection to the safety thickness of top plate, generally the safety thickness increases 10 %. It has no obvious affection to the safety thickness when fissures are far away from the top of a cave.

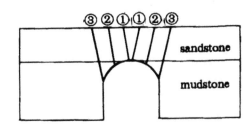

Fig. 14. The position of the fissure in the cave

Table 5. The result while there are fissures in the cave

	$S_f=0$	$S_f=0.5$	$S_f=1$
T_f(m)	5.7	5.5	5.2
T_0(m)	5.2	5.2	5.2
$(T_f-T_0)/T_0$	9.6%	5.8%	0

CONCLUSIONS

Summing up the studies of this paper, we can obtain following main conclusions:

(1)   The most important factors affecting the safety thickness of top plate is tensile strength of rocks of top plate (St), then the upper loading and the span;

(2)   The predictive model of safety thickness of top plate provided in this paper has not only considered multi-factors, but also high precision;

(3)   Only the fissures near the top of a cave have distinct affection to the safety thickness of top plate, no obvious affection if fissures are far away from the top of a cave.

REFERENCES

[1]   M. C. Wang, A. Badie, Underground cave affecting the foundation, *World Geology (Chinese)*, 1987, Translated by Zou Zhengsheng;

[2]   Shang Yuequan, et al, Numerical analysis method in engineering geology, *Publishing House of Chengdu University of Science and Technology*, 1991;

[3]   Probability and Mathematical Statistics, *Geological Publishing House*, 1981;

[4]   He Gaoyi, et al, Safety thickness selection method between top of artificial caves and foundation of constructions in Chongqing City, *Third National Engineering Geological Conference*, 1988;

[5]   Chen Shangming, et al, Numerical simulation analysis on stress changes in the foundation above shallow underground caves in Chongqing City, *Research Progress in Engineering Geology, Publishing House of Southwest Jiaotong University.*

Proc. 30th Int'l. Geol. Congr., Vol. 23, pp. 151-157
Wang Sijing and P. Marinos (Eds)
© VSP 1997

Fuzzy Dynamic Cluster Zoning of Stability and Its Application in Meishan Underground Mine *

GAO QIAN, SU JING, & SU YONGHUA
University of Science and Technology, Beijing, 100083 ,China

Abstract

According to the demand of the second phase design of development work for Meishan iron mine, this paper presents the fuzzy method for dynamic cluster zoning analysis of rockmass stability. The study includes the acquisition and analysis of rockmass condition, the selection and determination of zoning factors and parameters, and the results of its application.

Keywords: Underground mine, Fuzzy classification, Dynamic zoning

INTRODUCTION

Engineering geology   zoning is developed from the surface engineering, such as open pit slope, municipal works, and the reservoir   drainage area   planning.   And it is chiefly used in the stability evaluation of area,   the forecast and analysis of natural calamity. Although engineering geology zoning is still used a little in underground mine, it does not means that there is no necessary to take the engineering geology zoning, in underground mine,   on the other wise, engineering geology zoning is very important to the mine development design, the forecast and evaluation of   underground   water and even the selection of mining method, There are three reasons for engineering geology zoning is not used in underground mine:

The engineering geology zoning material is difficult to acquire
Unlike the earth's surface engineerings which can collect   many material directly from outcrop, the underground engineering is very difficult to obtain zoning material . Especially in the development design phase, the only   geological   material is the prospect borehole, because the   main object of prospect borehole is go prospecting, seldom touch upon other information such as dislocation, RQD parameters, it is difficult for   engineering geology zoning to acquire necessary geological material.

There is still no adaptable engineering geology zoning method
Although   many rockmass classification   method for underground engineering were put

* This paper is partly financial supported by the Engineering Geomechanics Laboratory, Institute of   Geology , Chinese Academy of   Sciences.

forward now(such as CSIR classification , NGI classification etc.), it is impossible for whole mining area to obtain detailed geological material. So it is infeasible to use this classification   method directly.

It is difficult to determine the zoning factors
Because there are many difference in the cause of formation of deposit, burred conditions and the geological structure effect between different mines, the factors affecting the rock stability are different. This makes it difficult to select engineering zoning factors with a same mode.

As a kind of attempt of   underground engineering geology zoning, this paper presents the underground engineering rockmass stability zoning method carried out by author in Meishan iron mine.

ENGINEERING GEOLOGY SURVEY OF THE MINIING AREA IN MEISHAN IRON MINE

Meishan iron mine is sited in   the ring of Huaiyang epsilon-type structure , the north segment of Ningwu basin, and the crossing location of Meishan-Fenghuangshan minerogenetic zone(North-East). Based on the cause of formation of rock, lithological characters alteration of engineering geological condition,   mining area rockmass are divided into 8 rock groups.   (1) block mica augite andesite tuff, tuff breccia rock group, (2) tuff siltstone rock group   (3) black mica andesite rock group,   (4) tuff breccia rock griouip, (5) quartz diorite andesite rock group, (6) quartz sand rock group, (7) diabase rock group,   (8) gabbro diorite rock group.

Meishan iron mine is a continental facies volcano rock and rich ore iron mine.   The deposit burred in the connect of the augite diorite and andesite. The deposit roof is augite andesite.   Rockmass alteration is violent. The alteration involves kaolinize, siliconize, carbonatize, pyrites, micarize. The deposit floor which is augite diorite rock , is also alterated, such as kaolinize, siliconize, sandinitize, augustitize, etc.   Because of the alteration and rock imbed, some deposits, especially these where ore-rock connecting zone is slacking, the stability is bad. According to the material analysis, the land block where alteration is violent, especially the block with kaolinize, chloritize, gypsumize, all contain bellied mineral, such as montmoril onite etc.

The structurre of   mining area is also developed.   It involves five groups : NE trend, NEE trend, NWW trend ,NNW trend and almost SN trend fracture group. The joint and crack is chiefly the NW trend and NE trend.

THE STUDY SCOPE OF MINING AREA ENGINEERING STABILITY ZONING AND THE ZONING MAP CONTENT

Mining area engineering stability zoning is serviced for the second phase project development design, the zoning results will provide foundation for the vertical shaft deepening, the engineering design of vertical shaft, haulage   roadway selection, and the stability evaluation of built haulage roadway.

Because the second phase project is related to the whole mining area rockmass from shallow to   depth, such   as chief   ramp project, north airshaft project etc., the zoning shows the engineering rockmass stability of whole mining area by 3 plan view and 3 sectional view. And these 3 plan view show -198m, -258m, and -330m level respectively, plan view control the rockmass stability layout statement of 3 haulage level.   3 sectional view is A, B, and C respectively. The A sectional view is the 302 exploratory line, it extends to north to auxiliary shaft, extends to west to the west-south airshaft. The B sectional view is the 401 exploratory line.   It extends to west to the west airshaft, extends to the east to the east-south airshaft, The C sectional view is the 408 exploratory line, it extends to the chief shaft.

In order to make the zoning map clear and easy to use, the zoning   map chiefly shows four contents as   follows.

Lithological character boundary.
The   lithological characters boundary is the boundary of 8 groups of rock in this mining area mentioned before. And the lithological character boundary chiefly shows the relationship of rockmass stability and lithological character.

Large fracture group or crushing zone .
Because the large fracture and crashing zone controls the rockmass stability and underground water trends around the structural features, in order to emphasize the effect of the fracture to engineering rockmass stability, it is considered as one of the content of engineering stability zoning map.

Excavated engineering
The   excavated   roadway, camber and other mine workings   are drafted on the engineering stability zoning map. And the goal is to show the relation between engineering   stability and excavated engineering. On one hand, to evaluate the reliability of engineering zone by the excavated engineering stable situation, on the other hand, based on the stability zone where excavated engineering is sited, to forecast the engineering long term stability, so as to investigate the engineering site in unstable rockmass at interval, and reinforce it if necessary.

The boundary is the chief element which affects the mine engineering rockmass stability. Here, we adopt   the fuzzy dynamic cluster analysis method.

THE SELECTION OF ENGINEERING ROCKMASS STABILITY ZONING FACTORS AND THE DETERMINATION OF ELEMENT INDEX.

The selection of engineering rockmass stability zoning factors

As we know, not only there are many factors affecting the rockmass stability, but also there are many differences between different mine. So, selecting several factors as the chief factors for engineering zoning from many mine elements is the key to mine engineering zoning.

Based on the investigation and analysis of 5030m roadway's stability of excavated -330m level, three factors are selected as zoning   factor.

① Lithological characters .   Because the mineral content and quantity varied greatly in the rock with different lithological character, and the alteration category and degree have great difference, the lithological characters factor contains rock intensity, alteration kind and alteration degree. Meanwhile, because alteration category reflects the mineral's developed degree and the content of bellied mineral on certain degree, the lithological characters reflect the effecting of bellied mineral to rockmass stability.

② Joint and crack.   The large fracture construction is drafted on zoning map directly. For many IV-V joints and cracks also affect rockmass stability on great degree, it is considering the development degree of joint and crack as one of the zoning factors.

③ Underground water.   Analysis of destroy of existing engineering shows that many destroy of engineering are related to underground water. So the underground water plays an important role to the rockmass stability of Meishan iron mine.   Study also shows that the underground water's effect to rockmass stability is related to the sensitive degree of lithological character to underground water, so we consider the sensitivity of lithological character to underground water. When we select this under-ground water factor. Only so the effecting of underground water to engineering rockmass can be showed correctly.

The determination of factor index

After selecting zoning factors, it is important how to quanititize the factor   indexes. Because this not only related to if the quanititize factor index reflects the effect of selected factors correctly, but also involves if the determinated index can be obtained. The principal of the zoning study is using the geological material as much as possible, to quanititize the qualitative elements by the subordinate degree of fuzzy mathematics. Therefore, the cluster analysis is carried out by the computer. Based on above study, follow zoning factor   indexes are determined:

① The uniaxial compressive strength Rc.   This index is obtained from the compressive strength lab test, the stochastic   results are shown in table1.

② Rock construction effect index Jc.   Here, the subordinate degree of fuzzy mathematics and expert accreditation method are adopted to determine the subordinate degree quantitatively expressing the crack developing degree. At the mean time, geological materials, such as crack rate, core adopt rate are adopted to determine the rock construction effect index Jc in zoning.   The adopted values are shown in table2.

Table 1. The stochastic results of ore uniaxial compressive strength in Meishan iron mine
(strength unit   Mpa)

ore kind	sample num.	Max.	Min.	Average	Ave.Variance	Varience
block aimant	40	275	25.6	154.3	58.9	0.382
dip-dye aimant	47	270	20.2	98.5	57.0	0.579
stain aimant	3	177	58.5	110.4	60.6	0.549
breccia aimant	12	220	16.0	86.4	62.2	0.720

Table 2. The adopted value of rock construction effect factor index

Crack development description	very develop	develop	quite develop	midi develop	no
crack rate (%)	>4.5	4.5-3.5	3.5-2.5	2.5-1.5	<1.5
core adopt rate (%)	0-20	20-40	40-60	60-80	80-100
crack indensity(zone/m)	>40	25-40	10-25	5-10	<5
zoning element index	0.0-0.2	0.2-0.4	0.4-0.6	0.6-0.8	0.8-1.0

③ Underground water affecting index Ws   Ws is determined by the product of underground water amount and the soften coefficient of rock. Ws =W•s. Because the underground water amount is often described quantitatively, it should be quanititivize firstly. The sensitivity of lithological character to water is expressed by the soften coefficient of rock block compressive strength, the value is showed in table3.

Table 3. The subordinate degree of underground water amount qualitatively description

Description	surging water	filter water	seeping water	dripping water	dry
subordinate degree	0-0.2	0.2-0.4	0.4-0.6	0.6-0.8	0.8-1.0

ENGINEERING   ROCKMASS STABILITY ZONING METHOD AND THE RESULTS

For the Meishan iron mine is a productive mine and needs second phase design, the study presents a engineering rockmass stability zoning method. The step of this method is as follows:

Collection and analysis of materials
Because the zoning scope is very extensive, the geological data should be collected as much as possible, including the quantitative and qualitative material, sketch map, such as excavated engineering ,prospect engineering, borehole columnar section etc. Then analyze the material, and quanitatize the factor index of qualitative described material with fuzzy math theory.

The selection of zoning factors

For any mine, there are many factors affecting the engineering rockmass stability. For most of mines, the factors affecting rockmass stability are varied because of the difference of deposit contributing factor, and the difference of geological construction. So it is very important for the reasonable of zoning result to select several most important factors from many mine factors as the zoning factors.

The determination of zoning factor index

After selection of zoning factors, the zoning factors must be quanitititize so as to dispose amount data by means of computer.

The delimitation and value adopting of zoning unit

Drawing necessary geological map for the zoning plan view or sectional view, and selecting the land block with same or similar engineering geology conditions as independence unit, determining the zoning factor index of every geology unit on zoning map, thus the zoning sample set is obtained. It is obvious that the bigger the sample set, the accurate the factors index, and the precise of the zoning result.

Fuzzy dynamic cluster analysis method

The step of analysis the obtained zoning sample set by using of dynamic cluster analysis method is as follows.

①Determine the zone number, it is 3-6 normally according to the zoning goal and precision.

②*Determine the initial zone.*

③Calculate the center of gravity of every zone, make the center of gravity as the standard of initial zoning.

④Calculate the distance of every sample(geological unit) to the center of gravity of every zone, and delimit the sample to the nearest zone according to the principle of distance is nearest.

⑤Calculate the center of gravity of every zone as the new standard of zoning again.

⑥By means of iteration, adjusting the zone of every sample should be in, and checking if the center of gravity obtained by twice calculation is same or if the error is in the acceptable range. If the iteration condition is contented , then stop the calculation, and output the final zoning result.

Above zoning algorithm had been composed into a computer program, and the zoning result can be quickly obtained by means of computer. The engineering rockmass stability zoning result of -330 level is showed in Fig.1.

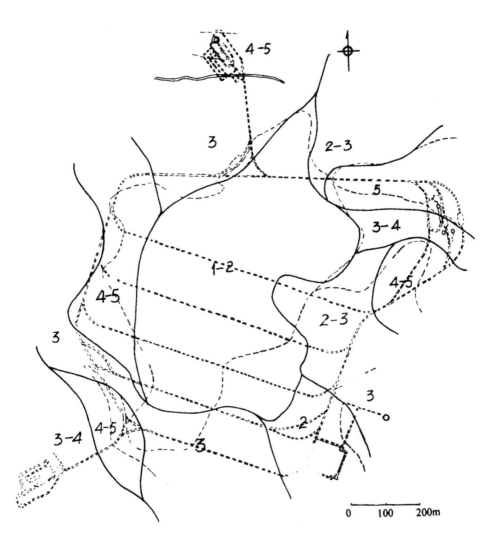

Fig 1. The Engineering Geological Zoning Plan View of -330 m Level

CONCLUSIONS

The research results have been applied in the design of development work for the second phase, and shown that the zoning results are reliable.

Proc. 30th Int'l. Geol. Congr., Vol. 23, pp. 159-178
Wang Sijing and P. Marinos (Eds)
© VSP 1997

A Study on Engineering Properties and Counter-measures of Soft Rock and Soil Tunnels of Wanjiazhai Yellow River Diversion Project

SONG YUE

Tianjin Investigation, Design and Research Institute of Ministry of Water Resources (TIDI), Tianjin, China.

Abstract

Wanjiazhai Yellow River Diversion Project is located at Shanxi Province in north part of China. Along the alignment of the diversion tunnel the strata are the loesses of Mid-Pleistocene (Q2) and Upper Pleistocene Series (Q3), red clay (N2) of Pliocene Series and swelling rock of Permian and Triassic Systems. The physical and mechanical properties of them as well as the design and construction counter-measures for tunnelling during the implementation of tunnelling work in these strata are experienced and studied. Furthermore, the main concerns of the methodologies in tunnelling by drill-blasting and TBM excavating are probed too. It is noted that special anti-seepage measures should be taken to the potentially collapsible loesses which will greatly affect the stability of the tunnel. The engineering properties of N2 red clay changes tremendously as the geological condition differs. At the same time, the swelling rock is also introduced in the aspects of methodologies in geological investigation and characteristics of its distribution. It is highlighted from the above that these soils and rocks are characterised by low strength, obviously theological property, and complicated properties upon the influence of water, which will control the self-stablising time and mechanism of failure of the tunnel, therefore, will affect the formulation of methods or plans for excavation and supporting.

Keywords: Collapsible Loess, Over-consolidated Red Clay, Swelling Rock, Argillaceous Rock, Squeezing Rock, Collapsibility Coefficient of Loess, Dry Density, Open-aerial Texture, Collapsibility Deformation, Anti-seepage Design, Drill-blasting Method, TBM Excavating Method.

INTRODUCTION

Wanjiazhai Yellow River Diversion Project is a large-scaled interbasin diversion project, by which the water will be taken from Wanjiazhai Reservoir in Yellow river to three bases of energy and heavy chemical industry, namely, Taiyuan, Shuozhou and Datong cities by three water diversion lines which include the General Trunk Line (hereinafter called GTL), South Trunk Line (hereinafter called STL) and North Trunk Line (hereinafter called NTL) as shown in Fig.1. The General Trunk Line (GTL) is 44.4km long capable of diverting a flow rate of $48m^3/s$, and then it is divided from Xiatuzhai in Pianguan County into two branch diversion channels, i.e. the South Trunk and North Trunk Lines, of which the former is 103km long with a designed diversion flow rate of $25.8m^3/s$ and the latter is 167km long with a designed diversion flow rate of $24.2m^3/s$.

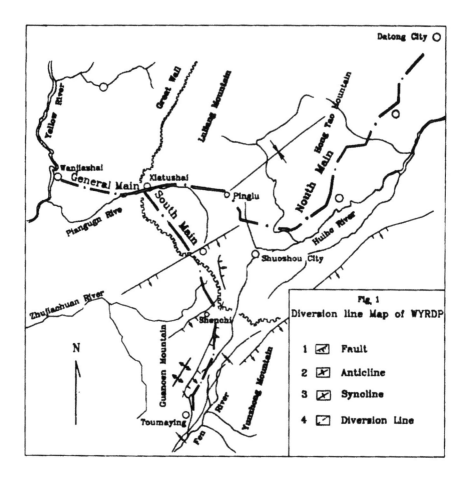

Fig. 1
Diversion line Map of WYRDP

1 [symbol] Fault

2 [symbol] Anticline

3 [symbol] Synoline

4 [symbol] Diversion Line

The total length of tunnels in Wanjiazhai Yellow River Diversion Project(hereinafter called WYRD Project) is 192km, including 23 separate tunnels among which the longest tunnel is 55km. The tunnel goes through sandstone, mudstone and shale formation of Mesozoic Era by a total length of 46.3km, while it runs 10.5km in red clay (N_2) of Pliocene Series of Tertiary System and loess of Mid-Pleistocene (Q_2) and Upper Pleistocene Series (Q_3) of Quaternary System. The loess of Q_3 has a moderate to strong collapsibility, while the loess of Q_2 is non- to weakly collapsible. N_2 formation is characterised by red clay. The strata of Upper Permian, Triassic and Jurassic Periods contain the argillaceous swelling and soft rocks. These rocks are characterised by low strength, high deformation, obviously rheological property, and strongly mechanical effects from surrounding water and pressure, which properties will control the self-stabilizing time of tunnels and mechanism of failure in rocks and therefore will affect the methodologies or plans to be employed in excavation and supporting. It is a difficult problem for the whole project to form a tunnel in such soft rock and soil, furthermore, sections with such rocks or soil are potentially weak upon the project operation in the future. Some effective engineering counter-measures to be exchanged hereinafter were practically taken by studying the physical and mechanical properties of the above-mentioned soil and soft rock and by the experiences in construction practice.

TUNNELS IN LOESS OF UPPER PLEISTOCENE SERIES (Q3)

It was investigated that the tunnel was aligned below such loess of Q_3 in places crossing the valleys. The tunnel is buried some 10~ 30m deep with an inner diameter of 5.0~ 5.40m.

Physical and mechanical property of Q3 loess

The Q_3 loess consists of alluvial-pluvial, dluvial-pluvial and aeolian loamy sand with a moderate to high collapsibility and compressibility, showing a lightly yellow colour and loess texture with typically columnar joints. It is shown by large amount of tests that the contents of different particles are 8~15%, 47.8% and 41.2% for their respective grains of clay(D <=0.005mm), silt(D=0.05~0.005mm) and sand(D=0.1~ 0.05mm), while the moisture content is w =4.27~ 24.8% varying from depths as shown in Fig.2. Void ratio is averaged to e=0.883~ 0.977 with a maximum value of 1.157.

The dry density ρ =1.23~1.56g/cm^3 of such Q_3 loess is relatively low and it obviously varies from depths as shown in Fig.2.

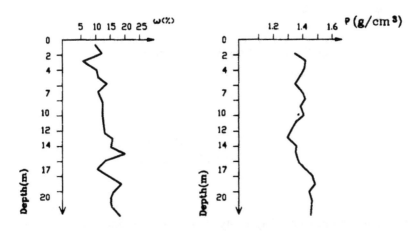

Fig. 2   The Variable Curves of Natural Water Content and
Dry Density of Q_3 Loess Varying With Depth

The collapsibility of such loess(Q_3), which is closely correlated to buried depth (h), dry density (ρ) and additional loading (p) of soil layer, is very complicated and characterised by :-

i) The collapsibility of loess tends to decrease as the depth increases, since the soil layer is gradually compacted under its self weight. Such tendency in changes of loess' collapsibility is harmonious with that of loess' dry density, i.e., the loess has a higher dry density, it will be of weaker collapsibility and vice versa as shown in Fig.3. As revealed by laboratory test results, if the loess has a dry density of higher than $1.5 g/cm^3$ the collapsibility of such loess becomes obviously weaker;

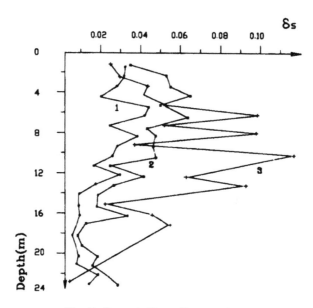

Fig. 3 Correlation Curves between

Collapsibility Coefficient and Buried

Depth of Loess at Different Pressures

1   • — •   0.2MPa Pressure

2   ⊠ — ⊠   0.6MPa Pressure

3   + — +   1.0MPa Pressure

ii) The relationship between the collapsibility of loess and the additional loading thereof is featured by a higher such pressure the loess has, a higher collapsibility the loess will be as shown in Fig.3;

iii) Initial pressure of collapsible loess is generally obtained under a collapsibility coefficient of ▷s=0.015. The initial pressure of such collapsible loess is smaller in a shallower depth where the loess has a smaller dry density as shown in Fig.4;

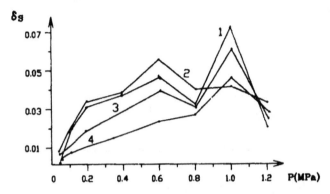

Fig. 4 Correlation curves between collapsibility coefficient
of Loess and Pressure at Different Dry Density

1 . —— .   ℓ =1.30-1.35g/cm³   Correlation Curve

2 . —— .   ℓ =1.35-1.40g/cm³   Correlation Curve

3 . —— .   ℓ =1.40-1.45g/cm³   Correlation Curve

4 . —— .   ℓ >1.45g/cm³   Correlation Curve

iv) Peak value of collapsible loess will indicate the sudden collapse failure of such loess in texture under certain pressure. It was disclosed by laboratory test that this value is closely related to the dry density and buried depth of the loess, the law it obeys is that in a deeper depth where the dry density is higher such value is higher too, on the contrary, such value becomes smaller if the loess is buried shallower and has a smaller dry density as shown in Fig.4.

v) Texture of loess is observed under an electron microscope, showing that the loess particles are accumulated and cemented with minor amount of clayey grains and soluble salt. The highly collapsible loess is mostly characterised by an open aerial-texture, while the weakly collapsible loess presents a partially aerial-texture or inlaid aerial-texture as shown in Table 1. The occurrence of collapsibility in loess is believed to have resulted from the physical and chemical actions of water that will destroy the texture of loess and reduce the void ratio of it too.

Table 1   Micro- Structure of Q_3 Loess

Formation	Sampling Depth (m)	Micro-Structure	Diameter of Aerial Void ($S_\mu m$)	Collapsibility Class	Remarks
Secondary loess(Q_3)	4	opening aerial texture	10~ 30	moderate to high	based on loess sample extracted from Da Foundation of Dalian Reservoir in NTL
Secondary loess(Q_3)	10	opening -partial aerial texture	-	moderate to high	
Secondary loess(Q_3)	15	partial aerial or inlaid aerial texture	5 ~15	weak	

Engineering Geological Problem for Tunnels in Loess
Stability of surrounding rock during tunnel excavation   The loess is classified as extremely unstable surrounding rock on the basis of loess' mechanical properties indices, including the internal friction angle $\emptyset$ =19~ 20°, cohesion c=0.01~ 0.02MPa, and solidity coefficient of loess (f=tg (19~ 23°)=0.35~ 0.42). Seeing from the tunnel excavation, the self-stabilizing period of surrounding rock is very short, especially in slope and shallowly buried sections, as a result, the tunnel is susceptible to deformation, furthermore, once it collapses it is very difficult to form a stable crown-arch, on the contrary, it will successively develop to the ground surface. An example is taken from the tunnel No.9 in General Trunk Line, which was excavated by manpower. During the excavation, the collapse occurred in a chainage of 0+030~0+046, where it collapsed from a height of some 20m above the tunnel floor,   forming a 16m long collapsed zone with a total volume of 6,000m^3 for such collapsed body as shown in Fig. 5. Another such collapse occurred in the outlet section of tunnel No.7 in a chainage of 48~ 30m in GTL. This tunnel was excavated by double   shield excavator of TBM type. The collapse also occurred in surrounding rock of Q_3 loess where the collapsed height is 35m.

Collapsing deformation of loess   The pressure of loess foundation caused by lining structures in loess tunnel and water mass through the tunnel is about 0.2MPa. If the rock pressure and ground stress are taken into consideration, such pressure will be higher than the initial pressure of loess in most cases and even higher than the peak pressure of collapsible loess. Therefore, during the tunnel operation, once seepage occurs which may saturate the loess, the collapsing deformation cannot be avoided. At the same time, the loess tunnel may be destroyed by soil flow and piping in loess foundation, which are caused by further development of cracks in lining and strong leakage of water. Such destroy of water diversion tunnel also occurred previously in other project in China. Owing to that the working period of the diversion tunnel for WYRD Project is up to 10 months per year, the attention should be duly paid to seepage in lining and to other engineering defects which may cause collapsing deformation or destroy of loess tunnel.

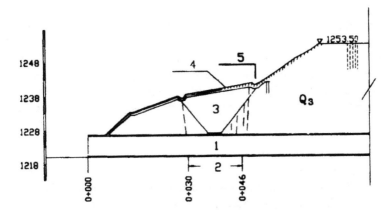

Fig. 5   The Cave-in Treatment Longitudinal
          Profile of Tunnel NO. 9 On the GTL
1   Tunnel
2   The Cave-in Section
3   Back-filling after Excavation
4   three-Seven Soil Layer
5   Drainage Ditch on the Ground

Differential deformation in loess   Owing to the significant difference in the strength of two different foundation media, such as loess/bedrock, loess/N_2 red clay and interface between loess tunnel and aqueduct, the loess tunnel is prone to differential and collapsing deformations which may result in the destroy of such tunnel under the adverse action of building pressure or earthquake forces.

It can be seen from the above that the loess should be kept strictly free from water during soil tunnel construction or diversion tunnel operation.

Design of Loess (Q_3) Tunnel
Based on the engineering geological characteristics of loess, the following counter-measures are envisaged for tunnel design :-

(1)   Type of tunnel section :
Standard horseshoe section or the approximately circular section tunnel to be adopted.

(2) Supporting design
The rockbolting and shotcreting combined with wire-mesh in tunnel crown and side walls are suggested in full section of the tunnel. The advance pipe frame work combined with grid steel arch truss is recommended to strengthen the primary supporting, while a 40~50cm thick reinforced concrete lining is envisaged for secondary supporting in inlet

and outlet portions of loess tunnel as shown in Fig.6.

(3) Anti-seepage design

In order to stop the water inside the tunnel from any leakage, rubber water seal strip was designed in circumferential expansion joints of each concrete placing section, while joints on water face were filled with water seal material, simultaneously, water seal strip was used in longitudinal construction joints. The high density polyethylene film was placed on the inside surface of the primary shotcreting support layer, at the same time, the waterproof agent   TMS(5%) was mixed into the secondary reinforced concrete lining so as to prevent any cracking and water leakage.

Methodology of Tunnel Construction

(1) Artificial excavation method

Full-face tunnelling with upper and lower benches method was usually employed for tunnel excavation. The primary supporting during excavation is about 1m long, which is more than 6m(tunnel diameter) away from the far excavated surface. A deformation monitoring section was practically designated at a 20m interval during tunnel excavation.

(2)   Construction by TBM Excavator

Excavation of tunnel was performed by using double shield TBM excavator in the outlet section of tunnels No.7 and No.8 in GTL where the loess (Q_3) is encountered. The hexahedron reinforced concrete precasted shell was adopted for tunnel lining. Serious collapse of such linings occurred, furthermore, disturbance of such precasted shell was also encountered due to the differential settlement of foundation. Anti-seepage measures are difficult to be taken. Therefore, the existence of quality defects arising from such method give rise to the conclusion that tunnelling method in Q_3 loess by TBM excavator is still under-mature.

TUNNELLING IN LOESS OF MID-PLEISTOCENE SERIES (Q_2)

The Q_2 loess is 20~40m thick light brownish red coloured sandy clay, with sand and gravel lens in its lower portion, and in its upper portion the loess is of loose texture and weak collapsibility with large voids and columnar joints.   Several layers of calcareous concretion were encountered in middle and lower portion of loess, which are relatively dense without collapsibility.

According to laboratory test results, the Q_2 loess is mainly composed of silt and sand, totalling 80~ 90%, and the clay content is 10~20% only. The dry density ρ =1.39~1.69g/cm^3 and its average value of 1.55~1.57g/cm^3 were obtained. The internal friction angle and cohesion are respectively achieved as φ =23~27° and c=10.5~24.5kPa,

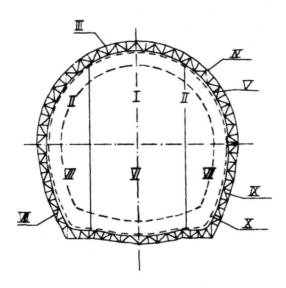

Fig.6 Transerval Section Excavation and the Primary
Circuit Support Consequence Diagrammatic Sketch

I   The Top Heading Excavation
II   The Top Enlarging-Excavation
III The Advance Bolt Bar
IV   Installation of Steel Framework on the Top Arch
V   Netting and Shotcrete and Pouring Concrete
     on the Top Arch
VI   The Medium Heading Excavation on the Lower Part
VII Overexcavation on the Lower Part
VIII Installation of Steel Framework on the Side Arch and
     Bottom Arch
IX   Netting and Shotcrete and Pouring Concrete on the
     Lower Part
X   Layout of HDEP Geotechnical Model on the Inside
     Surface of the Primary Circuit Support in Q_2 and Q_3
     Collapsed Loess Tunnel Sections

showing that the loess is a low to moderate compressibility soil.

As far as soil's physical and mechanical properties are concerned, Q_2 loess with a solidity coefficient of f=tg $\emptyset$=0.5~0.7 for surrounding rock is slightly superior to Q_3 loess. It is noted that attention should be paid to the contact zones between Q_2 and N_2 loesses nearby which minor ground water outcrops occasionally. The surrounding rock of tunnel is in a very poor stability condition since the soil there is in soft and plastic state, moreover, excessive water during tunnel excavation is difficult to completely drain out.

Q_2 and Q_3 loesses are similar in the aspects of tunnel design and methodologies for tunnel construction. The above-mentioned two methods, inclusive of tunnel by manpower and TBM excavator, were successfully applied in this project.

TUNNEL IN N_2 LATERITE(RED CLAY)

Engineering Properties of N_2 laterite

The strata of this kind distribute locally in the deeply down-cut valleys or fault subsidence basin. The underlying rock is limestone of Ordovician System, while the overlying strata are mainly loesses of Q_2 and Q_3. It is up to 210m thick and unconformity with aforesaid two strata. The N_2 strata are composed of red clay in its upper portion, and intercalation of red clay and gravel in its lower portion.

The difference between N_2 laterite and Quaternary clay under same moisture content and density is that the former has a higher strength(Rc=0.17~0.55MPa), and is classified as over-consolidated stiff clay though not yet formed to rock(Rc<1MPa). This N_2 red clay is characterised by follows:-

Cohesion c=0.012~ 0.062MPa, c value is twice of ordinary clay;
Bulk density ρ =1.85~ 2.05g/cm^3, relatively high;
Longitudinal wave velocity is 800~ 1400m/s;
Elastic modulus E =40 ~80Mpa;
Clay and montmorillonite contents are 30~50% and 10~20% respectively;
Free expansion ratio is 30~ 55%;
Specific surface area is 254~ 302m^2/g;
Fe_2O_3 content is 6.15~ 7.5%.

It belongs to weak expansibility soil with an expansion force of generally 0.3~ 0.5MPa and a maximum such force of 0.7MPa.

After site investigation, laboratory tests and construction exercise, the attention should be paid to the following problems:-
(1) Moisture content of N_2 loess greatly affects its mechanical properties as well as the stability of surrounding rock of tunnel. When its natural moisture content is higher than its plastic limit (w_p =25~ 50%), it presents a soft and plastic state. In such a case, the

surrounding rock has a poor stability condition and prone to deformation and collapse, hence, primary supporting measures on such sections should be taken. If TBM method is used for tunnelling, attention should be paid to the cutting tools which usually covered with clay or mud. The soil mass is relatively in the perfect stability condition when the moisture content is 15~25%.

(2) In localised tunnel sections, for example, Muguagou section in STL, the joints in soil mass are well developed, among which both tension and shear joints can be found. These joints are closely spaced about 50cm and can be classified as 3 to 4 sets. This will definitely affect the stability of surrounding rock of tunnel. However, there are diversities in opinion on the genetic classification of N_2 red clay. The main opinions can be summarized as tectonic, ancient landsliding and earthquake geneses.

When tunnel is buried deeper, such as tunnel No.5 buried 175m deep in STL in Liminbu area, the plastic deformation of 30~40cm occurred in 2m thick surrounding rock by way of definite element calculation.

Collapse of Tunnel in N_2 Laterite

The collapses occurred at a chainage of 0+30~0+46 in tunnel No.10 in GTL and at a chainage of 7+730~7+741 in tunnel No.3 in STL, where these two tunnels are respectively buried 50m and 30m deep. After occurrence of collapse, the settlement joints and depression can be found on the ground surface as shown in Fig.7.

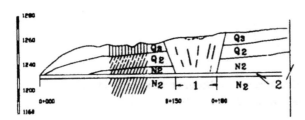

Fig. 7 The Cave-in Longitudinal Profile
of Tunnel No. 10 on the GTL

1   The Cave-in Section
2   Tunnel No. 5

The advance concrete grouting and the pipe jack framework supporting measures combined with concrete lining were applied in the treatment of collapsed body. Even though, the progress was delayed by 4~6 months. Seeing from the whole process of tunnel collapsing, much attention should be paid to the N_2 red clay that is of very poor stability and classified as Class V surrounding rock.

Design and Counter-measures on Tunnels in N₂ laterite

1) Horseshoe-shaped lining section, rockbolt supporting combined with reinforced concrete   lining are to be applied for tunnel design;

(2) Methodologies for Construction: excavated by using the upper and lower short benches, the former is prepared in advance and 6~ 8m far away from the latter; It is noted that the excavated surrounding rock should be timely supported at a 1m interval by rockbolt and shotcrete;

(3) Advance rockbolting and grid steel arch framework supporting should be increased in quantity for sections with a poor geological condition, such as the soil layer with a high moisture content or the joints well-developed belts as shown in Fig.8;

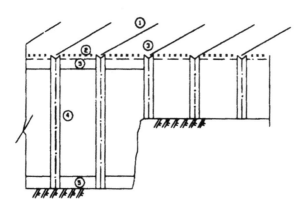

Fig. 8   The Construction Diagrammatic Sketch
         of N₂ Laterite Tunel Sections

          ① Advance bolt bar
          ② Steel Net
          ③ Shotcrete
          ④ Grid Steel Framework
          ⑤ Concrete Liner

(4) The deformation should be monitored after primary supporting;

(5) The secondary supporting of 30 40cm thick reinforced concrete lining should follow up.

ENGINEERING GEOLOGY OF SWELLING ROCK TUNNELS

There are 33km long tunnels aligned in south part of STL of this project, where the strata are sandstone, mudstone and shale of Carboniferous(C), Permian(P), Triassic (T) and Jurassic Systems. In order to identify the engineering properties and distribution of such swelling rock, site investigation was therefore carried out, in which some shallow test pits were excavated and argillaceous rock samples were taken for mineralogical analysis and specific surface area testing. Test results are listed in Table 2. Based on these results, the distribution of swelling rock along the tunnel alignment can be forecasted, and then the tests on physical and mechanical properties of argillaceous rocks extracted in boreholes which are proposed in typical sections will be performed.

Distribution of Swelling Rock

It can be seen from Table 2 that the minerals are mostly kaolinite or illite with minor amount of montmorillonite of Benxi Formation (C_2b) of Middle Carboniferous System and Taiyuan Formation (C_3t) of Upper Carboniferous System, showing that claystone in Carboniferous System contains no swelling rock.

Shanxi Formation (P_1s) and Xiashihezi Formation (P_2x) of Permian System are the oldest land facies of argillaceous strata. Montmorillonite content in moderately weathered rock samples is mostly less than 10%, though it increase in a few samples owing to the weathering action on ground surface, therefore, this formation can also be classified as non-swelling rock formation.

The strata in Shangshihezi Formation (P_2s) and Shiqianfeng Formation (P_2sh) are dominated by greyish white, greyish green, greyish black and purplish red coloured argillaceous rocks. The mono-layer of the above-mentioned rocks is thick to moderately thick. The rock on ground surface is highly weathered, showing the argillaceous and clastic states. The clay consists of mainly of montmorillonite, moderate amount of kaolinite and minor amount of illite. Owing to the higher contents of montmorillonite(mean value of 33.36%) and specific surface area(mean value is $361.17m^2/g$) in argillaceous rocks in Shangshihezi Formation (P_2s), rock of this kind is classified as the highest expansible formation in WYRD Project. The average content of montmorillonite in argillaceous rocks of Shiqianfeng Formation (P_2sh) is 18.1%, while the mean value of specific surface area is $254.1m^2/g$, therefore, they are of less expansibility than that of P_2s mentioned above. It is believed that the swelling rock occurs from Shangshihezi Formation (P_2s) of Permian System in northwest Shanxi province.

Liujiagou Formation (T_1l) and Heshanggou Formation (T_1h) of Trassic System are characterized by reddish brown color. Owing to little montmorillonite and dominant cementation in hematite (Fe_2O_3 =5.61~ 9.61%), the strata therefore become relatively stable argillaceous rock formation.

Table 2. Distribution of Swelling (or expansible) Rock in Niugwu Area, Shanxi Province
(Samples from Test pits) (to be continued)

Strata Symbol	Name of Strata	Montmorillonite Content(%)		Specific Surface Area (%)	
		Average	Range	Average	Range
C_2b	Benxi Formation	9.19	3.59	-	99.66
C_2t	Talyan Formation	6.76	4.72-8.61	109.1	80.1-169.69
P_1s	Shanxi Formation	8.42	4.4-12.44	-	88.1-265.73
P_2x	Lower Shihexi Formation	11.74	6.09-19.61	157.59	80.71-249.95
P_2s	Upper Shihexi Formation	99.36	10.99-59.56	961.17	185.5-365.9
P_2sh	Shiqlanfeng Formation	18.61	7.9-55.07	254.1	108.37-506.89
T_1t	Luijiagou Formation	9.95	8.67-18.26	179.66	129.91-202 99
T_2h	Heshanggou Formation	8.46	7.20-9.89	178.51	154.91-207.59
T_2er	Ermaying Formation	11.67	7.98-18.49	178.49	94.22-349.75
T_2t	Tongchuan Formation	11.95	6.22-19.65	169.99	72.92-246 76
J_1d	Datong Formation	11.95	6.22-19.86	169.89	72.92-246.76
J_2y	Yungang Formation	12.99	18.88-19.88	282.89	145.99-209.61
J_2t	Tianchi	9.24	6.17-18.98	132.6	132.6

Table 2. (continue) Distribution of Swelling (or expansible) Rock in Niugwu Area, Shanxi Province
(Samples from Test pits)

Expansibility and Numbers of Such Layer				Identification of Expansibility
High	Moderate	Weak	Non-expansible	
-	-	-	9	no expansible rock
-	-	-	4	no expansible rock
-	-	-	2	no expansible rock
-	-	-	4	no expansible rock
5	2	4	1	main formation with highly expansible rock
2	1	9	4	main formation with highly expansible rock
-	-	-	4	no expansible rock
-	-	-	5	no expansible rock
-	-	9	9	minor weakly expansible rock
-	-	5	1	with weak expansible rock
-	-	28	15	with many layers of expansible rock
-	-	9	4	with weakly expansible rock
-	-	1	1	minor weakly expansible rock

Ermaying Formation (T_2er) and Tongchuan Formation (T_2t) of Trassic System is dominated by sandstone with minor amount of thin argillaceous rock. Although the content of montmorillonite and specific surface area tend to increase in some argillaceous rocks, they have little adverse impact on deeply buried tunnel. Hence, this formation can be classified as weakly expansible formation.

Datong Formation (J_1d) of Jurassic System consists mostly of sandstone in its lower part, while in its middle and upper parts, 13 layers of coal beds or lines are identified. Underlying such coal beds, the rather thick argillaceous rocks can be found occasionally in which the clayey minerals are mainly illite and montmorillonite(11.95%) followed by kaolinite, giving rise to the weakly expansible formation too.

Yungang (J_2y) and Tianchihe (J_2t) Formations of Jurassic System are characterised by thick sandstone intercalated with minor amount of marlite. The mineral component and its characteristics are similar to those of Datong Formation(J_1d), therefore, it also belongs to the weakly expansible formation.

Engineering Properties of Swelling Rock
Unfavorable engineering properties of swelling rock, such as swelling, shrinkage, disintegration and softening, are formed by its complicated internal and external factors. Internal factors of swelling rock cover mineral component, nature of cement, etc., on the other hand, changes in moisture content, weathering degree, state of stresses, etc. in swelling rock are the external factors.

(1) Montmorillonite content and expansibility
Characteristics of swelling, shrinkage, disintegration and softening in swelling rocks are believed to have resulted from absorbing and loss of water in clayey minerals. Among various clayey minerals, montmorillonite has a very high expansibility and inner and outer surface area, which give rise to the great impact of montmorillonite on the engineering properties of argillaceous rock. Some experts regarded that no significantly adverse effect will arise from an argillaceous rock with montmorillonite less than 10%, however, adverse effect to certain degree will occur in an argillaceous rock with montmorillonite more than 10%. As such, in engineering practice the rock's expansibility is classified by the content of montmorillonite as follows:-

Montmorillonite	Expansibility
<10%	non expansible
10 20%	weakly expansible
20 30%	moderately expansible
>30%	highly expansible

In WYRD Project Area, the content of montmorillonite in most argillaceous rocks is less than 10%, while it is 10~20% in minor amount of such rocks, however, it is relatively high in Shangshihezi Formation (P_2s) and Shiqianfeng Formation (P_2sh), resulting in a moderately to highly expansible rock in these formations. The saturated expansion force reaches about 1.0MPa for weakly expansible rock under dry condition, unfortunately, no ideal test results achieved for that of highly expansible rock, but it is estimated to be higher than 3.0MPa.

(2) Relation between mineral component of cement and expendability of argillaceous rock
Much purple coloured argillaceous rock was found in Triassic argillaceous formation. Such purple rock has more hemertite (Fe_2O_3) content, featuring an insignificant expansibility and disintegration. Example can be taken from T_1l and T_1h formations, where the hemertite (Fe_2O_3) content is 5.61~9.61% and the rocks are non expansible formations.

(3) Relation between water absorption, natural moisture content and the expansibility of swelling rock
Another index adopted in WYRD Project for identifying rock's expansibility is the saturated water absorption after the swelling rock specimen is oven dried:-

Water absorption	Expansibility
< 5%	non-expansible
=5 20%	weakly expansible
=20 50%	moderately expansible
=50 100%	highly expansible

The test results on samples taken from borehole No.951 at Jiangzhuang in STL are in conformity with such classification as shown in Table 3.

Much attention was paid to the natural moisture content of argillaceous rocks in WYRD Project. Laboratory tests were carried out on samples extracted from boreholes, showing that the argillaceous rock's natural moisture content is smaller than that under saturated condition as proved by data listed in Table 3. As such, during tunnel excavation the groundwater in sandstone aquifers may flow into swelling rock formation, then this kind of rock will adversely affect the stability of the surrounding rock of tunnel if it is soaked by water.

(4) Relation between the expendability and the buried depth and weathering degree
Tunnel No.7 in STL of WYRD Project is generally buried at a depth of 50 400m. What is the difference between the slightly weathered argillaceous rock in upper portion and highly weathered such rock in lower portion? a) the rock extracted from lower part of borehole is mostly column-shaped. Though it will crack to failure if it is exposed in the air about several hours or more, it has a strength to some extent under its natural states. On the other hand, the rock in upper part of borehole is very fractured with a low strength, furthermore, cave-in and shrinkage of borehole occurs during drilling exercise;

Table 3 Test Results of Samples from Upper Shihezi Formation P_2s (to be continued)

Sampling Location	Strata	Sampling Depth (m)	Rock Type	Natural Specific Weight (G/cm^3)	Natural Moisture Content (%)	Saturated Absorption (%)
Borehole No.951 at Jiangzhuang in Ningwu County	P_2s Upper Shihezi Formation	32.7~32.9	grey sandy mudstone	2.32	5.45	23.83
		97.8~97.3	grey mudstone	2.27	18.30	20.76
		119.8~120	grey mudstone	2.08	20.16	55.67
		120~120.2	grey mudstone	2.89	15.86	31.44
		120.5~120.7	grey mudstone	2.87	15.81	35.58
		120.7~120.9	greyish green mudstone	2.51	5.81	7.81
		120.4~129.7	dark grey mudstone	2.34	5.84	9.75
		139.2~130.5	greyish purple argillaceous siltstone	2.46	5.87	6.90

Table 3 (continue) Test Results of Samples from Upper Shihezi Formation P_2s

Destruction Degree in Water	Montmorillonite Content (%)	Itlite Content (%)	Specific Surface Area(m^2/g)	Expanding Percent without Loading (%)	Expansibilty Identification
strongly soften and disintegrated	11.48	-	122.05	5.2	weakly to moderately expansible
argillaceous or clastic state	19.39	14.78	437.96	0.4	weakly to moderately expansible
argillaceous state	40.76	25.76	466.44	-	highly expansible
argillaceous state	29.47	18.61	425.79	-	highly expansible
argillaceous state	31.98	17.22	429.46	-	highly expansible
softening	12.21	26.44	274.43	-	non-expansible
clastic state	12.10	18.56	245.55	2.45	non-expansible
clastic state	10.39	20.88	226.04	2.30	non-expansible

b) it is disclosed by tests that montmorillonite content at argillaceous rock in upper part of borehole is higher than that in lower part of borehole. The above shows that the expansibility of argillaceous rock in shallower part is stronger than that in deeper part. This may be caused by rock's weathering action which will reduce the degree of cementation of such rock and caused by montorillonitization of the rock. In overall viewpoints, the engineering properties of swelling rock in deeper portion of tunnel is superior to that in shallower portion.

Influence of Swelling Rock on Tunnelling Work
(1) Shangshihezi Formation (P_2s) and Shiqianfeng Formation (P_2sh) are the main rock formations with swelling rock in tunnel No.7, while Datong Formation comes second. The tunnel in these formations is shallowly buried where the swelling rock is relatively active. The mono-layer of swelling rock in Ermaying Formation (T_2er) and Tongchuan Formation (T_2t) is thin, at the same time, few such layers were found, furthermore, the tunnel was deeply buried, therefore, little adverse influence on tunnelling work arisen from these formations. It is worthy to mention that the presently confirmed alignment of tunnel No.7 passes through the moderately and highly expansible rock formation by 2.5km and the weakly expansible rock formation by 20km.

(2) The rock mass in which the tunnel excavated is characterised by multi-layer artesian water. Though the flow rate is small, overflowing and flowage of such groundwater after tunnel excavation will further deteriorate the stability of surrounding rock that is originally on a very poor stability condition, therefore, to which particular attention should be paid.

Design and Construction Counter-measures of Tunnel in Swelling Rock
Types of supporting and methods of construction for tunnels in swelling rock should be designed on the basis of the classification of such swelling rock. Circular section tunnel type will be adopted for tunnelling in moderately and highly expansible formation, simultaneously, totally-enclosed water-proof lining has to be employed to barricade the access of water.

During the construction, blasting or other mechanical disturbance of swelling rock should be minimised, at the same time, the exposed surrounding rock should be sealed as soon as possible to prevent any losses of moisture in swelling rock. Monitoring of deformation in surrounding rock of tunnel cannot be neglected. In addition, the tunnel section can be properly enlarged as a provision for deformation during the construction of tunnel.

DEFORMATION OF SOFT ROCK TUNNEL UNDER HIGH GROUND STRESS

Geological Setting of Tunnel Deformation
The tunnel No.7 in STL goes through the watershed of Huihe and Fenghe Rivers and is buried 300~400m deep with a total length of some 7km. According to the ground stress

testing results in two boreholes, the directions of maximum horizontal principle stresses ($\acute{\sigma}_H$=15~19MPa) are respectively NE87 and NW285 , intersected by tunnel axis at an angle of 50~70° . The strata the above said tunnel crosses are sandstone and argillaceous shale of Ermaying Formation(T_2er) of Jurassic System. The compressive strength of sandstone under dry condition is 18.6~70.5Mpa. It is noted that such sandstone has a higher porosity of 3.98~4.1% and a very low softening coefficient of 0.16~ 0.36, featuring a very strong softening character in water. Observed by a scanning electron microscope, cement and clastic feldspar in sandstone are believed to have undergone the montmorillonization, resulting from the long-term alteration of alkaline groundwater which has abundant of magnesian ion(Mg^{2+}), hence, the sandstone is sub-classified as altered sandstone. In addition, the argillaceous rock along the tunnel alignment has a very low compressive strength with Rc=7~25MPa. As judged by the proportion of ground stress and uniaxial compressive strength, i.e. $\acute{\sigma}_H/R_{c>=}$ 0.5, deformation of surrounding rock will occur in these swelling rocks which are under a high ground stress. Such deformation is 5~20cm exactly as determined by definite element calculation.

Deformation Design and Construction Counter-measures for Soft Rock in Tunnels
(1) The circular section tunnel is adopted to strengthen the lining in the aspect of resistance against external forces, furthermore, the compressible material will be filled into the 5cm-wide   deformation outage outside the lined section;

(2) Both flexible and rigid supporting types are adopted to suit the deformation of surrounding rock for tunnelling by drilling and blasting, while the optimal time for backfill and grouting should be carefully studied for tunnelling using TBM method.

(3) Deformation monitoring in some typical tunnel sections is also needed so that the radial shrinkage deformation and pressure thereof in surrounding rock will be measured to determine the optimal time and strength of supporting.

CONCLUSIONS

(1) The tunnel sections excavated in Q_3 and Q_2 loess and N_2 red clay are some 10.5km long which is 5.0% the total length of tunnels in WYRD Project. They are the most difficult sections for excavation and construction. It was shown by investigation, testing and experiences that Q_3 loess is of loose texture with low physical and mechanical property indices, furthermore, such soil is of relatively high collapsibility which is greatly hazardous to the   stability of tunnel. As such, special anti-seepage measures should be taken, at the same time, deformation monitoring during tunnel operation is also required. Some measures such as short excavation and strong supporting have been successfully taken during the construction of tunnels in Q_3 and Q_2 loess and N_2 red clay.

Excavating in Q3 loess by double shield TBM excavator is still under-mature and further study on subject is needed;

(2) As investigated, the swelling rock does exist in argillaceous rock formation of Permian to Jurassic Systems, among which the moderately and highly expansible rocks can be found in Shangshihezi Formation (P_2s) and Shiqianfeng Formation (P_2sh). Tunnels with such expansible formations are the sections which are unfavourable for excavation and construction. However, the counter-measures for such sections were taken as the adoption of totally-enclosed circular structure, minimised disturbance of surrounding rock and timely implementation of supporting works.

(3) According to the study of ground stress and rock's properties, the attention should be paid to the tunnels which are deeply buried in soft rock that is regarded to has deformation.

(4) Owing to that the water diversion tunnel is very long and deeply buried, and takes a long period for construction, employment of multi-functional excavator appears to have a broad prospect in WYRD Project. However, the special properties of soft rock and soil results in fresh difficulties and confrontations in choosing of excavation methods. Therefore, study on engineering properties of Q_3 and Q_2 loesses, N_2 red clay and swelling rock is believed to be play an important role in boost of the development and suitability of excavator.

Acknowledgements

This paper is translated from Chinese into English by Mr. Li Jiameng and Huang Xiangchun. The acknowledgements are hereby paid to Mrs. Luo Fuyun and Du Liren who gave their kind help on the compilation of this paper.

REFERENCES

(1) Latest Progress in the Study of Swelling Rock in Eastern China, « 90's Geoscience » , Qu Yongxin and so on, 1992;

(2) Rapid Identification of Weak and Altered Sandstone, « Engineering Geomechanics Research», the 1st issue, Wu Zhilan, Qu yongxin, Song yue, Liu jishan and so on, 1995;

(3) Settlement Deformation of Loess in Daliang Reservoir Dam Foundation and its Impact on the Project, « Engineering Geomechanics » , VoL.3, the 3rd issue, Song Yue, 1995.

ROCK AND SOIL PROPERTIES

Proc. 30th Int'l. Geol. Congr., Vol. 23, pp. 181-197
Wang Sijing and P. Marinos (Eds)
© VSP 1997

THE CAUSE OF FORMATION AND ENGINEERING GEOLOGICAL PROPERTIES OF HYDROTHERMAL ALTERED ROCK IN GRANITE MASS FOR GUANGZHOU PUMPED STORAGE POWER STATION

CHEN YUNCHANG, WEI BINGRONG, & WU ZHAOWEN
Guangdong provincial Investigation Design and Research Institute of Water Conservancy and Electric Power China

Abstract

The clay-grouting altered granite which formed due to hydrothermal alteration is rarely to be seen in the engineering construction of water conservancy and hydropower.It is understood that the altered granite formed under special geological condition based on the study of the cause and distribution regulation of the altered granite body at the site of Guangzhou pumped storage power station(GPSPS).There are several alteration types in the granite mass, one of them is montmorillonization.The montmorillonite in montmorillization altered granite can swelling when it absorbs water that makes the compact-structure granite loose,calving and disintegrate into sandy-gravel soil which is a kind of soft belt in the granite mass.The support of altered rock and its impact to the substructure work,especially to the large span underground powerhouse and high pressure tunnel and manifold are analyzed and discussed with respect to the integrity and strength of the rock mass and based on the engineering geological characteristics of the altered granite in this paper.

Keywords: Granite Hydrothermal alteration, Altered belt Clay grouting granite, Montmorillonite, Engineering geological character, Support of surrounding rock

INTRODUCTION

The Guangzhou pumped storage power station with a total installed capacity of 2400 MW is the first large-size pumped storage power station in China which is characterized with high water head and deep burying. The station will be constructed in two stages, each stage is 1200MW in the capacity. The two construction stages will use the same upper reseroir, diversion tunnel, high pressure tunnel, high pressure manifold, underground powerhouse,tail-water tunnel and the lower reservoir.The maximum static water-head is 610 m and the maximum dynamic water-head is 750m at the manifold and their diameters are 8.5-3.5m after lining. The rockbolt/shotcrete is used for permanent support of the under-ground powerhouse and the main transformer cavern and rock-anchoring crane bean is used.

Guangzhou pumped storage power station is located about 90 km on the northeast of Guangzhou, northern side of Nankun Mountains.The upper and lower reservoirs have natural reservoirs basin and natural water resource respectively. The distance between the upper and lower reservoirs is more than 3 km and the fall between the reservoirs is over 500 m.The underground structures are place in a slightly weathered-fresh granite mass. Six groups of faults and fractures developed in the rock mass, among them, the mostly developed is NW, NNW, and NNE directions.The main engineering geological problem in this power station is the clay grouting altered granite which caused by hydrothermal alteration developed along the faults and fractures after postmagma stage.

GEOLOGICAL BACKGROUND

In regional geological tectonics,Guangzhou pumped storage powerstation lies at the southeastern flank of joint area of E-W Fugang-Fengliang fault belt and NE-NNE Guangzhou-Conghua fault belt, within ground-thermal water active area (Fig.1). The fault is the main geological tectonic type. sedimentary rocks distribute in the northern   flank of the area where is Zheng'an basin extending in E-W direction and in the northeastern flank where is Xiaosong basin extending in N-E direction.   The basin are lower lands on the top of granite mass.The main rock body forms the southern margin of E-W Fugang complex rock body. In lithology, the rock is Yenshanian Stage 111(Jurassic) medium-coarse grained porphyritic granite ($\gamma52(3)$). As the Nanling granite body, in the early time of the rock forming, gas-liquid replacement autometamophysim formed albitization or potassium replacement granite,tourmaline-postassium feldspar, weakly sericitized and chloritized granite. Therefore, the rock is easy to be weathered and softened when it absorbs water. The mean wet-compressive strength is 95.4 MPa and softening coefficient is 0.80 for the fresh and slightly weathered medium-coarse grained granite. It does not reach the strength which granite should has,however, the rock mass is complete and has integrate or massive structure. Hence, ingeneral,the engineering geological condition of the rock mass is quite good.

GENESIS ANALYSIS OF ALTERATION

The main alteration in the granite mass of Yenshanian Stage III in the area is clay grouting. The clay grouting process is described as follows based on field survey,comprehensine analyses of various test results and the information concerned:

In the early stage,k+-rich alkli hydrothermal solution replaced plagioclase of the wall rock when it moved upwards along the tectonic fissure and formed clay grouting belt characterized with illite-sericite-kaolinite complex on the sides of fissures. when the hydrothermal solution continuously moved upwards and reached ground-water table, the temperature, pressure, PH value of solution lowered and K+ activity decreased because of the Ca2+-and Mg2+-rich down wards-leaked surface water entering the solution. The mixed solution interacted with the wall rock and formed the clay grouting belt characterized with montmorillonite-illite-kaolinite type.The hydrothermal solution from deep part of crust interacted with the wall rock which is above the groundwater surface. Even though the interaction was weaker than the preceding mentioned above, the kaolinite-illite clay grouting belt formed. In other words, the clay grouting results from

K+-rich thermal solution coming from the deep part of the crust interacting with Ca2+- and Mg2+-rich surface water.That is called the hydrothermal alteration or mixed hydrothermal solution alteration. It was clearly found based on field survey,lithological and mineralogical determinations of drilling cores and excavation of the underground works that in 250-300m deep and bellow, K-replacement reaction prevailed between medium-low-temperature solution and the wall rock, in 150-250 m deep the reaction existed between the medium-low-temperature solution with ground-surface infiltration water and the wall rock, in the depth within 150m the filtered surface water mainly led to weathering of the rock.

The altered minerals forming during hydrothermal alteration are mainly illite hydromica, sericite, and kaolinite,secondary montmorillonite, hydromica montmorillonite irregular mixture,quartz, calcite, pennine, leuchtenbergite and hydrobiotite, ect. content of the altered minerals is 2-70%, usually 10-30% of the whole rock.

There were two kinds of thermal solution, thermal groundwater and magma, in the hydrothermal alteration. After the main rock mass of Yenshanian Stage III formed, multiple stage of magma activation appeared, such as the medium fine grained granite of Yenshanian Stage IV(γ53(1)) and granite-porphyry of Yenshanian Stage V(γ53(2)) . Brought in F-rich fluid of silicion. A great quantity of Ca- and Mg-rich solution, which derived from the granite surface or from the carbonate strata on the top of Upper Paleozoic System, was heated and mixed with the post-magma thermal solution during it moved downwards along the faults and fractures, then the mixed thermal solution replace and reacted with the granite of Yenshanian Stage III(γ52(3)) when the solution moved upwards along the faults and fractures and periodicity manner

The alteration temperature is 132-250^{0}C which belong to medium-low temperature, is inclined to low temperature(200^{0}C)based on the typical altered minerals.

ALTERATION TYPE AND ITS DISTRIBUTION REGULATION

Alteration type
The hydrothermal alteration is controlled by the nature,component, temperature and pressure of the hydrothemal liquid and by the nature and component of the country rock. The thermal liquid moved upwards and outwards along the faults and cracks,frome main faults-secondary faults-fissures-micro fissures-pores,finally into the replacement place and the placement appeared in manner of filling, filtering and diffusion. The composition of the thermal liquid changed, the temperature and pressure decreased during replacement. Altered rocks with different degrees and different types formed in different tectonic parts.

Depending on characteristic altered minerals, the alteration includes silification, carbonization, illitization, kaolinization, montmorillonization, fluorization and chloritization, ect.

During alteration process, decomposition of original minerals and formation of new minerals   were a chain reaction which formed a mineral series. The altered minerals can hardly individually form and exist, contrary, they were an assemblage. Based on mineral assemblage, the altered minerals can be summarized into three types:

(a) Quartz-calcite assemblage(main vein of alteration);
(b) Illite-microcrystal quartzassemblage(illite-silified rock,microcrystal quartz, illite rock)and
(c) Montmorillonite-kaolinite-illite assemblage(clay mineral belt).

Distribution and changing trend
It was observed in the field that the altered minerals or the minerals assemblages mentioned above have obvious horizontalzonal distribution, a altered belt from the center of a main vein outwards can be approximately divided into:

(a)Microcrystal quartz-calcite belt(replacement-filling main vein);
(b)Leuchtenbergite-hydromuscovite-silification belt(leuchtenbergite-silification granite);
(c)Clay grouting alteration belt(montmorillonite-type altered granite,kaolinite-type altered granite);
(d)Week hydration belt(weakly altered granite)and
(e)Normal granite.

The zonal pattern can be seen in Fig.2 and mineral component is listed in Table 1. The chemical component is in Table 2. The clay mineral content in the clay grouting altered granite determined with X-ray diffraction and differential thermal analysis is shown in Table 3. The content of clay minerals in the altered granite usually is not more than 24 of the whole rock.

There are transition belts among the altered belts mentioned above and the width of the transition belts is not same each other,usually are 1-2 m and their length is meters to hundreds of meters.

The important property of atlitude of altered rock belts in the area are controlled strictly by the main fault and its secondary faults and fissures, therefore the belts have zonal distribution.The   appearances of the belts are NW,NNW and NNE,in general, there are two groups of NE and NW, the NW is the dominate. The altered rock arrange into cross, complex, parallel and oblique distribution and formed parallel zonal, gridiron, trapezium(ladder-shaped)and feather-shaped patterns(Fig.3). Further cutting the granite body into smaller sub-bodies that enhanced thermal solution replacement and made the altered rock belts become meters to tens of meters wide.

The vertical distribution of clay minerals is shown in Fig.4.Four mineral-assemblage types are distinguished in this area depending on content of the altered mineral(Fig.4):

(a)Microcrystal quartz-hydromica-kaolinite assemblage
The feature of the assemblage is that the altered main vein is silicified rock (microcrystal

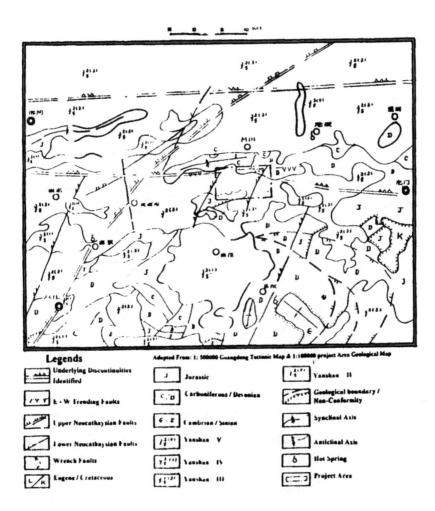

Fig. 1. Guangzhou Pumped Storage Project Regional Structure System

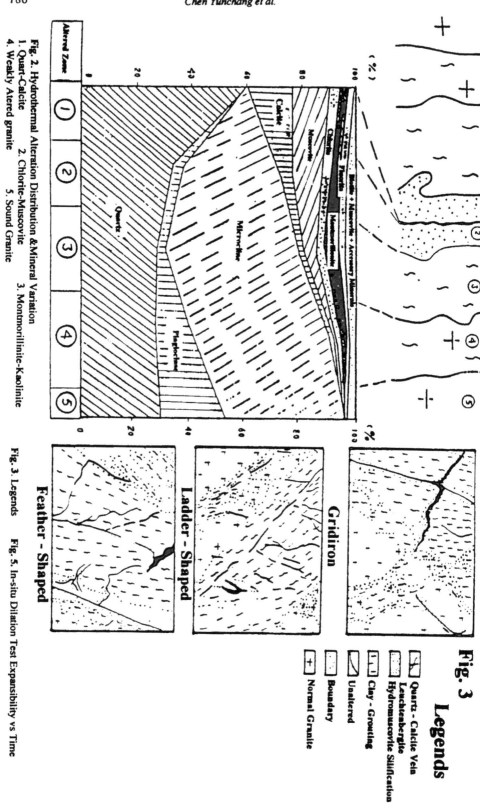

Fig. 2. Hydrothermal Alteration Distribution &Mineral Variation
1. Quart-Calcite   2. Chlorite-Muscovite   3. Montmorillinite-Kaolinite
4. Weakly Atered granite   5. Sound Granite

Fig. 3. Legends   Fig. 5. In-situ Dilation Test Expansibility vs Time

Fig. 3
Legends

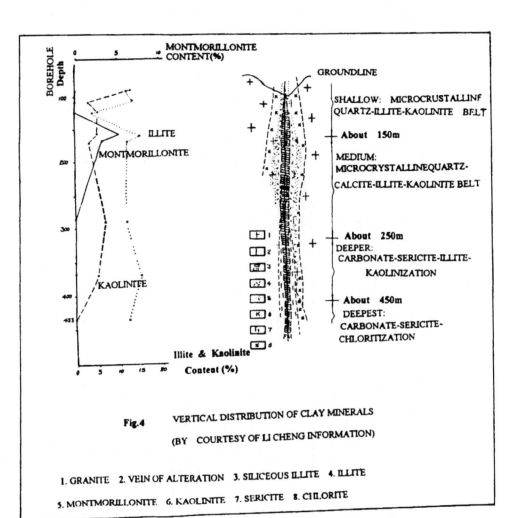

Fig.4   VERTICAL DISTRIBUTION OF CLAY MINERALS

(BY   COURTESY OF LI CHENG INFORMATION)

1. GRANITE   2. VEIN OF ALTERATION   3. SILICEOUS ILLITE   4. ILLITE

5. MONTMORILLONITE   6. KAOLINITE   7. SERICITE   8. CHLORITE

Table 1 the minerals component of altered rock (%)

Items	Silifiation & quartz vein ①	Quartz-calcite vein	Leuchtenbergite hydromuscovite silification belt ②	Montmor-illonization ③	Kaolin-ization	Weak hydration belt ④	Normal granite ⑤
K - Feldspar	5~8	12~40	42~52	46~58	40~56	45~53	48~53
Quartz	55~87	27~47	28~38	25~31	28~34	25~28	25~33
Biotite Muscovite Chlorite	<3	<3	2~4	2~4	2~4	3~4	2~5
Plagioclase	0~2	0~2	0~8	0~4	0~1	10~17	15~18
Kaolinite	2~3	2~4	2~5	5~7	5~13	0.5~1	
Hydromuscovite Sericite	10~30	9~10	5~10	2~4	2~3	2	2
Montmorillonite	trace	trace	0~2	4~12	0.5~3	trace	trace
Clinochlore	trace	0~4	1~5	0~12	0.5~1	0.5~1	trace
Calcite	1~2	10~70	0.5~8	0.5~4	0~1	0.5~4	trace
Trace Minerals	Total conten	is less than	1%, including	flu orite, zircon,	magnetic,	apatite, m	onzonit

Table 2   The Chemical Component Of Altered Rocks (%)

Items	SiO2	Al2O3	CaO	FeO	Fe2O3	MgO	MnO	TiO2	K2O	Na2O	P2O3	SO3	H2O+	H2O-	LOI
silification-quartz vein	87	2~3	2~3	1.73	0.49	0.46	0.10	2.63	0.12	0.009	0.06	2.94	△	△	△
quartz-calcite vein	55.61	7.93	15.68	2.06	0.64	1.58	0.06	0.15	5.56	0.26	0.02	0.11	3.39	2.35	14.2
leuchtenbergite silification granite	64.17	14.77	1.95	1.57	0.43	0.95	0.04	0.15	5.05	0.37	0.02	0.05	3.39	4.92	4.92
montmorillonite or kaolinite altered granite	69.4	13.49	1.34	1.36	0.73	0.18	0.03	0.14	5.15	0.02	0.02	0.05	2.94	2.77	3.60
weakly altered granite	73.52	12.11	1.31	1.36	0.73	0.18	0.03	0.12	3.08	2.94	0.02	0.04	0.74	1.02	1.28
normal granite	73.28	13.12	1.50	1.93	0.74	0.19	0.07	0.26	4.83	2.89	0.07				1.28

quartz); in the clay grouting granite,illite content is 12-30% of the whole rock; kaolinite content increases near the groundsurface(Fig.4). Depth of the assemblage(belt) is from the surface to 150m deep. Where montmorillonization rock is not developed.

(b)Quartz-calcite-illite-montmorillonite irregular mixture -montmorillonite-kaolinization assemblage

The assemblage(belt) mainly is montmorillonization altered rock belt, it is buried from 150-250m deep.

(c) Carbonate-sericite-illite-kaolinization assemblage

The assemblage(belt) is buried in depth from 250-350m.Illite and and sericite obviously increases in the depth deeper than 350 m. it means that K-mica content is increasing with depth. Content of clay minerals in the altered rock decreased bellow 450-500m deep, which indicates that the alteration reaction range of medium-low-temperature solution is from ground surface to the depth of about 450m. From 450m downwards, sericitization and chloritization was intensified because of weakening of clay grouting, therefore there could be the fourth assemblage,i.e.carbonate-sericite-chloritization assemblage there in the depth(Fig.4)

PROBE INTO CONDITION OF ALTERATION

Carbonization, leuchtenbergitization and montmorillonization in granite area are seldow and they can form only under very special physical and chemical condition.

Fault tectonic condition
The faults and fissures are the main passages of hydrothermal solution and the main place water-rock materials exchange and thermal solution alteration. Therefore the fault tectonics is the essential prerequisite of thermal alteration in the area.

The direction, range and degree of the fault tectonics directly control the developing direction, range and degree of thermal solution alteration. The larger the fault tectonic range is, the larger the alteration range is. Conversely, the smaller the fault tectonic range is, the smaller the alteration range is.The alteration belt or complex alteration belt distribution in this area is mainly controlled by the faults and fissures.

In regional geological tectonics, Guangzhou pumped storage power station is located at southeastern flank of joint area of EW Fugang-Fengliang fault belt and NE-NNE Guangzhou-Conghua fault belt where three groups of faults of NE, NW and EW directions developed. After the faults formed, in the later time of Yenshanian period(Neocathaysian tectonic period), a group of NE faults appeared and the original two groups of NE and NW faults further developed, especially the NW fault group obviously extended in large scale, such as faults of F1, F2, F4 and F6. Meanwhile NW faults became of extension that was advantageous for hydrothermal solution moving, alteration and replacement. Therefore, the thermal solution alteration in this area mostly developed along NE and NW faults and fissures, especially along NW one.In general trend,the main alteration belt distributed also along NW faults.

The hydrothermal solution moved along the main tend-shear-faults and diffused to the secondary faults. The diffusion appeared from main faults-secondary faults-fissures-microfissures -pores and finally into replacement places. Therefore, the degree of the alteration belt from main fault center outwards is getting weaker. Rocks in different alteration degree demonstrate that tectonic faults position obviously controlled alteration type, degree and distribution of altered rock belt.,In addition,the nature of faults frequently controlled alteration type. the permeation replacement mainly appeared in NNE compress-shear faults and formed montmorillonization and leuchtenbergitization altered rock,whereas the filling replacement mainly appeared in NNW extension faults that was a good passages for thermal solution moving upwards in the early time and formed silification, carbonization and hydromuscovitation altered rock and that also was the good places for basically material, alkaline earth, Ca and Si deposition which were superfluous or extracted during replacement in the later time of thermal solution, such as those appeared in fault f7012 of the powerhouse of the first stage. The tectonic unit of different scale has different feature of alteration replacement.Thermal solution filling replacement was dominant and diffusion replacement and permeation replacement were secondary in larger faults,whereas diffusion replacement and permeation replacement were dominant and thermal solution filling replacement was secondary in smaller fissures.

Magma and wall-rock condition
Magma autometamorphism and gas-thermal alteration (potassium feldspar, albitization and tour malinization, ect.) found in granite rock mass indicate that the gas-thermal, which was separated from the magma in the processes of intrusion and crystallization, was a direct factor of   alteration in early time in this area. The clay grouting alteration in the late time was also related to silification, carbonization and fluorization that indicates that heat quantity of the remanent gas-thermal in the late time of magma crystallization somehow impacted to the appearance and distribution of the clay-grouting alteration.

The alteration in this area presents not only in medium-coarse grained biotite granite of Yenshanian Stage III, but also in the fine or medium-fine grained porphyritic biotite granite of Yenshanian Stage IV. Further more,the alteration belt vertically cut diabase vein and made it altered that shows that the late alteration appeared after the diabase vein formed and there was a long time gap between crysto-differentitation and the late alteration. During the time gap, the heat quantity of the remanent gas could mix with deep circulate groundwater and formed mixed hydrothermal solution.

In addition, the high contents of Si, K and Na of granite provided the materials for the medium-low-temperature alteration that made the alteration easily. The replacement alteration could be more possible if the rock is coarse structure, high permeability and groundwater has good circulation condition.

Groundwater condition
This area is the place where the alteration, especially montmorillonization alteration relatively concentrated and developed. The area is a rolling country below an elevation of 500 m above sea level. Faults well developed and cut the area into valleys and smaller hill-bodies with medium-gentle slope. The alteration is rarely to be seen in the granite

area where the landform is very steep that indicates that the alteration could also be related to landform(especially to paleoerosion surface and paleogroundwater level)besides related to tectonic fault. it could be because the steep landform is disadvantageous for watercatchment and preservation, water-rock reaction time is short,water quantity less, degradation is fast in this kind of landform.On the contrary, flat landform has large catchment area(if there are rich rainfall and well-developed faults and fissures), good groundwater circulation and good exchange of water-rock materials that provides good space, enough time and substance conditions for alteration development.

The change of assemblage type of alteration minerals vs. depth also shows that alteration is related to erosion surface and paleogroundwater surface. hydrogeological logging of drilling holes indicates that groundwater mainly is vertical flow within150 m deep; groundwater is relatively static below 150 m deep,but the static water pressure is not large. The heat quantity of the remanent gas rose up to the depth,and was cooled and mixed by Mg-rich groundwater, then formed montmorillonite. It could be the cause why the montmorillonization alteration concentrated in the depth of 150-200 m in this area.

There, the development type and degree of the alteration belt are directly   related to the difference   of chemical component of groundwater based on the difference of distribution regulation in the space and development degree of the alteration belt.

Physical and chemical conditions of montmorillonization
Montmorillonite[$(NaCa0)0.33(AlMg)2(Si4O10)(OH2)nH2O$] is a kind of water-bearing Al-silicate minerals which forms mainly when plagioclase in granite is replaced. The plagioclase turns to montmorillonite that needs hydrolysis, additive Mg, neutral to weak basicity pH, insufficient discharge and weak circulation system of groundwater.

Hydrographic information shows that the surface water of this area belongs to HCO_3-Ca-Mg type, Mg content reaches to 0.174 gram-equivalent/liter. Therefore, it is proposed that Mg, where is necessary for montmorillonite forming, mainly comes from surfacewater permeating down to the depth along the faults and cracks.The forms process of montmorillonization altered rock is that the medium-low-temperature liquid moved upwards along the faults and fissures to groundwater table and mixed with Mg^{2+}-rich surfacewater which permeated down to groundwater table, then the mixture liquid decreased temperature, pressure and K^+ activity and increased Mg^{2+} content, the mixture liquid reacted with plagioclase in granite and finally formed montmorillonzation altered rock.

Formation and preservation of montmorillonite needs suitable temperature and pressure. Test indicates that the temperature of montmorillonite forming runs from 160-190 C with a mean value of 172 C;the pressure (depth, where montmorillonite is stable)changes from 150 to 200 m; when the   temperature and pressure increase,the montmorillonite turns to illite-chlorite or non-water or less-water Al-silicate minerals. Based on the regulation obtained from the test , to set the underground structure deeper lot can reduce the impact of montmorillonization altered belt of granite as possible.

The test also proved that under certain temperature and pressure conditions and if the proportions of [Ca2+]/[H+] and [K+]/[H+] in the liquid system are suitable, montmorillonite, illite and kaolinite can coexist. The coexistence of them found in the depth of 150-250 m in this area(Fig.4) could be a good evidence of the proper proportions of [Ca2+]/[H+] and [K+]/[H+] during the thermo solution evolution. The experiment pH value is 7.47 (neutral-weak basicity) for the coexistence.

The montmorillonite expansibility when montmorillonite absorbs water is an important factor of impacting stability of surrounding rock. It makes altered rock with a close texture be disintegrate and loose. The higher, the montmorillonite content is, the severer the texture damage of the rock will be. The tests and measurements show that montmorillonite content in the whole rock usually is <1-13%, at some strongly altered sections can reach 25-26%. The total amount of clay minerals is <20% in the area. When the montmorillonite content is ≤3%, the strength of the rock decreases but rock dose not disintegrate if is exposed in open air for a short period; when the montmorillonite content is 4-7% and the total quantity of clay is<25%, the rock can early be hammered broken and can automatically disintegrate in open air;When the content is > 7% and the total quantity of clay is >25%, the rock can quickly disintegrate in open air and can completely destroy within a week

ENGINEERING GEOLOGICAL PROPERTIES OF ALTERED GRANITE

The physical and mechanic features of the altered rock belt of different alteration type and degree are listed in Table 4 and 5.The indexes of the slightly weathered-fresh granite and clay grouting altered granite are quite different each other. The mechanic index of clay grouting altered rock is similar to that of strongly weathered medium-coarse granite. Depending on the typical load displacement curves of kinds of rocks, all the rocks belong to elastoplastic deformation or plastoelastic deformation except the medium altered granite which belongs to plastic deformation.

Acoustic test shows that the acoustic velocity in altered rock and fault belt is obviously lower than that weakly weathered or slightly rock. It reduced the elasticity and deformation modules of granite rock. the permeability of altered granite is weak,water pressure test of drilling hole indicates that Lugeon values 1-2 and there is no water directly permeating out from altered rock belt in underground opening.

Expansibility of montmorillonization altered rock is distinguished in the clay grouting altered granite. The result of the site swelling test is listed in Table 5 The expansibility is mainly caused by montmorillonite, the swelling quantity is closely related to montmorillonite content. The larger the montmorillonite content is, the larger the swelling quantity is.the swelling force is only 0.072 MPa when the montmorillonite content is 4%; the swelling force is 0.483 MPa when the content is 8%.It is also possible that there is weak swelling in the altered rock when illite content is high enough and the swelling force is only 0.01-0.03 MPa.

Table 5 The Physical And Mechanic Indexes Of Altered Granite (Field Test)

items / rocks	cohesion C MPa	internal friction angle Φ (°C)	static young's modulus ×10³MPa	defomation modulus ×10³ MPa	poisson's ratio μ	swelling depth cm	test swelling value mm	swelling stress KPa
montmorillonite	0.25	25	1.4 (0.26~2.4)	0.25 (0.02~0.34)	0.30	0 5 10 15	8.9~2.1 < 1 < 0.4 < 0.05	482 ~ 72
kaolinite	0.40	30 ~ 34	9.0 (5.9 ~ 14.8)	2.5 (0.98~5.27)	0.25	0		< 72
weakly altered	0.70	40 ~ 45	20.0 (14.8~ 22.2)	5.0 (2.7 ~ 8.5)	0.22	0	0.47 ~ 0.23	15 ~ 5

Note:Denotes range value in bracket.

Table 3 The Mineral Content Of Clay Grouting Altered Granite

Zones / Minerals	② Hydromuscovite	①Clay-grouting altered Montmorillonization	Kaolinization	①Sandy-Gravel Soil Of clay-grouting altered
Kaolinite	4.37 (52.4)	0.76~2.30 (10.0~23.5)	4.2~5.4 (80.9~85.7)	4.94 ~ 13.34
Hydromuscovite	3.08 (37.0)	0.40~0.34 (5.3~4.4)	0.7 ~ 2.1 (14.3 ~ 26.0)	1.54 ~ 4.12
Montmorillonite	0.88 (10.6)	6.45~7.11 (84.7~72.1)	0 ~ 0.65 (0 ~ 4.8)	3.36 ~ 10.87
Content Of Clay Minerals	15 ~ 24	12 ~ 21	10 ~ 24	15.5 ~ 19.9

Note: Indexes in bracket denotes the percentage in total amount of clay minerals.

Chen Yunchang et al.

Table 4 Collect Tables Of Physical-Mechanical Indexes Of Altered And Unaltered Granite

Rocks \ Values	Items	Specific Gravity (VS)	Bluk Density (g/cm³)	Void Rate (%)	Water Absorption (%)	Single Axial Strength (Mpa) Saturation (Rw)	Single Axial Strength (Mpa) Stoved (Rd)	Softening Coefficient	Elasticity Modulus ×10³ MPa	Deformation Modulus ×10³MPa	Poisson's Ratio μ
Clay Altered Granite	Range	2.62~2.69	1.92~2.59	6.02~9.33	1.46~3.15	2.5~14.0	7.0~27.4				
Clay Altered Granite	Mean	2.66	2.2.35	9.02	2.48	8.1	20.1	0.40			
Weakly Altered Granite	Range	2.62~2.65	2.42~2.56	4.48~9.33	0.53~2.75	11.9~43.3	25.5~52.3		8.82~14.0	6.99~12.69	0.12~0.28
Weakly Altered Granite	Mean	2.63	2.52	6.97	1.81	25.9	41.1	0.63	12.57	9.78	0.21
Slightly Weathered and Fresh Rock	Range	2.61~2.66	2.55~2.63	1.13~3.79	0.14~0.80	71.1~129.0	92.0~159.0		29.5~46.3	23.8~42.9	0.05~0.31
Slightly Weathered and Fresh Rock	Mean	2.64	2.60	2.22	0.31	95.4	119.0	0.80	(39.1)	(35.65)	0.15
Weakly Weathered Rock	Range	2.61~2.67	2.49~2.64	1.89~6.08	0.13~1.19	33.2~71.7	44.7~92.8		20.6~43.8	12.5~38.3	0.08~0.33
Weakly Weathered Rock	Mean	2.64	2.54	2.90	0.48	51.0	67.3	0.77	(35.4)	(31.0)	0.15
Strong Weathered	Range	2.62~2.65	2.11~2.52	4.49~14.0	1.63~8.74	1.8~15.2	6.8~26.1		1.51~7.78	1.09~5.40	0.14~0.40
Weathered	Mean	2.64	2.37	10.46 7	4.2	8.0	18.2	0.44	4.85	2.9	0.25

Note: Indexes in bracket denotes the **mean** values of little indexes

Swelling test shows that the swelling quantity of montmorillonization altered granite is rapidly decreased from the tunnel-wall surface to the inner of rock body when the surface is the only free and sock surface. The swelling quantity is very weak at the depth of 0 cm (Fig.5). Figure 5 also shows that at the begging of swelling there is a stage of high swelling, but the stage is very short, usually not longer than 48 Hs. Since then, the swelling is getting reduced and finally turns to a stable stage with low swelling. The stable time is usually 400-500 Hs. To understand the swelling feature of altered rock is most helpful for the support of surrounding rock.

SUPPORT OF ALTERED ROCK

Altered rock belt in granite is a kind of weak belt.The belt irregularly distributes in granite in this area and its exposed width is larger than that of fault belt. It damaged rock mass and reduced strength and deformation modules, further, influenced classification and stability of the surrounding rock of underground structure. Altered belt is main factor for sites election of the underground structure complex and optimum design. In classification of surrounding rock, a percentage of the exposed width of fault and altered rock vs. the length of underground opening (10 m used) is taken as the macroclassification index. Certainly, underground structure complex should be set at the place where the fault and altered belt are as less as possible.

Using the rockbolt/shotcrete to support the altered rock is available based on rockbolt-shotcrete mechanic test. Usually, the mortar-rockbolts with a length of over 0.5 m can stably support the altered rock. The rockbolt made with $\Phi 25mm$ and 16Mn screw-threaded steel bar can undergo 14-15 tons pulling force, maximum18-19.5 tens. The cohesive force between shotcrete and altered rock mass for the micro-weakly altered mass can reach 0.53-0.57MPa that is similar to that for slightly weathered rock and fresh rock; the cohesive force of slightly altered rock mass and medium altered mass is about 50-25% lower than that of micro-altered rock and fresh rock. To sum up, for the altered rockmass, the rockbolt/shotcrete(grid reiforcing bars, rebar) support are a kind of reliable measure. The shotcrete is to close the altered rock body, prevent weathering and deliquescence from developing. The grid rebars are used to reinforce the shotcrete layer to prevent the layer from breaking and falling down when it is drying. The mortar-rockbolts can directly reinforce surrounding rock and provide fixation or elasticity of the shotcrete layer.

After excavating and cleaning loose rock on the rock surface, a immediate guniting mortar is necessary for close the rock mass and preventing the rock mass from absorbing water. The strength in the inner of altered rock mass is quite high, and the mass will not swell. Before guniting mortar, the rock surface should be cleared with pressured air or water for removing loose or deliquescent material. The groundwater from fissures should be led out with drilling drain-holes before guniting mortar. Two days later, to gunite a more than 5 cm-thick layer of mortar is necessary as a protection layer, then dryly to drill holes for dowel. The dowel, in the places where the altered rock belt is wide or the belt is concentrated, should be plunged through the altered belt into rock mass 1-2

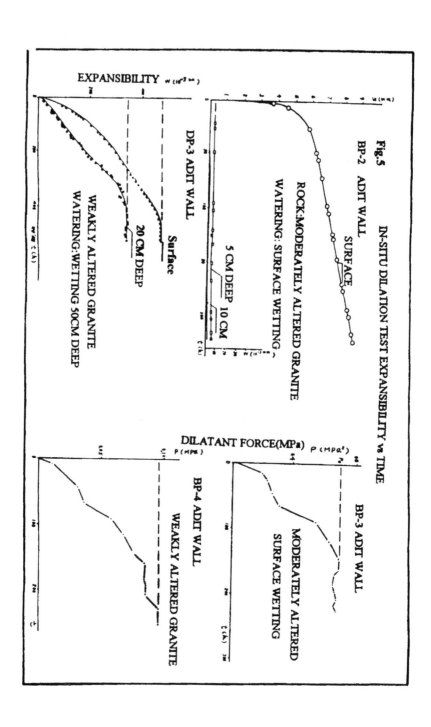

Fig.5 IN-SITU DILATION TEST EXPANSIBILITY vs TIME

m deep and soleplate and hanging net should be used. After 15-20 days, the mortar should be gunited again to make the layer to the designing thick. The gunited mortar-layer should be so thick that the surrounding rock can recover to the state before excavation and montmorillonite expansion can be really controlled. There, the thickness is determined not only depending on the designing request of rockbolt/shotcrete ete, but also depending on expansion force of montmorillonite

Acknowledgments

Authors heartily thank Senior Engineers Zhang Fugang and T uXixian for checking and approve this paper.

REFERENCES

[1] D.Dudoigon, D.Beaufort, And A.Meunier, 1988,Hydrothermal and supergene alteration in the granite cupola of montebras, Creuse,Clays and Clay Minerals, Vol. 36, No.6.1988, France.

[2] Guangdong Provincial Investigation Design and Research Institute of Water Conservancy and Electric Power, 1989,Engineering geological report of tech-construction designing stage of Guangzhou pumped storage power station.

[3] Li Cheng, Zhang Yanzhu, Du Haiyan and Huang Linghui, 1991, A study on the type and degree divisions of mixed thermal solution alteration in the granite body at the working section of Guangzhou pumped storage power station

[4] Tongji University, 1992, An investigation of rational support type and mechanics character of rockbolt/shotcrete technique used in altered rock in the construction stage II of Guangzhou pumped storage power station.

[5]Guangdong Provincial Investigation Design and Research Institute of Water Conservancy and Electric Power,1994, The engineering geological report of tech-construction designing stage in the construction stage II of Guangzhou pumped storage power station.

[6] Guangdong Provincial Investigation Design and Research Institute of Water Conservancy and Electric Power, 1994, Engineer geology, in Chapter III of Design Specification of tech-construction of Guangzhou pumped storage power station.

[7]Zhang Fugang and Tu Xixian, 1994, Engineering geological feature of clay grouting altered granite, Guangdong Geology.

Proc. 30th Int'l. Geol. Congr., Vol. 23, pp. 199-206
Wang Sijing and P. Marinos (Eds)
© VSP 1997

Mineralogical, Chemical and Geotechnical Characteristics of Basaltic Soil in Niksar, Northern Turkey

ATİYE TUĞRUL
Geological Eng. Dept., Faculty of Engineering, Istanbul University, 34850 Istanbul/TURKEY
ERSİN AREL
Civil Eng. Dept., Faculty of Engineering, Sakarya University, 54188, Sakarya/TURKEY

Abstract

The weathering of the Eocene aged basalts originates typical basaltic soils, at a site in Niksar in the northeast of Turkey. Assessment of the engineering behavior of the soils is of great importance for the design and construction of site formations and foundations. Various analyses were carried out to determine the clay mineralogy, major elements geochemistry, physical properties and finally the strength parameters of the extremely and completely weathered basalt and the residual soil. Along with chemical analyses, X-ray diffraction analysis of the whole rock and clay fraction were carried out. Traditional classification analyses such as sieve analyses and Atterberg limits were also carried out. Finally, consolidated undrained triaxial compression tests were performed on some of the samples to determine the shearing resistance and the pore-water pressure coefficient of the soils at different degree of weathering. The interpretations, depending on the test results, have shown that, besides its high resistance and low compressibility, the extremely weathered material has less activity and low swelling potential than the completely weathered material and the residual soil.

Keywords: weathering, basaltic soil, geotechnical characteristics

INTRODUCTION

Differences in structure, mineralogical, geochemical and geotechnical characteristics of soils are often caused by weathering. The aim of this study, is to analyze the engineering behavior of the soil derived from basalts pertaining to the Eocene aged Hasanseyh formation. The investigation was carried out with the help of X-ray diffraction methods. For each weathering state major chemical elements were also determined. Where the weathered rocks had lost the original structure due to weathering, they were considered as soil and soil mechanics principles were used for the determination engineering behavior of the weathered materials.

Geology

The studied area is located within the boundaries of Tokat province in the northeast of Turkey. The oldest rock stratum is the Upper Cretaceous aged Vezirhan formation in the study area. Vezirhan formation comprises limestone, clayey limestone, marl and tuff.

Middle Eocene aged Kusuri formation consists of sandstone, sandy clayey limestone, claystone and marl. These deposits disconformably locate at the top of Vezirhan formation. In the area, at some places lateral and at the others vertical transitions exist between the Kusuri formation and the Hasanseyh formation. From a petrographic point of view the Hasanseyh formation presents some diversity from basalt, largely predominant, to agglomerate, autoclastic breccia, tuff, tuffite, sandstone and marl. The relationship between the Hasanseyh formation and the others are shown in geological map (Fig. 1). Depending on the fault tectonics of the region; Hasanseyh formation shows partly unconformed and partly faulted relation with the overlying Yolüstu formation [11].

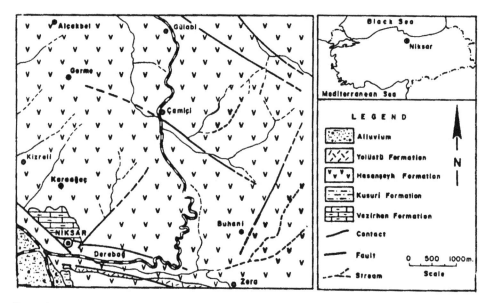

Figure 1. Location map for the study area, showing the generalized geology

Basalts are composed of mainly plagioclase phenocrysts and pyroxene, chlorite, epidote, calcite, volcanic glass, opaque minerals and rarely olivine crystals. Chlorite, iddingsite and epidote are the main weathering products.

Weathering of the basalts
The eight weathering grades determined in the basalts are given in Table 1. The effects of weathering are seen to increase upwards from fresh basalt to almost residual soil. The study were performed on the extremely and the completely weathered materials and the residual soil. Along the soil profiles at the site, the boundary between the zones was not always observed to be distinct and some zones were clearly absent in some profiles.

SOIL MINERALOGY AND CHEMISTRY

The mineralogical composition of the clayey fraction is influenced by many factors, among which the grain size, that controls in particular the kind of predominating clayey

Table 1. The eight weathering grade in the basalts

Weathering Class	Weathering Grade
Residual soil	IV
Completely weathered	IIIC
Extremely weathered	IIIB
Highly weathered	IIIA
Moderately weathered	IIC
Slightly weathered	IIB
Faintly weathered	IIA
Unweathered (Fresh rock)	IV

minerals [5]. The engineering properties of soils are partly related to the types and distribution of the clay minerals present [9]. The mineralogical composition of the soil was determined with the X-ray diffraction analyses. The mineralogical data indicate that; extremely and completely weathered materials contain clay size materials that are montmorillonite as the primary component, lesser amounts of kaolinite and plagioclase. The silt size fraction contains significant quantities of plagioclase, although with relatively less montmorillonite and more mica and goethite than in the clay size fraction. The clay size fraction for residual soil contains a systematically higher percentage of montmorillonite and kaolinite and a lower percentage of plagioclase than in the lesser weathered zones.

According to the chemical analysis results (Table 2); the amounts of the SiO_2, CaO, Na_2O and K_2O have diminished while the amount of MgO rose with chemical weathering in these rocks. The pH values of completely and extremely weathered basalts and the residual soil were also determined. The results show that, pH values are between 6.6-7.8 and less basic fresh basalt become either acidic or neutralized when it is weathered. This phenomenon is believed to be connected with the increase of the clay minerals by weathering.

Table 2. Chemical analyses result. (IIIB: Completely weathered basalt, IIIC: Extremely weathered basalt, IV: Residual soil)

Major Element Oxides	IIIB	IIIC	IV
SiO_2	48.56	49.43	48.21
Al_2O_3	18.68	19.27	19.33
Fe_2O_3	8.71	8.85	10.23
TiO_2	0.81	0.991	1.04
MnO	0.15	0.16	0.17
CaO	7.53	5.13	4.50
MgO	3.42	3.53	5.17
Na_2O	2.68	1.38	0.92
K_2O	2.02	1.67	0.58
Loss of ignition (1000°C)	7.56	9.70	9.86
Total	100.12	100.03	100.01

GEOTECHNICAL CHARACTERISTICS OF THE SOILS

To evaluate the engineering properties of the extremely and the completely weathered materials and the residual soil, bulk and tube samples were collected at various depths from the studied area. The bulk samples were used to determine the grain size distribution, consistency limits and natural water content according to ASTM [1,3]. The undisturbed tube samples were used to determine the mechanical properties. The shearing resistance of the basalt weathered to a residual soil has been determined in the laboratory by means of the consolidated undrained triaxial compression test (CIU) with pore-water pressure measurements in accordance with ASTM [2].

Particle size
Physical disintegration causes finer particle size in the soils and finer the more particle size higher the surface area. The intensity of the chemical weathering   increases with the increment of the surface area (Caroll, 1970). According to the calculation of Baver (1956) the surface area of the clay size material where the chemical reaction takes place is 100 times higher than the surface area of the sand material. Concerning the all the above mentioned findings, priority was given to determining particle size properties of the soils derived from basalts. To evaluate the soil texture, the grain size distribution of the extremely and completely weathered rocks were determined. The test results show that (Fig. 2), the extremely weathered basalts contain gravel size materials that are more than 50%. The completely weathered materials consist of mostly sand and silt size material with little or no clay. The residual soils derived from basalt contain slightly or no clay, but are also generally silty in nature.

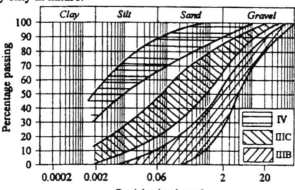

Figure 2. Particle size distribution of the basalts at different weathering grade

Soil density
Along the weathering profile, specific gravity of the soil derived from basalt ranges between 2.71 and 2.92, bulk density is between 17.76-21.26kn/m3, dry density is between 13.61-17.81 kn/m3 and moisture content is between 18-34%.

Void ratio and porosity values for the completely weathered and highly weathered basalts were computed as well as the residual soils. According to the results, porosity values are between 41.46% and 51.49%, void ratios between 0.51 and 1.06. The results indicate that the more weathered materials have high void ratios and high porosity values.

Plasticity characteristics and classification of the soil
Besides the particle size distribution of the weathered materials, the Atterberg limits were also determined. The range of the liquid limit for fine grained part of the extremely weathered materials is between 25% and 45%, for completely weathered materials between 35% and 52%, for residual soil is between 50% and 70%. The variation with depth is important. There is a noticeable decrease in values with decreasing weathering.

The plasticity index ranges between for fine grained part of the extremely weathered materials 5% and 20%, completely weathered materials 18% and 30%, and residual soil 30% and 45%. These index values also decrease with decreasing weathering.

The soil samples were also evaluated in accordance with USCS (Unified soil Classification System). The extremely weathered materials are generally in the SM group according to this classification system. The completely weathered materials are SM-SC type soils and the residual soils, which have mostly fine materials are partly in the SC group. In view of the plasticity, the weathered materials can be classified as follows (Fig. 3): Fine grained part of the extremely weathered materials can be classified as CL-ML and ML-OL and fine grain part of the completely weathered materials as CL but rarely CL-ML classes and the residual soils as CH. Obtained values for the Atterberg limits usually agree with the soil textural composition. This fact is generally in accordance with the results obtained in previous works [5, 8].

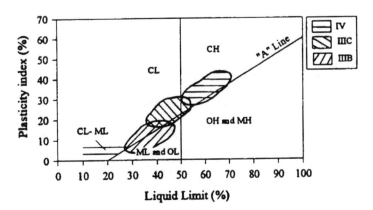

Figure 3. Plasticity chart

The determined value for the activity parameter, Ac (Ip/%<2μm) allowed us to classify these soils, using the criteria of Skempton [10]. According to the classification (Fig. 4), the extremely weathered materials mostly show low activity, the completely weathered materials give low to moderate activity and the residual soils generally give extremely high activity. In Figure 5, the materials are classified according to the Van Der Merwe diagram [12], which defines the swelling potential on the basis of the relationship between activity and clay fraction. The classification shows that the extremely weathered

materials

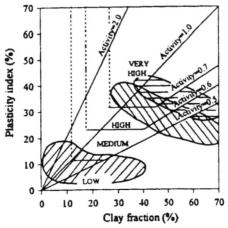

Figure 4. Determination of activity of the weathered materials (Legeng as in Fig. 2)

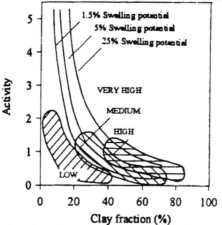

Figure 5. Evaluation of swelling potential of the weathered materials (Legend as in Fig. 2)

have mostly low swelling potential. The completely weathered materials generally have low to high swelling potential and the residual soils have high to extremely high swelling potential. The high to extremely high swelling potential of the residual soils is related to the abundance of the montmorillonite in the clay component.

Mechanical behavior of the soils
The shearing resistance of basalts from extremely weathering stage to residual soil has been determined in the laboratory by means of the consolidated undrained triaxial compression test with pore-water pressure measurements. Pore-water pressures were measured with a pressure transducer. The stress-strain curves for a few of the samples and pore-water pressure measurements of the undrained compression tests are shown in Figure 6. As it is demonstrated in these diagrams the less weathered materials have higher strength than the others. In addition, the deformability of the samples increases with the dropping of the pore water pressures.

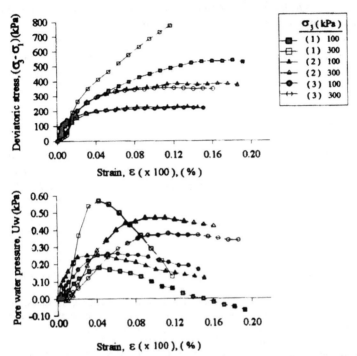

Figure 6. Stress-strain and pore water pressure-strain curves for basalts at different weathering stages (1: Completely weathered basalt, 2: Extremely weathered basalt, 3: Residual soil)

CONCLUSION

The following conclusions can be drawn from this study.

1) Samples relating to the some place show substantial differences in their mineralogical composition, depending on the degree of weathering.

2) The extremely weathered basalts contain more than 50% gravel size particles. The completely weathered basalts consist of mostly sand and silt size material. The residual soil is mainly clay and silts including some fine sand and, not so often, a coarser fraction composed essentially of medium to coarse sand. Gravel is also rarely present.

3) Attending to mineralogical composition of the clay fraction (montmorillonite and kaolinite) a higher plasticity would be expected for the fine grained part of the extremely and completely weathered basalts. However we may not forget that the clay fraction only in residual soils exceeded 30%, therefore, the obtained values can be considered as normal.

4) It was found that, the extremely weathered materials mostly show low activity and low

swelling potential. The completely weathered materials generally give low to moderate activity and low to high swelling potential while the residual soils mostly show extremely high activity and high to extremely high swelling potential.

5) Consolidated undrained triaxial test results indicate that the more weathering processes effect the rock, the more the rock losses its strength and the compressibility increases.

Acknowledgment

This work was supported by the Research Fund of The University of İstanbul. Project number:Ö-III/4/0300496. The authors would like to thank to Prof. Okay GÜPINAR, Prof. Akin ÖALP and Eymen AREL for their encouragement and support during this study.

REFERENCES

1. A. S. T. M. (422-69). Particle size analysis of soils
2. A.S.T.M. (4767-88). Triaxial test in cohesive soils.
3. A.S.T.M. (4318-93). Determining the liquid limit, plastic Limit and plasticity index of soils.
4. L.D. Baver. *Soil Physics*, 3rd. Ed. John Wiley and Sons, New York (1956)
5. R. Bertocci, P. Canuti, E. Pechioni and C. A. Garzonio. A contribution to the geotechnical and mineralogical characterization of the clayey and shaley complexes involved in mass movements affecting historical towns in the Tosco-romagnolo Apennines, *Proc., 7th. Int. IAEG Cong.*, 983-992 (1994)
6. D. Carroll. Rock Weathering, Plenum Press. New York-London, 203p. (1970).
7. A. Cassagrande: Classification and identification of soils, *Trans. ASCE*, 113, 901-992 (1948).
8. L. M. Duarte, I. M. A. Almeida and S. Prates. Mineralogy, crystallinity and activity of Lisbon's basaltic soils, *Proc., 7th. Int. IAEG Cong.*, 945-950 (1994).
9. S. A. Ola. Mineralogical properties of some Nigerian residual soils in relation with building problems, *Engineering Geology*, 15, 1-13.
10. A. W. Skempton. The colloidal activity of clays, *Proc. 3rd Int., Conf. Soil Mech., Found. Eng.*, Switzerland, 1, 57-61 (1953).
11. A. Tugrul. *The Effects of Weathering on The Engineering Properties of Basalts in The Niksar Region.* Unpublished Ph.D. Thesis, Department of Geological Engineering, Istanbul University, Istanbul, Turkey, 168 p. (in Turkish) (1995).
12. D. H. Van Der Merwe. The prediction of heave from the plasticity index and the percentage clay fraction of soils. *Civil Engineering in South Africa*, 6, 103-107 (1964).

Proc. 30th Int'l. Geol. Congr., Vol. 23, pp. 207-220
Wang Sijing and P. Marinos (Eds)
© VSP 1997

Dynamic Properties of Weak Intercalation Encountered in Xiaolangdi Project on Yellow River

XUE SHOUYI
Shijiazhuang Railway Institute , Shijiazhuang 050043, China

WANG SIJING
Institute of Geology , Chinese Academy of Sciences , Beijing 100029, China

Abstract

The deformation and strength of rock mass are dominated to a considerable extent by its soft and weak intercalations. The dynamic properties of the weak intercalations in the left hill mass of the Xiaolangdi Damsite, which is located in the downstream of the Yellow River, China, and which is mainly composed of sandstone with clay intercalations of Traissic time, have been studied through laboratory shearing experiments. The samples used in the experiments are from the site and contain gouge, clastic sediments, and materials with slickenside. Furthermore, the dynamic shear modulus and damping ratio are studied by a resonant column method using moulded samples. The experimental results under multi-cyclic loading on the model with soft and weak intercalations show that a non-linear relationship emerges between dynamic shear stress and strain, and that the strain is of obvious hysteresis. In case of increasing the total stress level, the dynamic strain will become bigger and bigger with the increase of the number of cyclic loading, and finally appears as plastic deformation. The experiments also show that the dynamic strength is closely related to the duration of dynamic loading or the number of cycles. The realtionship between the dynamic shear strength and the initial normal stress follows the Mohr-Coulomb Law. The dynamic friction coefficients are larger than the static ones. Another important fact is that the exsistence of smooth slickenside can reduce 50% of the cohesion and 30% of the dynamic friction coefficient. The method and results in this paper can reasonably be used to evaluate the dynamic stability of the left hill mass of the Xiaolangdi Damsite.

Keywords: Weak intercalations, Dynamic properties, Xiaolangdi Project

INTRODUCTION

The deformation and failure of rock mass are dominated to a considerable extent by its weak intercalations, and therefore the mechanical behavior of weak intercalations is one of important factors for the evaluation of rock mass stability. Particularly, the study on dynamic properties of intercalations under cyclic loading might be a critical problem to the engineering design in the areas of high seismic intensity. The static behaviors of weak intercalations are well known, but their dynamic ones have rarely been reported so far. This paper presents the test results of the dynamic properties of weak intercalations ocurred in the left hill mass of the Xiaolangdi Damsite on the Yellow River.

As a major project, the main structure of the Xiaolangdi reservoir is designed to resist the damage of seismic intensity 8. This designing code is also suited to the left hill mass because of its close relation to the main structure. Moreover, the influence of slab vibration on the left hill mass stability caused by flood discharging must also be considered. Thus the dynamic characteristics of the weak intercalations in the Damsite must be carefully investigated. Some experimental studies have been carried out and the results achieved have been used in the stability analysis and design of the project (Wang Sijing et al. 1990, Xue Shouyi et al., 1997).

GEOLOGICAL CONDITIONS AND WEAK INTERCALATIONS

The bedrock formation outcropped at the Damsite of the Xiaolangdi project consists of clastic rocks of Triassic time and belongs to continental or shallow water deposits. The rocks mainly contain sandstone, clayshale, and claystone, and occur as interbeds. Owing to tectonic movements some thin layers of weak rock were altered into shear zones. Seepage water perhaps played a role in this altering. Slightly dipping clay intercalations are common in all kinds of rock in the left hill mass. A typical profile is shown in Fig.1.

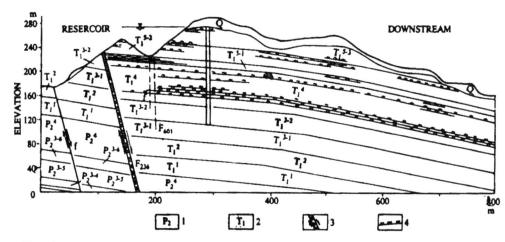

Fig. 1 Geological profile of the left hill mass in the Xiaolangdi Damsite
1. Upper permian deposits; 2. Lower Triassic deposits; 3. Fault; 4. Clayey intercalation

It can be seen from Fig.1 that the layers dip downstream and contain weak intercalations on the slope of the river. According to the principles of engineering geomechanics, downward sliding along the intercalations is expected to produce strong deformation and failure if a greater earthquake or strong slab vibrating caused by flood discharging takes place. Therefore the qualitative and quantitative study of dynamic behavior of the intercalations is a prerequisite for the stability analysis of the hill mass. According to their compositions, the intercalations at the hill mass may be divided into three groups as follows: Group AB-intercalations consisting of either silt or silt with fine debris; Group CE-intercalations consisting of silt sand with silt; Group D-intercalations consisting of debris with silt. In the tests only AB intercalations were investigated because of their low strength and controlling role on the slope stability.

SAMPLING AND TEST METHOD

Dynamic direct shear test

The field stress conditions of the intercalation unit is schematically shown in Fig.2. The dynamic direct shear test in this study can approximately represent the stress conditions. The samples were collected mainly from layer T by means of cutting ring sampling. Except samples of normal intercalations, rock slices with mirrorlike planes and slickensides were carefully collected because they may cause a significant reduction of the strength of the intercalations. To meet the requirement of the experimental equipment and procedure, the raw material was cast by expoxy resin to make a test specimen, which is 5.1cm in diameter and 2.1cm in height. However, in oder to study the influence of mirrorlike planes on the dynamic properties of intercalations, a group of test specimens was prepared that contain no such a plane. Fig.3 shows a sketch of the dynamic shear test specimens used.

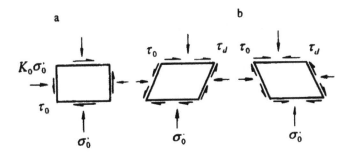

Fig. 2 Stress conditions of the intercalation
a. Initial stresses ; b. Initial stresses plus cyclic stresses

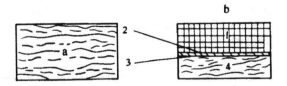

Fig. 3 Typical sedtion of the samples of intercalations
a. Full clayey intercalation; b. Intercalation with mirror planes or slickensides;
1. Epoxy resin; 2. Rock slice; 3. Mirror plane of slickensides; 4. Clay and debris of intercalation

Before the test, the specimens were treated by air pumping and water saturating. Then they were seated on the instrument and applied by initial normal and shear stresses ($\sigma`_0$ and τ_0) so that they were consolidated under the constant initial stresses. After the consolidation, they were applied by a dynamic shear stress (τ_d) and thus began vibration till failure occurred. To simulate seismic loading, the dynamic shear stress is of the form of a sinusoid curve with a frequency of 1 Hz.

Dynamic triaxial shear test
The core samples used in the dynamic triaxial shear test were obtained by using the three-pipe drill. The diameters of the samples range from 5.16cm to 5.2cm, with a height of 12cm. In the rock block on the two sides of the intercalations of some samples, cleavages and joints developed, which makes the blocks break and the samples unused in the test. When the other usable samples were placed on the apparatus, they generally still needed some processing work. Fig.4 is the sketch map of the triaxial samples. The diameters (d), intercalation thickness (ts) and angles between the intercalations and the end planes (θ) of the samples, are given in table 1.

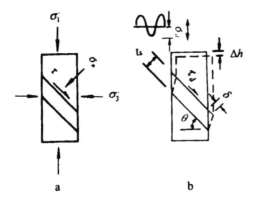

Fig. 4 The Sketch map of the dynamic triaxial samples

The triaxial shear test was performed on the electromagnetic type dynamic triaxial shear apparatus made in China. The test conditions can be described as follows. Firstly, the samples are saturated by means of the air pumping method. Secondly, the initial stresses are loaded and the samples are permitted to consolidate to be stable. Lastly the cyclic stresses are applied under the condition of undrainage to simulate the influence of the vibration generated by flood discharging.

Through comparative tests by adopting two kinds of loads with different frequencies (2.5Hz and 0.2Hz), it was discovered that the results under the two conditions are very approximate. Therefore most of the samples were first shaken 240 loops under loads with a frequency of 0.2Hz, and then they went on being shaken 1500 or 3000 loops under loads with a frequency of 2.5Hz. And some samples vibrated 3000 loops or more under loads with the same frequency of 0.2Hz.

Table 1 Some parameters of the samples and test conditions

No.	d cm	ts cm	θ deg	σ'$_3$ kPa	σ'$_1$ kPa	σ'$_2$ kPa	τ$_0$ kPa	α$_0$
1	5.16	3.8	45	98	144	121	23	0.19
6	5.16	3.1	45	98	144	121	23	0.19
7	5.19	3.8	40	98	144	125	22.7	0.182
10	5.18	2.7	45	98	144	121	23	0.19
16	5.20	2.6	45	98	143	121	22.5	0.186
2	5.20	3.5	50	294	430	350	67	0.191
4	5.18	3.3	42	294	431	370	68.1	0.184
5	5.18	3.5	40	294	431	374	67.5	0.18
8	5.18	2.8	45	294	431	363	68.5	0.189
12	5.20	0.8	45	294	429	362	67.5	0.187
18	5.20	2.1	45	294	430	362	68	0.188
19	5.20	2.2	45	294	430	362	68	0.188
11	5.20	2.5	45	98	189	144	45.5	0.316
15	5.19	2.2	45	98	189	144	45.5	0.316
9	5.20	2.1	45	294	566	430	136	0.316
13	5.20	2.4	45	294	566	430	136	0.316
17	5.20	1.6	45	294	566	430	136	0.316

DEFORMATION BEHAVIOR OF THE INTERCALATIONS

Results obtained from the dynamic direct shear test
The typical relationship between the shear strain and the number of cyclic loading is shown in Fig.5. Here the total shear stress (τ) denotes the sum of the dynamic shear stress (τ$_d$) and the initial shear stress (τ$_0$), the total shear strength (τ$_f$) denotes the sum of the dynamic shear strength (τ$_{df}$) and the initial shear stress (τ$_0$), and the ratio of over is defined as a stress level.

The curve shown in Fig.5-a gives results of a test speicmen without mirrorlike planes at a stress level of 85%. It can be observed that the shear strain hardly increases when the number of cyclic loading exceeds 20. The results of a test specimen with mirrorlike planes at a stress level of 80% are shown in Fig.5-b. It can be seen that the shear strain is almost proportional to the cyclic number. Obviously, the kinds of test specimens are very different in deformation behavior, and the test specimen with slikensides seems to be less resistant to cyclic loading. But the reason why this phenomenon occurs is not clear so far.

The experimental results show that the intercalations behave not only non-linear deformation, but also strain hysteresis. This illustrates that the intercalations have similar deformation characteristics to soils. Because of the similarity of such intercalations to soils, an equivalent visco-elastic model which was proposed by Seed et al. (1969), and which is still widely used in soil dynamics has been introduced into the study of the intercalations. According to this model, the stress -strain relationship at a given strain amplitute can be approximately expressed as a hysteresis loop shown in Fig.6. To link the

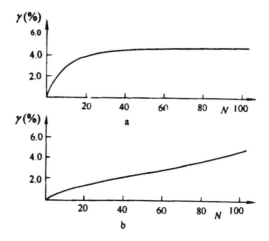

Fig. 5 Relationship between the dynamic shear stress strain and the number cyclic loading

    a. Test specimen without mirrorlike planes or slikensides (σ_0'=400kPa, τ_0=60kPa, τ_d=74kPa)

    b. Test specimen with mirrorlike planes or slikensides   (σ_0'=200kPa, τ_0=30kPa, τ_d=25kPa)

original point and the top points obtains a characterized segment. The inclination of the segment denotes the dynamic shear modulus at the given strain amplitude. This secant modulus is defined as the equivalent modulus G. The stress-strain curves for the intercalations generally show the following features(Fig.6).

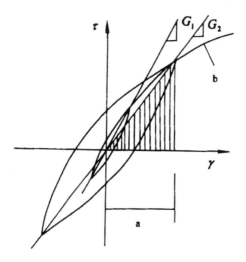

Fig. 6 Stress-strain hysteresis loop
a. Shear strain amplitude; b. Trajectory of the end points of hysteresis loops

a. The inclination of the segment linking the two end points of a hysteresis loop decreases with the increase of the cyclic strain level;

b. The area embraced by a hysteres loop increases with the increase of the cyclic strain level;

c. The end points of hysteres loops with different strain levels, namely the points which indicate the maximum and minimum values of stresses and strains, are located on the stress-strain curve gained by the initial loading and unloading.

If the intercalations are assumed to be of visco-damping, the damping ratio can be expressed as follows.

$$\beta = \frac{A_L}{4\pi A_t} \tag{1}$$

where A_L -area of the hysteresis loop; A_t -area of the triangle shaded.

Obviously, the equivalent modulus G and the damping ratio   vary with strains because of the non-linearity of the intercalations. Although the top part of the hysteresis overlaps the initial loading curve, the dynamic shear modulus determined by the inclination of the secant segment can be obtained as long as the initial loading curve is known.

The experimental results also show that the relation between the dynamic shear stress and the dynamic shear strain   can be described by a hyperbolic curve. The stress-strain curve for the cyclic loading number of 20 is shown in Fig.7 and can be formulated by

$$\upsilon_d = \frac{\gamma G_{max}}{1+\gamma/\gamma_\gamma} \tag{2}$$

where, G_{max} is an initial modulus for a small strain; τ_{dmax} is maximum dynamic shear stress; γ_γ, i.e. τ_{dmax}/G_{max}, is reference shear strain.

From the stress-strain curves obtained actually, G_{max} has been estimated to range from 30 to 100 Mpa for the test specimen with mirrorlike planes and from 30 to 150 Mpa for the one without mirrorlike planes. But the estimated results are not very reliable because the shear tests were conducted in the case of large strains.

Results obtained from the resonant column test
The resonant column test performed in a moulded test specimen is to study the dynamic shear modulus and damping ratio of intercalations. The material making up the test specimen has a dry density of 2.0g/cm³ which is compatible with that of the real intercalations. Before experiments the test specimen is treated by water saturating, and then is consolidated at a confining pressure. The maximum shear modulus of the test specimen are expressed by the following formula

$$G_{max} = 7181\left[\left(\sigma'_1+\sigma'_3\right)/2\right]^{0.49} \tag{3}$$

where G_{max}, σ'_1 and σ'_3 all are calculated in Kpa. The relation curves between the shear modulus ratio G/G_{max} , the damping ratio β and the shear strain γ are given in Fig.9. Generally speaking, the maximum shear modulus of the moulded specimen are 100%-200% larger than those of the intercalations in situ.

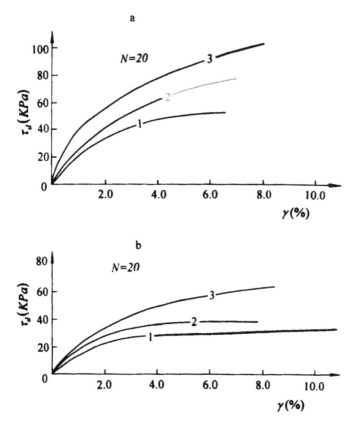

Fig. 7 Stress-strain Relationship Curves
a. Test specimen without mirrorlike planes or slickensides;
b. Test specimen with mirrorlike planes or slickensides;
1. $\sigma'_0 = 100kPa$, $\tau_0 = 15kPa$; 2. $\sigma'_0 = 200kPa$, $\tau_0 = 30kPa$; 3. $\sigma'_0 = 400kPa$, $\tau_0 = 60kPa$

Results obtained from the dynamic triaxial shear test
Fig.10 to Fig.12 show the relationships between the shear strain and the cyclic number of three typical samples, and three typical curves are given out in each figure. It is known from the curves that:

a. When the total shear stress ratio (the ratio of the total shear stress τ over the initial normal stress σ'_0 of the intercalation) was smaller, the shear deformation of the intercalations tended stable with the cyclic number and mainly developed within two or three hundred loading loops. Only when the ratio was approximately greater than 0.4, did the deformation with the cyclic number take a destructive trend.

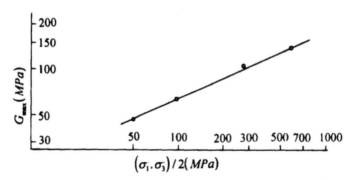

Fig. 8 Relationship between the maximum shear modulus G_{max} and the mean effective consolidation stress (σ'_1 +σ'_3)/2

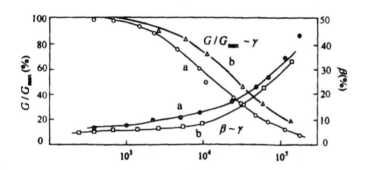

Fig. 9 G/G_{max} ~γ and β~γ relation curves
a. σ'_1=49.0kPa; b.σ'_3 =588.4kPa

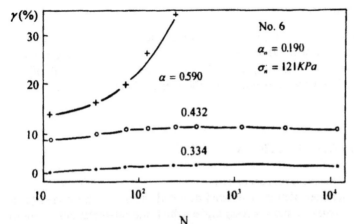

Fig.10 The relationship between the shear strain and the cyclic number(1)

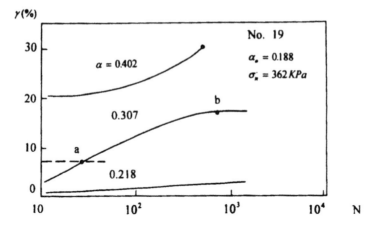

Fig.11 The relationship between the shear strain and the cyclic number(2)

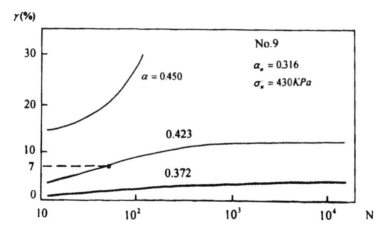

Fig.12 The relationship between the shear strain and the cyclic number(3)

b. The test in which cyclic loads were applied interimly was carried out on a few samples, with interim period from several hours to several ten hours and on the conditions of drainage or undrainage in the interim period. It was found that the trend of the deformation with the cyclic number did not interimly change. Point b in Fig.11 is the record of a interim test.

STRENGTH OF THE INTERCALATIONS

Results obtained from the dynamic direct shear test
The dynamic strength of the intercalation is an axial deviatial stress or a shear stress(sum of initial shear stress and dynamic shear stress) under which the intercalations come to failure. Here are two uncertain aspects, namely, the criterion of failure and number of cyclic loading. In this study, the test specimen is considered as being of failure when the

dynamic strain reaches 5%. This is a criterion widely used in soil dynamics. On the other hand, the strength is closely related to the number of cyclic loading. An equivalent cyclic loading number Neq is usually adopted to determine the strength. Neq is concerned with the earthquake duration and the number of main peaks on the accelerogram.

According to the basic seismic motions at the Damsite of the Xiaolangdi project, Neq for nearby and distant events can take 8 and 20, respectively. Thus using the experimental data for the intercalations, the relation between the dynamic strength and the initial normal stress are obtained, as shown in Fig.13.

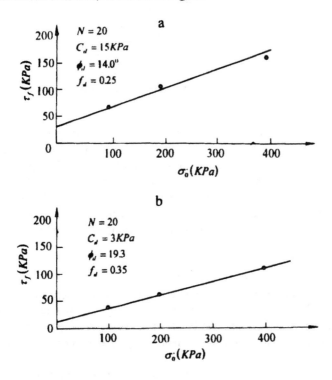

Fig. 13 Dynamic strength of the intervcalations
a. Test specimen without mirrorlike planes or slickensides;
b. Test specimen with mirrorlike planes or slickensides

It can be found that the relation is in accordance with the Mohre-Coulomb criterion, which is expressed as follows.

$$\tau_f = C_d + \sigma'_0 \, tg\varphi_d \, or C_d + \sigma'_0 \, f_d \tag{4}$$

where, C_d is a dynamic cohesion; φ_d is a dynamic friction angle; f_d is a dynamic friction coefficoent.

When Neq = 20 , C_d=30kPa and f_d = 0.35 have been obtained for the intercalations without mirrorlike planes, and C_d =15kPa and f_d =0.25 for the intercalations with mirrorlike planes. The existence of the mirrolike planes significantly reduces the strength because C_d decreases 50% and f_d decreases 30%. When Neq=8, f_d is 0.263 for the intercalations with mirrorlike planes. This indicates that the selection of the earthquake equivalent cyclic number Neq has a larger effect on the strength(Fig.14).

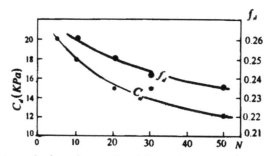

Fig.14 Relationship between the dynamic strength and the number of cyclic loading for the test specimen with mirrorlike planes

Because of the importance of the intercalations in the Xiaolangdi project, their static mechanical properties have been studied widely and thoroughly. According to the results provided by 11 groups large scale in situ shear tests, the mean values of the peak frictional coefficients(f_d) and the cohesions (c_d) are 0.23 and 5 kPa respectively. As mentioned above, the corresponding dynamic values for the intercalations are 0.25 and 15 kPa, respectively. Therefore, the dynamic parameters are larger than the static ones. This illustrates that it seems to be conservative to use static parameters in the dynamic stability analysis of rockmass.

Results obtained from the dynamic triaxial shear test
Assigning Poisson's ratio of the intercalation of 0.4 and the failure shear strain of 7%, we can obtain the relationships between the dynamic strength(τ_{df}) and the cyclic number(N) shown in Fig.15 and Fig.16. It is found that the dynamic strength decreases together with the cyclic number and tends to a certain value when the cyclic number is to one thousand.

According to Fig.15 and Fig.16, the upper limit and the lower limit of the dynamic strength to every stress condition may be adopted to strengthen out the total shear strength parameters(c_d and f_d)shown in Fig.17. The stable values of the dynamic strengthes with the cyclic number were used in Fig.17 to accord with the condition of the long time vibration by the flood discharge . The total strength parameters corresponding to the two initial stress conditions are given out in table 2.

Because the test condition that the initial shear stress ratio α_0 equals 0.187 is very close to the working condition of the intercalations (tg10^0 =0.176), and the number of the samples in this group (group 1) is larger than the other group(group 2) ,the determination of total shear strength parameters is mainly on the test result of group1. The total strength parameters obtained by using the least square method are c_d =8kPa , f_d=0.278. For the sake of safety, it is recommended that the average values of the obave and the lower limits should be used as the total strength parameters of the all siltized intercalations, namely, c_d

= 5kPa and f$_d$ =0.27.

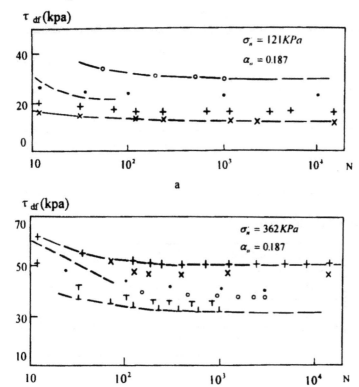

a

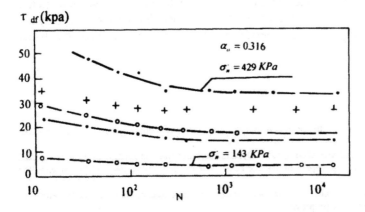

b

Fig.15 Relationship between the dynamic strength and the cyclic number(1)

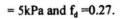

Fig.16 Relationship between the dynamic strength and the cyclic number(2)

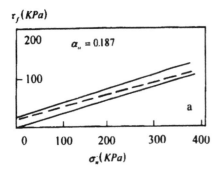
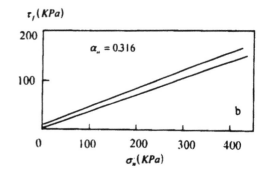

Fig.17 The upper and lower limits of the total strength curves

Table 2 The total strength parameters of the intercalations

Initial stress ratio α_0	Number of samples	The upper limits		The lower limits	
		C_d(kPa)	f_d	C_d (kPa)	f_d
0.187	12	20	0.274	2	0.270
0.316	5	7	0.378	0	0.346

CONCLUSION

The existence weak intercalations in the Xiaolangdi project is a key factor controlling the left hill mass stability. The experimental study shows that the weak intercalations under the action of cyclic loading behave as a non-linear hysteresis body, and that the equivalent visco-elastic model can be adopted to simulate their behavior. The dynamic frictional coefficient for the intercalations is higher than the corresponding static one. The dynamic stability analysis and design for this project, therefore, should be based upon the dynamic test results.

REFERENCES

Seed, H. B. and Idriss, I.M. 1969,Infuence of soil foundations on ground motions during earthquake, Proc. of ASCE, Vol.95,No.SM1

Wang Sijing, Zhang Jingjian and Xue Shouyi, 1990, Dynamic soil modulus and strength of weak intercalations contained in rockmass, Selected Papers of Hydroelectric Science and Technology, Hebei Science Press, pp.318-329

Xue Shouyi, Wang Sijing and Yu Peiji, 1997, Dynamic triaxial test investigation on core samples with siltized intercalations, Proc. of Geot. Eng.(in press)

Proc. 30th Int'l. Geol. Congr., Vol. 23, pp. 221-231
Wang Sijing and P. Marinos (Eds)
© VSP 1997

Engineering Property of Seafloor Sediment in Huanghe Subaqueous Delta Since Late Pleistocene

SHEN WEIQUAN, YANG SHAOLI, FENG XIULI, LIN LIN& WANG HAIPENG
College of Marine Geosciences, Ocean University of Qingdao, China

Abstract

According to a great number of bore holes and comparison of surveying data of shallow stratum profiles, the seafloor soil in Huanghe subaqueous delta since Late Pleistocene is divided into five groups in terms of the sedimentary character. The difference in their engineering property and the engineering geologic problems possibly met are given. The suggestion of selecting feasible bearing stratum of pile foundation and pile penetrating depth are proposed on the basis of engineering practice for more than ten years.

Keywords: subaqueous delta, sedimentary character, static and dynamic feature, pile penetrating depth

INTRODUCTION

We had successively undertaken the compiling work of marine engineering geology in "Chengdao oil explorating and developing marine environment" and "Dongying harbor natural environmental data Compilation" in request of Shengli oil field managing

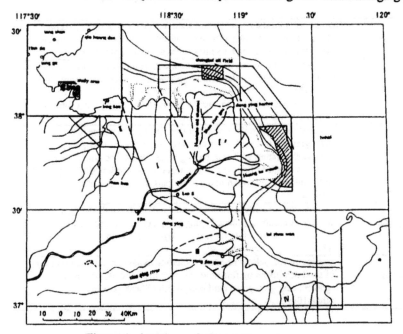

Figure 1   location of the study erea

Ⅰ. mordern Huanghe delta   Ⅰ'. contemporary Huanghe delta bureau, government of Dongying city and important engineering office of Shandong province. We systematically selected and arranged the research results of our unit and brother units during this period, moreover we have much engineering practice in Chengbei area, So we divided the strata which have important influence on the engineering building since Late Pleistocene into five groups and proposed different requirement for pile penetrating depth to different building. It has been proved feasible and has gained good economic benefit through engineering practice examination of Chengdao oil field, Kendong oil field and Dongying bureau extension. The study area in this article is shown in Figure 1.

SEDIMENTARY CHARACTER AND STRATUM DIVISION SINCE LATE PLEISTOCENE

Marine transgression and marine regression which had happened many times in Huanghe subaqueous delta since Late Pleistocene together with the deposit of a large amount of silt and frequent swinging of tail river course each year after Huanghe entered the sea from this area made the seafloor engineering geologic condition much complex. We first found that the seafloor strata since Late Pleistocene could be divided into five groups according to their sedimentary character from engineering geologic surveying and soil dynamic analysis of Chengbei borehole 20-1. Compared with other boreholes and shallow stratum profile surveying data, it had been proved that this kind of division had general law in the whole Huanghe subaqueous delta and could be the basis for marine engineering geologic research. The sedimentary environment and character of each soil layer are shown in Table 1.

Table 1   Sedimentary environment and character of each soil layer
            in Huanghe Subaqueous delta since Late Pleistocene

division of soil stratum	sedimentary environment and character
first group	Hanghe delta frontal accumulation sediment , depth4.6-10.7m, average 8-9, mainly containing silty sand and silt, intercalating with silty clay and organic form, brown yellow without bedding structure
second group	Shallow sea and seamarsh facies sediment, depth7-9m, mainly including sludge of muddy silty clay, brown yellow in the upper, grey -grey black intercalating with peat layer in the middle , grey yellow in the bottom, with oibvious bedding, containing much marine living thing form, soft plastic-flow plastic
third group	Continental deposit dominated by river-lake facies before maine transgression in Late Pleistocene, original Holcene epoch, depth 17-18m, coarse alternating with fine, having much obvious bedding structure, grey-brown silt or fine sand in the upper, brown yellow silty clay or silt in the middle, yellow silt in the lower part, sand in moderately dense state
forth group	Continental -oceanic interactive facies sediment, dominated by littoral-delta river mouth sediment, having fairly developed fine bedding, silt alternating with silty clay, single layer depth 1-3mm, grey brown plastic state, depth 28m or so
fifth group	marine facies, yellow fine sand including muddy gravel , composition of sand dominated by fine sand and very fine sand, dense, muddy gravel being grey-green hard clay, depth of layer possibly more than 10m

Table 2. Chengbei 20-1 gathering table of distinguishing results to liquefaction.

group number	soil type	thickness of layer(m)	research depth(m)	natural unit weight	content of clay pebble	e	D_{50}	blow count of SPT	[7] standard of building seismic design	[7] standard of water transport engineering seismic design	[7] Yanqi method	[7] seed method	[8] standard of building seismic design	[8] standard of water transport engineering seismic design	[8] Yanqi method	[8] Seed method
1			0.01	1.99	6.65	0.665	0.038	10	x	x	x	O	x	O	x	x
			1.68	1.99	6.65	0.665	0.38	10	x	O	x	x	O	O	x	x
			2.54	2.03	12.44	0.662	0.027	10		O			O	O		
	silty sand	5.04	2.54	2.03	12.44	0.662	0.027	8	x	O		x	x	O		x
			3.41	2.03	12.44	0.662	0.027	8	x	O	x		x	O	x	
			4.15	1.99	6.95	0.687	0.043	8	x	O				O		
			4.15	1.99	0.95	0.687	0.043	8	x		x	x	x	O	x	x
			4.88	1.99	6.95	0.687	0.044	10					O			
			5.04	1.99	5.01	0.687	0.028	<1	x		x		O	O	x	O
	silty	3.47	5.04	1.99	5.01	0.687	0.028	<1	O	O	x	O	O	O	O	O
			0.68	1.99	5.01	0.687	0.004	<1	O	O	x	O	O	O	O	O
			6.68	1.99	5.01	0.687	0.004	<1	x	x	x	x	x	x	x	x
			7.71	2.00	4.67	0.697	0.034	30	x	x	x	x	x	x	x	x
			8.51	2.00	4.67	0.097	0.034	30	x	x	x	x	x	x	x	x
2	silty sand	7.15	10.28	1.96	17.55	0.757		7	x				x			
	clay	2.23	15.66	1.69	29.34	1.408		5	x				x			
	silty sand	2.23	17.89	1.95	41.59	0.772		5						x		
3	silty sand	2.51	19.12	1.95	4.36	0.633	0.075	37	x	x	x	x	x	x	x	x
	silty clay	1.51	21.93	2.07	35.79	0.574		10	x	x	x	x	x	x	x	x
	silt	5.2	22.23	2.14	22.93	0.490	0.039	17	x	x	x	x	x	x	x	x
			23.96	2.14	22.93	0.490	0.049	19	x	x	x	x	x	x	x	x
			25.93	2.02	12.20	0.569	0.058	37	x	x	x	x	x	x	x	x

Note: O represents possibly liquefy, x represents impossibly liquefy   Revise according to data from earthquake bureau of Shandong Province

STATIC-DYNAMIC CHARACTER AND MAIN ENGINEERING GEOLOGIC PROBLEM OF EACH SOIL LAYER

Main character of the first soil layer is that the soft intercalates with the hard, having soft in hard, having hard in soft. lithology changes largely, physical-mechanical feature changes a lot, blow count of SPT, shear strength and compression coefficient have not certain law. The compactness of sand is low, water content of clay is high with flow-plastic state. Compression coefficient in the harder part: $\alpha_{1-2} = 1.1 \times 10^{-4} KPa^{-1}$, moderate compression, unstrained shear test: u=26.9°, Cu=2.94KPa, over consolidation ratio is less than unit. There is no test data in the further soft soil layer because of having no way to get sample, but we can know from the autofalling of the bore rod during the course of drilling and the small blow count of SPT (less than unit)that the compressibility is very high , the shear strength is possible to reach zero. Because of the sediment resource from yellow river, content of silt is high. Thus the coefficient of permeability is high (among $10^{-4} \sim 10^{-5}$ cm/sec), coefficient of consolidation is large(10^{-3} cm²/sec), speed of consolidation is very fast, the degree of sensitivity is high. Generally, it is from 2 to 6, sometime it is larger than 6. This kind of soil belongs to moderate to high sensitivity. In addition, there is obvious thixotropy in this layer, initial strength turns to decrease after disturbance, it can be recovered passing a period of time. It is known from triaxial dynamic test that dynamic shear modulus is very small, dynamic strength is not high (Table 4). This layer is easy to cause liquefaction and in a large area under the storm loading because of its high silt, low clay content(Table 2). It also has the possibility to happen the revival phenomena of seafloor slides and to form siltflow gully, breakdown and pits following the liquefaction of silt. All these factors would have large influence on ocean engineering.

Table 3   Chengbei 20-1 allowable bearing capacity of foundation of each layer

group number	number of layer	name of soil layer	buried depth (m)	water content (%)	void ratio (e)	water-plasticity ratio (I_L)	blow count of SPT $N_{63.5}$	allowable bearing capacity (Kpa)
1	1	silty sand	0-5 04	28 9	0 786		7	120
	2	silt	5 04-8 51	30 4	0 777	0 40	6	100
2	3	silty clay	8 51-14 22	38.0	0 930	0.85	2	85
	4	mucky silty clay	14 22-15 56	50 8	1 403	1 05	<1	60
	5	clay	15 56-17.89	32 3	0 872	0.68	3	140
3	6	silty sand	17 89-20 42	22 4	0.637		22	200
	7	silty clay	20 42-21 96	26 4	0 674	0.71	6	180
	8	silt	21 96-27 13	21 0	0 656	0 62	12	200
	9	silty fine sand	27 13-33 03	24 4	0.680		13	210
4	10	silty clay	33 03-35 13	25 6	0 700	0.55	11	200
	11	silty sand	35 13-37 60	25 6	0 747		9	190
	12	silt intercalating sity silty sand	37 60-44.89	24 4	0 672	0 66	12	210
	13	silt	44 89-65 19	24 0	0 680	0 70	14	220
5	14	fine sand with muddy gravel	65 19-68 99	20 3	0 621		45	300

Table 4   Reference shear strain, maximum damping ratio, maximum shear modulus and unit weight after consolidation (20-1 borehole)

number of soil layer	sample number	depth of sample (m)	name of soil sample	natural water content %	natural unit weight KN/m³	unit weight after consoli-dation KN/m³	maxim-um shear modulus MPa	Maxim-um damping ratio	confin-ing pressure σ_3 MPa	referen-ce shear strain
1	20-2B	2.71-2.87	silty sand	22.0	20.3	20.9	34.3	0.262	0.094	3.59×10^{-4}
1	20-3B	4.26-4.51	sandy loam	24.2	20.3	20.9	34.3	0.246	0.110	9.241×10^{-4}
3	20-6B	8.83-9.01	loam	26.0	19.5	20.4	30.4	0.251	0.153	2.45×10^{-4}
3	20-7B	10.82-11.00	loam	40.6	18.1	19.6	23.0	0.268	0.196	2.51×10^{-4}
3	20-9B	14.04-14.22	loam	28.9	19.4	21.0	31.7	0.242	0.199	2.23×10^{-4}
5	20-12B	17.14-17.32	loam	26.4	19.7	20.6	35.5	0.258	0.229	3.40×10^{-4}
6	20-14B	20.09-20.27	silty sand	20.6	18.4	20.2	92.3	0.248	0.253	1.17×10^{-4}
7	20-15B	21.75-21.93	loam	19.9	20.8	21.1	39.4	0.228	0.276	2.10×10^{-4}
8	20-17B	25.55-25.73	sandy loam	18.8	19.4	20.9	46.7	0.240	0.310	7.57×10^{-4}
9	20-19	28.93-29.00	silty sand	22.3	18.5	20.2	113.4	0.240	0.340	9.63×10^{-4}
9	20-21B	32.85-33.03	silty sand	21.9	20.5	20.5	64.94	0.273	0.301	1.43×10^{-4}
10	20-22B	34.95-35.13	loam	20.1	20.9	21.3	78.74	0.225	0.320	1.04×10^{-3}
12	20-24B	38.88-39.06	silty sand	25.7	19.6	19.7	104.93	0.303	0.36	8.40×10^{-3}
13	20-27B	44.95-45.13	sandy loam	23.2	20.3	20.7	83.33	0.240	0.42	1.23×10^{-3}
13	20-29B	48.90-49.00	sandy loam	25.6	19.6	19.8	100.91	0.304	0.46	1.22×10^{-3}
13	20-32	53.18-53.36	sandy loam	27.1	19.9	20.3	83.33	0.310	0.50	1.85×10^{-3}
13	20-34	57.13-57.31	sandy loam	25.1	20.0	20.6	99.01	0.380	0.54	2.00×10^{-3}
13	20-36	60.38-61.26	loam	22.0	20.8	21.2	78.74	0.248	0.58	1.43×10^{-3}
14	20-38A	65.01-65.19	sandy loam	14.3	21.6	22.3	142.25	0.207	0.62	1.48×10^{-3}
14	20-39	67.18-67.36	fine sand containing muddy gravel	21.2	20.3	20.5	150.38	0.246	0	1.43×10^{-4}

Main character of the second soil layer is its high compressibility, low coefficient of permeability ($2\sim3\times10^{-6}$ cm/sec)and shear strength, high sensitivity (larger than 6), big water content(generally 40-50%, the highest 80-90% often larger than liquid limit), small blow count of SPT(Table 3)and low dynamic shear modulus (Table 4). Coefficient of consolidation is small($3.0\times10^{-3} \times 5.9\times10^{-4}$ cm²/sec under the loading of 2-8 kg/cm²) it has the feature of soft soil. The over consolidation ratio is from 0.57-0.68, belonging to under consolidation soil. The strength is further low in the muddy clay with black peat intercalation of the middle part of this soil layer, coefficient of compression is fairly large, unconsolidated and Unstrained test: u=3.5~7.0°, Cu=2.94~6.86KPa, triaxial consolidated and unstrained strength: u=12°, Ccu=14Kpa. It is easy to produce depression under the earthquake loading. The main engineering geologic problem is its low bearing capacity and poor engineering feature. Soil layer of this group can not be acted as the bearing stratum of pile foundation.

The third soil layer is the stratum of sand intercalated with mud. The compressibility, shear strength and other mechanical indexes change largely following the alteration of lithologic character. It is possible to gain different evaluating result because of the different soil condition disclosed in different bore hole. Thickness of sand layer is large in the area of No.5 pile harbour, the first silt-fine sand layer appears in the depth of 30-35m under the seafloor with thickness of 5-10 m, having high relative density (Dr=0.76), fairly large blow count of SPT, low coefficient of compression, good permeability, coefficient of permeability is from 1.54×10^{-4} to 9.29×10^{-5} cm/s. It is the good bearing stratum for pile foundation in the harbour area. Most of the soil layers are sand layers in No.14 borehole of Chenbei oilfield, soil layers in chengbei 20-1 bore hole and 11-A bore hole are sand layers intercalated with mud layers, dominated by sand layers . Soil layers in the three boreholes of number one manmade island are mainly mud layers. We take the example of 20-1 borehole in chenbei area, Coefficient of compression of grey-brown silt in the upper part of this soil layer is from $6.0\sim9.0\times10^{-5}$ KPa^{-1}, the unstrained shear strength index: $u = 35°$, Cu=15.68Kpa, natural angle of repose is 36°. In the brown-yellow silty clay of the middle part: $\alpha_{1-2} = 1.0\times10^{-4} \sim 1.9\times10^{-4}$Kpa, triaxial unstrained consolidated test: cu= 32°, Ccu=39Kpa, unconsolidated unstrained shear test: u=22-30°, Cu=13-24KPa. Coefficient of consolidation under the loading of 2-8 kg/cm^2 is from $4.6\times10^{-3} \sim1.18\times10^{-3}$ cm^2/sec, overconsolidated ratio OCR is about 1, belonging to the normal consolidated soil. The yellow silt layer in the lower part of this layer: $\alpha_{1-2} = 8.0\times10^{-5} \sim 2.3 \times 10^{-4}$ KPa^{-1}, u=31°~37°, Cu=5.8~17.6KPa, cu=34°, Ccu=4.8KPa, natural angle of repose is 43°. It has the good engineering property in this layer.

The forth soil layer has low water content, low coefficient of compression ($\alpha_{1-2} = 7.0\times10^{-5} \sim 1.4\times10^{-4}$ KPa^{-1}, mostly with low compressibility), fairly high shear strength(u=20-30°, Cu=13~19KPa^{-1}, cu=26°, Ccu=47KPa), high coefficient of permeability (K=$5.39\times10^{-5} \sim 8.26\times10^{-5}$ cm/sec), low coefficient of consolidation($2.09\times10^{-4} \sim8.74\times10^{-4}$ under the loading of 2-8kg/cm^2.) The time required to permeable consolidation is short , it is fast to compress to stable state. Blow count of SPT is very large and its dynamic property is also fairly good. Thickness of the soil layer is large (Table 3 and Table 4), it is the considerately ideal bearing stratum of pile foundation.

The fifth soil layer is dominated by sand with high blow count of SPT(Table 3), small compression coefficient factor($\alpha_{r2} = 7\times10^{-5}$ Kpa^{-1}) ,shear strength high to 143-150Kpa. Sand is in dense state, muddy gravel is hard with bearing capacity high to 300Kpa(Table 3 and Table 4). This group is the ideal bearing stratum for pile foundation.

It is suitable to apply pile foundation in each engineering for having the soft, loose soil layer in the upper part and hard, dense soil layer in the lower part of the area . Considering the frist and second soil layer in the area is sediment lately deposited or underconsolidated soft soil, it is possible to appear negative skin friction during the course of penetrating the pile.Especially in the soil layer with very low blow count of SPT(Table 3), it is probable to appear fairly large negative skin friction. We should take enough consideration on this condition in the file foundation analysis.

DEPTH OF PILE INSERTING

It is not difficult to find that soil layers except the second group can be considered as the bearing strata of foundation and also easy to gain the buried depth of each bearing statum according to the borehole data(Figure2). Adopting the pattern of inserting piles for all kinds of ocean engineering is relatively good   because engineering feature of the first and second soil layers is relatively poor, but engineering character becomes better and better and bearing capacity becomes higher and higher from the thrid group. The penetrating depth of piles is decided by the requirement of engineering. It is proposed as follows according to the engineering practice in this area:

(1) The frist group can be conditionally selected as bearing stratum for general industry-civil building or small platform with basement pattern, generaly its penetrating depth is not more than 3-5m. The precondition is as follows: soil in the building area is relatively uniform; depth of silt layer is big enough (not less than 5m); there is not soft underlayers in the compressible layers; upper loading (selfweight of platform)of the structure is comparatively small; applying period is short; the soil layer can not lose stability because of the scouring, hollowing or penetrating through of pile foot. This kind of soil condition ordinarily only appears in several sandspits of new-old river mouth of yellow river area on the basis of engineering practice for many years.

(2) It is good to choose the third group as the bearing stratum of pile foundation for general engineering, penetrating depth of pile could be about 20m (such as extension engineering of yellow river harbour and small oil-drilling platform of Chengdao oil field), but it is required to examine if there is soft underlayers within 30m depth.

(3) The upper part loading of large ocean engineering(e.g. man-made island and steel-concrete gravity platform)is rather large, together with high security factor used for considering dynamic loading of wind wave and flow et al., bearing capacity of the third soil layer can not satisfy the requirement. The forth layer should be adopted as bearing stratum of pile foundation, depth of penetrating should be decided in line with the analysis of pile foundation engineering (Figure 3). Penetrating depth is suitably consider to be 45-50m according to the design requirement of chengbei man-made island.

(4) It is better to use the fifth group as bearing stratum of pile foundation for major projects which have special requirement in safety such as offshore oil tank and yellow river bridge. Ultimate bearing capacity of pile can be increased through two paths: improving pile end resistance and pile side fiction resistance. In such case the considered penetrating depth is about 70m.

(5) The problems of sand liquefaction and soft soil depression under earthquake force should be considered during the designing course when adopting the first and second soil layers as the bearing strata, necessary antiseismic measure should be used. We think that soil layer above 9m is slightly liquefied at 7 seismic intensity, moderately liquefied at 8 seismic intensity in the light of research for sand liquefaction problem of the four man-made islands in chengbei area. It should be emphatically considered the problems of scouring, hollowing, sand liquefaction and seafloor slope instability caused by wave loading when adopting the first soil cayer as the bearing stratum.

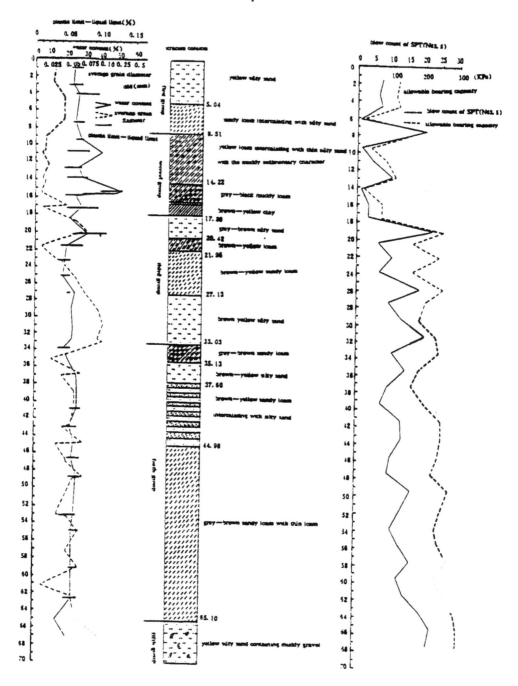

Figure 2 Chengbei 20-1 comprehensive Column

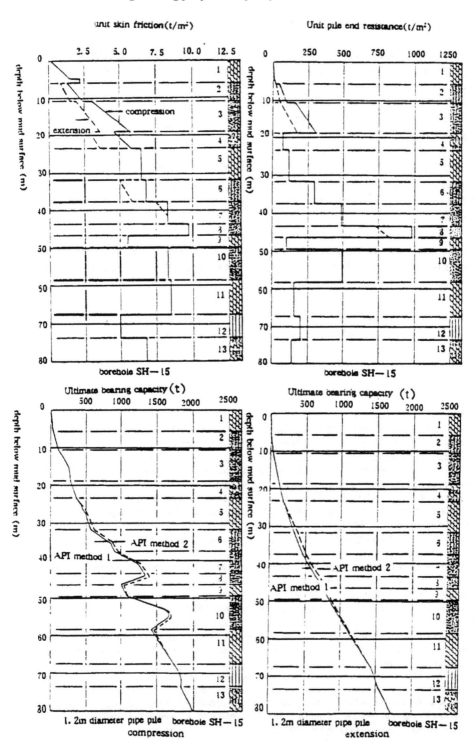

Figure 3 Pile foundation engineering analysis of borehole SH-15.
(from offshore engineering company, Bohai oil company)

CONCLUSION AND SUGGESTION

(1)The engineering property of different deposit facies of seafloor soil depositted within 70m depth in yellow river subaqueous delta since Epipleistocene has fairly big difference because of the change of depositional environment during the sedimentary course. It is relatively optimistic and easy to obsolve the evaluation of engineering geologic condition when we confine the influence of yellow river on this area within soil layers above 10m using scheme of five group division.

(2) We think that it is rather good to apply pile-inserting method for each kind of ocean engineering construction based on the engineering property of five groups. The penetrating depth is decided by the requirement of different structure to reduce the investment of projects.

(3) The multi-changeble and uneven property of yellow river sediment should be sufficiently considered when adopted as the bearing stratum, especially the possible alternation caused by dynamic loading.

(4) It can be extended to the whole Bohai area after further study of analysis of the engineering property of seafloor soil in yellow river subaqueous delta.

REFERENCES

1. Chengdao oil field explorating and developing marine environment, Shengli oil managing bureau, group compiled by ocean university of Qingdao, publishing house of Ocean University of Qingdao, 1993.8.
2. Dongying harbor natural environmental data compilation, goverment of Dongying city, group compiled by shengli oil managing bureau and improtant engineering office of shandong province, 1994.4.
3. Shengli oil field man-made island engineering geophysical prospecting paper, Service company of ocean research equipment of China , science college ocean research institute, 1990.12.
4. Engineering geologic explorating and evaluation paper on the man-made islandsite test 1# in Chinese Bohai gulf, second marine geologic exploration team, geomineral department Guangzhou marine geologic explorating bureau et.al., 1991.
5. Marine engineering geologic exploration, Shenli man-made island number 2 and number 4, offshore engineering company of Bohai oil company, 1991,10.
6. Paper on earthquake intensity check and dynamic factor prediction of engineering site number 1 man-made island in chengdao oil field, Geophysical institute of shandong province and China, 1990.12.
7. Test paper of engineering geologic exploration and soil dynamics of chengbei borehole 20-1 shengli oil field. Estuarine and Coastal studies. Ocean University of Qingdao. 1989.12.
8. River mouth sedimentary dynamic research papers( )Yellow river subaqueous slope instability, major author: Yang zuosheng, Shen Weiquan. publishing house of Ocean University of Qingdao 1991.8.
9. Yang shaoli , Sheng weiquan and Yang zuosheng, The Mechanism Analysis of Seafloor silt Liquefaction under Wave Loading, China Ocean Engineering, Vol.9.No.4.
10. Yang shaoli, Shen weiquan and Chang ruifang, The Response of soft soil under wave loading and its unstable Mechanism, Second international conference on soft soil engineering, Nanjing, May27-30,1996.
11. Active slope destruction, sediment collaps and silt flow of modern yellow river subaqueous delta, D.B.prior, Yang zuo sheng et.al. Weight redeposition and slope instability on coastal-river mouth area, Publishing house of Ocean University of Qingdao, 1990,12.

12. Yang zuosheng, Chen Weimin et.al. System of subaqueous slide in Yellow river. Ocean and limnology. Vol.25.No.6,1994.

Proc. 30th Int'l. Geol. Congr., Vol. 23, pp. 233-239
Wang Sijing and P. Marinos (Eds)
© VSP 1997

Engineering Geological Study and Evaluation of Red Soil in China

WANG QING, CHEN JIANPING, TANG DAXIONG, & LIU YING
Changchun University of Earth Sciences, Jilin, China

Abstract

Red soil is one kind of regional special residual soil with its own unique engineering geological properties, which is widely distributed in the south of Changjiang River in China. Based on the study of engineering geological properties of many kinds of red soil, the evaluation of the properties of swelling and shrinkage of red soil was done, The main reason for lower or higher strength of different red soil is determined by the type of free oxides, its state and the existing manner in the red soil. In this paper the value of subsoil bearing capacity of different red soil is studied.

Keywords: Red Soil, Engineering Geological Properties, Evaluation

INTRODUCTION

Red soil is one kind of special regional soil with its own unique engineering geological properties. Red soil has being widely used as a foundation of different kinds of building, man-made slope and fillings of constructions. Because of the different forming condition of red soil, there are many type of red soil in the nature, the properties of red soil is obviously different. According to the achievement in our research, a brief introduction to the engineering geological properties and evaluation of red soil in south of China is presented in this paper.

ORIGINAL TYPE OF RED SOIL

Red soil is one kind of red cohesive soil in which contains more clay fraction, more free Fe and Al and it was subjected to a fairly laterization under the conditions of humidly and warmly climate. Laterization is an important effect of chemical weathering in the tropical region. Under the laterization, most of the rocks and soil can be changed into red soil with more Fe and Al and lower ratio of Si-Al, more kaolinite and higher activation. According to the research on the laterization and coefficient of residual and ratio of weathering ect[1,2], red soil in the nature can be classified into in-situ residual red soil and transported (or sedimentary) red soil. Red soil is widely distributed in southern part of China, according to the bedrock of red soil there are mainly five type of them that is introduced as follows:

1. Granite residual red soil: Granite is widely distributed in southeast of China in the age of Yanshan period, and now, it is a formation of very thick weathered crust. weathered zone of upper strata is called residual soil; according to the composition and construction, residual red soil can be divided into homogeneous red soil, netty red soil and netty cohesive soil. The first two of them are named residual red soil.

2. Basalt residual red soil: Basalt is widely distributed in the Leizhou Peninsula and Hainan island in the age of Quaternary Period, and now, it is the formation of different thick of weathered crust. Upper strata red cohesive soil is called residual red soil.

3. Red layers residual red soil: Red layers is largely distributed in the center area and southern part of China. it is originated from sedimentary sandstone, siltstone and mudstone in the age of Tertiary Period and the Jurassic-Cretaceous Period. The cohesive soil that forms the weathered crust is one kind of residual red soil.

4. Red clay: Red clay is a cohesive soil with the color in brown and red, it is originated from carbonate under the condition of subtropical zone and subjected to the residual and slope accumulation, red clay is widely distributed in Guangzhou Yunnan and Guangxi province in southwest of China. The thick of red clay layer is different. The origination of red clay is also from alluvial or diluvial besides residual.

5. Alluvial netty red soil: Netty red soil is oddly distributed in the center and south of China and it is originated mainly from sedimentary sand, gravel and cohesive soil in the Pleistocene Period(Q^a_{12}).

Table 1. Varied characteristics of element of residual red soils

	AR		Value of cumulative(+) or eluviat(-)						CE		CR		RWRS	
BR	EA	N	SiO_2	Al_2O_3	Fe_2O_3	FeO	CaO	Na_2O	RS	BR	RS	BR	RE	RR
G	JX	3	-20.3	+45.9	+767	-84.1	-48.4	-91.5	13.08	6.28	6.05	1.23	2.08	4.29
G	FJ	8	-13.0	+5.29	+394	-56.9	-56.5	-97.8	-25.2	6.43	11.21	1.28	4.16	8.88
G	GD	8	-23.7	+97.3	+456	-86.9	-71.2	-97.3	27.97	5.66	18.0	1.16	4.88	15.29
B	QL	19	-17.9	+51.0	+201	-84.7	-98.9	-97.4	29.92	2.03	24.75	0.80	4.39	30.94
C	GZ	7	+112.	+89.4	+85.5	-	-99.5	-	10.23	0.086	7.36	0.072	119.0	102.2

Note:

BR   Bed rock; G   Granite residual red soil; B   Basalt residual red soil;C   Carbonate rock residual red soil; N   Frequency number; JX   Jiangxi province;FJ   Fujian province; GD   Guangdong province; QL   Qionglei area;GZ   Guaizhou province; A(+) OR E(-)   Accumulate(+) or eluviat(-); CE   Coefficient of eluviat; CR   Coefficient of residual; RWRS   Ratio of weathering of red soil; RS   Red soil; RE   Ratio of eluviat; RR   Ratio of residual

According to chemical composition of red soil (table 1), we known that a lot of FeO, CaO, Na_2O was eluviated, MgO and K_2O was also eluviated. Some of the SiO_2 in the magma rock was partly eluviated, but some of the SiO_2 in the carbonate rock was accumulated,a lot of Al_2O_3 and Fe_2O_3 was accumulated, so the ratio of weathering of red soil is very high, this makes known that each kind of red soil has a high degree of laterization, and the weathering stage is in the stage with more Si-Al, Fe and more Al. The eluviation of carbonate and basalt is stronger,the ratio of weathering of granite residual soil is gradually increased from lower part to upper part in the layer and also from north to south geographically in China , and the phenomenon is clearly observed.

ENGINEERING GEOLOGICAL PROPERTIES OF RED SOIL

The indices of mineral, chemical composition, physicochemistry, physico-mechanics properties of red soil in China is shown in the table 2 and 3.

Table 2. Composition of red soils (average value)

		GC(%)		FO(%)									FO(%)	
T	A	>0.1	<.005	SiO_2	Al_2O_3	Fe_2O_3	Total	I(%)	PH	SA	SA	K(%)	SSA	CEC
G	JX	33	26	2.24	2.30	4.62	9.16	70	4.84	5.06	2.36	75	130	20.0
G	FJ	43	23	5.32	3.38	3.27	11.97	69	6.04	5.57	2.17	79	105	15.4
G	GD	51	19	6.41	4.97	3.45	14.83	80	5.33	4.10	2.15	85	80	14.0
B	QL	5	50	2.66	3.97	9.67	16.30	55	5.37	2.41	2.14	95	113	13.2
B	YN	20	57								2.05			
R	JX	9	46	3.49	1.72	5.36	10.57		4.68	6.67	2.66	48	110	17.4
R	GX	17	58							5.61	2.31	57		
R	GZ	10	58	3.63	2.12	5.47	11.22	58	5.88	5.10	2.31	57		
R	YN									3.48		65		
Q_2	JX		32	1.00	1.33	5.10	7.4	52	5.69	8.53	2.75	76	16.7	18.3
A	HN	11	41					45	5.20	6.47		87		
RL	JX	6	39	2.70	2.02	5.28	10.00	87	4.81			58	139	17.5

Note:

T Type of soils; A Area; GC Gravel content(mm); FO Content of freeoxide;I Free iron oxide to total iron in the soil; SA Ratio of silica to aluminum;K kaolinite; SSA Specific surface area(m^2/g); CEC Cation exchange capacity(meq/100g); G Granite residual red soil;B Basalt residual red soil; R Red clay; A Alluvium natty red soil; RL Red beds residual red soil; HN Hunan province;YN Yunnan province; GX Guangxi province; JX Jianxi Province;FJ Fujian rovince;GD Guangdong province;QL Qionglei area;GZ Guizhuo

The clastic minerals is mainly quartz and less in fresh feldspar; the content of clay fraction is some more, the clay minerals is mainly kaolinite, and the content of illinite is less, there is some red hematite and needle ironstone, and still some of red soil contains gibbsite; the major chemical composition is SiO_2, Al_2O_3, Fe_2O_3, the content of RO, R_2O_3 is less, the ratio of silica to allumine is not high. The PH value of soil is lower, and the content of organism and the soluble salt is lower, too, the specific surface area and the cation exchange capacity are not high, but the content of free iron oxide is about 50%-80% in total iron of red soil. So red   soil is one kind of red cohesive soil that the main content   is kaolinite and quartz,the main cement material is iron. The engineering geological properties of some main type of red soil is expounded in this paper.

Table 3 . Physic-Mechanics properties of red soils (Mean value)

T	A	$W_{L(\%)}$	Ip	W(%)	I_L	e	$a_{(1-2)}$	Es	C		Pa	VH	Pp	Vs	es	Fs	Ves	CWs
		(%)		(%)			MPa⁻¹	MPa	KPa	(°)	KPa	(%)	KPa	(%)	(%)			
G	JX	45	18	28	0.11	0.83	0.29		36	25		1.5		2.7			2.5	
G	FJ	46	16	30	0.26	0.93	0.37	5.5	27	27	2.56	1.4				13	18	16
G	GD	43	15	27	0.16	0.83	0.38	5.5	25	32	30.7	1.0	<10	1.8	1.8	13	18	19
B	QL	56	19	31	0.21	1.25	0.43		32	20		<1.0		4.3		22	34	29
B	YN	53	23	38	0.35	1.17		6.7						14.5				
RL	HN						0.38											
RL	JX	44	18	25	0.0	0.78	0.21	9.0	59	21		1.3		2.6			12	
RL	GD	30	12	21	0.21	0.61	0.23		35	22		2.6		1.54				
R	JX	53	23	24	<0	0.88	0.14		70	24				10.8				
R	GX	68		38		1.11						0.9	38		2.5	34	21	33
R	GZ	73		47		1.45					230	0.5	14	8.4	8.0	36		
R	YN			44		1.30			50	21		0.5	20		6.0	33	2.9	53
Q₂	JX	40	17	23	<.4	0.70	0.16		67	25		3.0		1.8		33	2.9	24
A	HN			24	0.10	0.78	0.19		63	26	115							

Note:

G.B.RL.R.A.JX.FJ.GD.QL.YN.HN.GX.GZ. have the   same   meaning as   in table 2; A   Area; T Type of read soils; AV   average value; Pa   proportional limit strength value; VH   non-loading swelling ratio; Pp   swelling pressure;Vs   Volumetric shrinkage; es   linear shrinkage; Fe   free swelling; Ves   volumetricshrinkage ratio at the liquid limit; Ws   shrinkage index.

All of the red soil is overconsolidation soil. pre-consolidation pressure is higher, the ratio of overconsolidation is about 3-10 or more, but it is not   influenced   by self-weight of soil in the history, in fact, it is influenced by the major cementation of free oxides.The ratio of overconsolidation is reduced along with the increasing of depth. This is related to the degree of weathering and the content   of free oxides (table 4).Due to the large amount of free oxides, which   is the main characteristics of red soil that is different from general cohesive soil. Although the red   soil has more porosity, more clay, higher plasticity, it has higher strength, middle compressibility, lower swelling-shrinkage.

Table 4. The relationship between the ratio of overconsolidation and   free oxides in granite residual red soil

Type	S			Free oxides(%)			
	SN	V	R	SiO₂	Fe₂O₃	Al₂O₃	Total
HS	17	SV	1.8-18.0	1.5-7.8	1.4-5.8	1.5-5.3	7.3-16.3
HS	17	A	9.0	5.0	3.7	3.2	11.9
NR	12	SV	1.1-6.4	2.3-8.8	0 9-5.9	1.4-4.8	6.2-17.1
NR	12	A	3.4	4.8	2.8	2.9	10.5
MCS	5	SV	1.1-3.7	2.1-3.1	0.9-2.6	0.5-1.0	3.4-6.5
MCS	5	A	2.3	2.3	1.6	0.7	4.6

Note:

SN Sample number;S Statistics;V Value;R Ratio of overconsolidation; SV Scope value; A Average value;HS Homogeneous Soils; NR Netty red soils; Mcs Motley cohesive soil

We know that the red soil is one kind of loam and clay, the general state of the soil is plasticity and hard plasticity. According to the origination and the properties of engineering geology, the red soil in the nature can be classified into in-situ residual red soil and transported or sedimentary red soil. The properties of residual red soil is related to the composition of grain and minerals, one of them has coarse grain, more quartz, lower plasticity, higher strength,lower swelling-shrinkage, more loam, for example, granite residual red soil, sand-gravel rock residual red soil; the other one has fine grain, less quartz, higher plasticity, more clay ,lower strength, lower-middle swelling-shrinkage and some of them is belong to expansion soil, for example carbonate, basalt, mudstone, tuff residual red soil. The properties of transported or sedimentary red soil are related to the cementation material and the grain , the one is paleo-red soil, for example, netty red soil in Q_2-Q_3 period[[3], this soil has higher cementation, higher strength, no swelling-shrinkage, the properties of this soil is close to paleo-cohesive soil; the other one is new-soil, for example, the secondary red soil with lower degree of laterization and secondary sedimentary, this soil has weak cementation, lower strength, lower swelling-shrinkage, the properties of the soil is close to general cohesive soil.

RESEARCH OF SWELLING-SHRINKAGE OF RED SOIL

The swelling-shrinkage of red soil is determined by the composition , the structure, the state of soil , the   properties of   bedrock, and degree of laterization and condition of laterization(table 2, 3). We known that the granite residual   red soil and netty red soil have less in porosity , lower plasticity, lower activation,   lower swelling-shrinkage. It is no need to consider the influence of the swelling-shrinkage of these soil. The red clay and basalt residual red soil have more in porosity,higher plasticity, lower liquid limit (W_L=40%-55%) and middle liquid limit(W_L=55%-70%)and higher liquid limit(W_L 70%) shrinkage(location is below the A line),lower to middle swelling(location is above the A line, W_L 45%). When liquid limit is beyond 45%, we must consider the properties of swelling-shrinkage of soil.

THE EVALUATION OF THE STRENGTH AND SUBSOIL BEARING CAPACITY OF RED SOIL

The subsoil bearing capacity is depend on the properties of deformation and strength of soil.This kind of red soil is belong to overconsolidation soil and middle to lower the compressibility, the modulus of compressibility is 5-7MPa,the problem of deformation isn't important except the soft plasticity red soil or important building, the subsoil bearing capacity of red soil is mainly depend   on the strength. It is known that the strength of soil is determined by grain composition and   the   content of free oxides and its state and compactness and moisture content, Alluvial netty red soil in Q_2 has coarse grain with higher compactness, so C and   is larger.Granite residual red soil has more in the content of sand and   gravel, better in cementation, so   is very   high, C is middle,the strength   of this soil is higher. Red clay has fine grain, middle   cementation, so C   is higher although   it is higher in   void ratio, this soil has   higher strength. The strength of basalt residual   red soil is   very different, the strength of this soil is less than that of red clay(about 75%-85%). The standard value of subsoil bearing capacity of red soil in red soil district is proposed to use the proportional limit strength value, for the important building, it is proposed   to use bearing test and other field measurements, for general building it is proposed to use theoretical formula, standard penetration   test and static cone penetration test, for some of unimportant building it is proposed to use routine soil test for the primary estimation of subsoil bearing capacity. According to the properties of red soil and some data, standard value of subsoil bearing capacity in red soil district is shown in Table 5,6,7.

Table 5. Bearing   capacity of red   soil in the light of   number of standard penetration

T	N	5	10	15	20	25
Q_2 A			280	380	600	760
G		150	200	250	300	350
B		120	160	200	240	280

Note:

N   standard penetration test blow count; T   type of soil; A. G. B   have   same meaning as in the table 2

Table 6.   Bearing capacity of red soil in the light of Ps

T	P	100	150	200	250	300	350	400	450	500	550	600
Q_2 A							350	420	460	510	550	590	630
G				120	135	150	170	185	200	220	235	250
R		175	210	245	280	315	345	385	415			
B		125	170	200	230	260	285	305	325			

Note:

P   specific penetration resistance; T.A.G.R.B. have same meaning as above

Table 7.   Bearing capacity of red soil according to laboratory

Ia		Plastics state	Hard plastic state
		I_L=0.25-0.75	I_L=-0.00-0.25
T		a_w=0.70-0.85	a_w=0.55-0.70
	Q_2 A(e<0.8)	350-450	450-550
	G(e=0.8-1.0)	165-210	210-325
OR	W_L/Wp<1.7	165-210	210-325
OR	W_L/Wp>2.3	120-160	160-240
	SR	120-150	150-220
	B(e=1.0-1.5)	140-200	200-260

Note:

1. OR   original red soil; SR   secondary red soil; A. B. G. have   same   meanings above; I=IL; a=a(1-2); T=type of soils
2. If red soils has   higher density or looser, you   can   increase ordecrease the value of 10%-20%.
3. If red soil has harder or softer in the   state, you can increase   or decrease the value of 10%-20%.
4. If the content of gravel grade of granite   residual soil is more than 20%,this value can be increased 50KPa.

REFERENCES

[1.]Tang Daxiong. The Engineering Geological Properties of Red Soil in The South of China (1989).

[2.]Wang Yuhua. The Properties of Red Soil and Red Clay in China (1988).

[3.]Xiang Chunrao. The Engineering Geological Properties and The Evaluation of Subsoil Bearing Capacity of Netty Red Soil (1985).

Proc. 30th Int'l. Geol. Congr., Vol. 23, pp. 241-245
Wang Sijing and P. Marinos (Eds)
© VSP 1997

Research on Consolidation of Saturated Hard Clay

YAN SHIJUN, & CHEN GE
Institute of Hydrogeology and Engineering Geology, MGMR, China

Abstract

The study has carried out on the variation of pore graduation and compression volume (s) of hard clay with time (t) based on the Terzaghi Theory. We classified the pores in hard clay and the porewater in them into three types. The compression volume (s) shows periodicity with the time of consolidation. The boundary of the periods is a critical point. The consolidation rate is high at the beginning of each period, but getting slower with time, which we call it a consolidation cycle. The compress volume (s) shows correlatively with time by the form of Terzaghi Theory. But the s-t curve obtained by omitting the cycles does not fit with Terzaghi's theoretical curve well. The practical observation data of subsidence show that the consolidation parameters obtained by consolidation cycles are more approximate to actual conditions, with 10-20 times precision than that got from traditional soil tests. By the analysis of hard clay structure and evolution environment, we found that the occurrence of pores which results from the difference of evolution or environment.

Keywords: pore water, pore graduation, consolidation cycle

INTRODUCTION

Geological hazard control and engineering survey indicate that the consolidation parameters gained through the consolidation test used at present are quite far away from the observations, sometimes one or two magnitude orders of the error. The difference of hard clay is much bigger. In order to solve the problem, we have done much study on the consolidation characters of the clay mass based on the evaluation of foundation stability and the control of land subsidence in Tianjin, Haikou, Changzhou and Xiamen. The study began with the geological origin of soil, especially the evolutionary environments after being accumulated, and focused on the correlation of quantitative variation of the soil microtexture elements and the consolidation deformation (i.e., the volume of pore water released from the soil). It has the discussion on the inner relations between the pore grade in the soil and consolidation deformation under the stationary and additional pressure and pays great attention to the structural limits related to the consolidation rate variation, therefore, to describe the mechanism of the consolidation deformation (Fig. 1).

This paper just tried to make some understanding of characters of consolidation deformation of hard soil in Tianjin.

The soil mass to be studied is buried 500-650m deep, that is the Pliocene (N_2) fluviolacustrine deposit. The clay particle content is over 30%, the clay minerals mainly the montmorillonite and secondarily the illite with nodule and ooid. The microfissures are

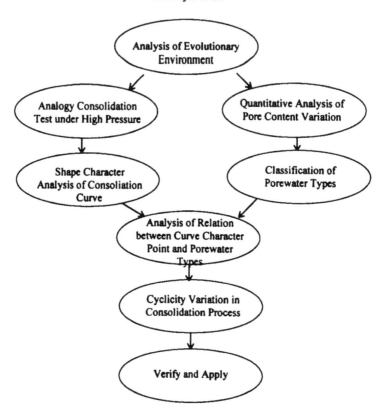

Figure 1. Flow chart of research on consolidation of saturated hard clay

well developed. There are many polished surfaces and scratches. The microtexture unit is much of multiple-film body with partial directive arrangement. The pore content is 0.1248-0.1623 ml/g and the majority of pore diameter is 0.01-0.07μm. The moisture content is about 20%, the permeability coefficient 10-9 cm/s. It is the ultra-consolidation and hard clay.

PORE WATER TYPES

Before and after consolidation, the mercury porosimetry measurements show that, under the common additional pressure, the uncompressive upper limit of the pore diameter is 0.02μm (Fig.2). That means, the pore water in the pores less than or equal to the pore size of 0.02μm can't move and be released out under the pressure. This is called the firmly bound water, but the loosely bound water has the upper limit determined by Domanskii and Geliiazin of 0.08μm. Therefore, the limit of the pore size is 0.08μm and 0.02μm. The pore water in the pores is divided into three types:
a. The gravitational water and capillary water occur in the pores bigger than 0.08μm. So it is called the unbound water. It is mainly effected by gravity, occupying 18% of the total pore water. The movement of the water can be expressed by Darcy's law.

b. The loosely bound water mainly occurs in the pores bigger than 0.02μm and smaller than 0.08μm. The consolidation particles have some absorption on the pore water besides the effect of gravity. The loosely bound water can move until it overcomes the absorption. The content is about 48% of the total pore water.

c. The firmly bound water occurs in the pores smaller than or equal to 0.02μm mainly effected by the absorption. It is uneasy to migrate under the mechanic force. The content is 34% of the total pore water.

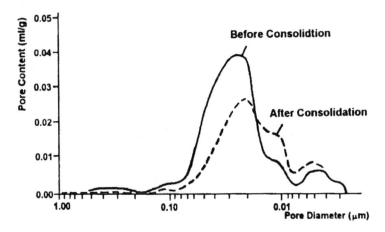

Figure 2. pore distributed curve

ANALYSIS OF CONSOLIDATION PROCESS

The soil is consolidated under the pressure that is bigger than the self-weight effective pressure. In the initial stage, the unbound water is first released out and the ultra-porewater pressure reduces. The consolidation rate becomes slow. The consolidation presents correlation with time to power function. The consolidation law is correspondent to Terzaghi Theory.

While the ultra-porewater pressure disappearing, the effective stress becomes increasing. As the value is close to the texture strength of the loosely bound water pores, these pores are compressed instantaneously and the pore water becomes increasing as well, with releasing the loosely bound water, which makes the consolidation rate to be quickened. The consolidated curve has left a sudden point. From then on, the consolidation rate again becomes slower and slower. In this stage, the consolidation with the time has the correlation of power function. The consolidation law is correspondent to Terzaghi Theory. In the whole process of consolidation, the correlation curve with the power curve of the consolidation has somewhat poor analogy, rather far from the Terzaghi Theory (Fig.3). We name the half periodic variation the cyclicity of the consolidation process of the fast consolidation rate and the slow consolidation rate. Each cycle includes the process from the fast to the slow consolidation. The consolidation parameters from the traditional consolidation cycle are 20 times higher in precision than the parameters got from the

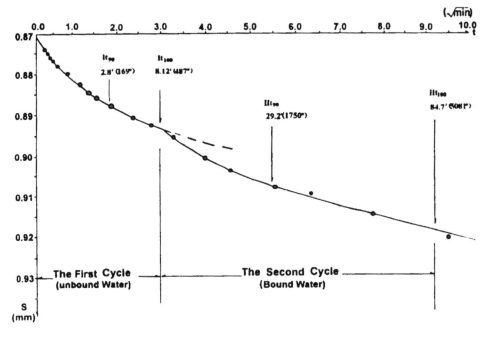

Figure 3. consolidation settlement curve

traditional consolidation test. It is proved by the data from the long-term observation of the land subsidence stratification scale.

GEOLOGICAL ORIGIN

Soil is the historical product of nature. Its physical-mechanical features are controlled by the composition and texture of the material that is constricted by the evolutionary and original environments. The writers consider that the original environment controls the quantitative variable of consolidation feature and the evolutionary environment controls the qualitative variable of consolidation feature. From the microview, the fluviolacustrine environment is homogeneous and stable. The microtexture of its accumulation is also homogeneous and continuous. It can be regarded as the homogeneous soil that is correspondent to Terzaghi Theory. So, the consolidation law can be described by the theory. The difference of consolidation features in different homogeneous soil lies in the consolidation rate. The external conditions --- evolutionary environments are various after soil accumulation. The different mechanic state of the soil makes the changes of the soil various. The evolutionary stress made the effects on the soil of homogeneous, continuous and unchanged or the homogeneous little changed, just as the syngenetic evolutionary environment and inner-genetic evolutionary environment. The consolidation law won't be changed, but the consolidation process of soil will present cyclicity as the evolutionary stress is alternatively strong and unstable, or the soil in the hard state suffers the strong endokimatic action that destroys the homogeneity of the soil texture and form the texture sequences in different type and texture strength such as the supergene evolutionary environments and endogenetic evolutionary environments. Tianjin lies in the vulnerable area of active structure. The soil presents hard effected by the tectonic stress, which produces many microfissures linking to the big ones, made up of a new microstructural sequence. The pore diameter, continuity and the structure strength are quite different from the original structure --- the microfissure sequence and super microfissure sequence. In the different structure sequences, the pore water has different activities. Therefore, in the process of consolidation, the structure sequence corresponding to the other sequences presents individual consolidation cycle.

Proc. 30th Int'l. Geol. Congr., Vol. 23, pp. 247-252
Wang Sijing and P. Marinos (Eds)
© VSP 1997

On the Internal Factors for Controlling the Swelling-shrinkage of Expansive Soil

LI SHENGLIN, SHI BIN, DU YANJUN, LI QI, & WANG BAOJUN
Department of Earth Sciences, Nanjing Univ., China

Abstract

In this paper, expansive soils in China are described including the distribution type of hazards and economic losses caused by expansive soil. Four internal factors for controlling the swelling-shrinkage of expansive soil are discussed in detail. The research shows that some expansive soils contain a certain amount of mixed-layer clay mineral, that the expansive soils mainly with the exchange cation Mg^{2+} and laminar flow microstructure have stronger expansion and shrinkage and that most expansive soils in China are natural highly dispersive soil with a clay content greater than 30%.

Key words: expansive soil, mineral composition, exchange cation, dispersibility of clay grains, microstructure

INTRODUCTION

Expansive soil is a natural, highly dispersive and plastic one, which contains mainly clay mineral and is very sensitive to dry or wet environment. Expansive soil's main features are: 1. Clay content (<2um) is more than 30% of all fractions; 2. Clay minerals includes mainly illite, montmorillonite, kaolinite and their mixed-layer minerals with strong water affinity; 3. When the moisture content of soil increases (or reduce), the volume of soil expanses (or shrinks) to form swelling pressure (or shrinking fissures); 4. As moisture content of environment changes, swelling and shrinkage of soil happen alternately, as a result, the strength of soil reduces; 5. Its liquid limit is more than 40%, See Fig. 1.; 6. It is a overconsolidated clay. The rock with the above features (2,3,4) is called expansive rock.

Expansive soil is widely distributed in the world and scattered over more than 40 countries and regions of six continents. China is one of the countries with larger scale and distribution of expansive soil which has successively been found in more than 20 provinces and regions in China. Its geological origins are mainly lacustrine, residual, slopewash, alluvial and diluvial ones.

The hazards of expansive soil in China are very serious. It is estimated incompletely that the area of house destroyed by expansive soil is up to 10 million m^2 per year in China, and the economic loss reaches to 0.5~1.0 billion RMB.

Types of expansive soil hazards in China include: (1) The foundation deformation; (2) Landslide; (3) Frost soil and mud pumping of roadbed; (4) Debris flow.

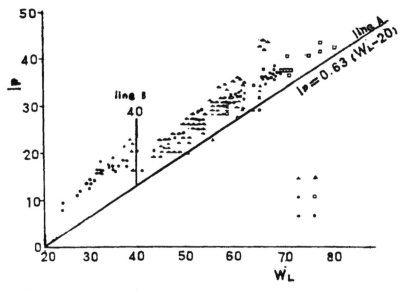

Fig. 1. Distribution of expansive soils in China on plastic chart

In order to mitigate these hazards, the first task is to understand these internal factors for controlling the swelling-shrinkage of expansive soil. These factors include (1)mineral composition, (2)Cation exchange, (3)Dispensability of clay grains, and (4) Microstructure.

MINERAL COMPOSITION

Expansive soil consists of more than 20 mineral species, among them clay minerals have a great influence on soil's engineering properties.

Most expansive soils in China (Except Guangxi) are natural, highly dispersive ones, with a clay content greater than 30%. Minerals composition includes mainly illite, montmorillonite, kaolinite, chlorite, and their mixed-layer minerals which form the substantial basis for strong swelling and shrinkage of the expansive soils. From table 1, it is found that a great amount of mixlayer clay minerals exist in some expansive soils. Such as location 3 and 4. Therefore, the swelling-shrinkage of expansive soil is not only determined by the content of montmorillonite, but also by mixlayer clay minerals and illite.

Table 1 Clay mineral composition of soil samples

LOCATIONS		1	2	3	4	5	6
	I	58-61	59.6	28.9	–	22-51	65-80
clay	M	7-13	10.4	–	38-48	13-18	–
mineral	K	24-29	22.3	13.3	10-15	45-71	10-20
composition	X	–	8.2	–	–	–	–
(%)	I-M	–	–	–	–	–	10-25
	M-I	–	–	49.2	40-49	–	–

NOTES   1--Ninmin 2--Nanning 3--Ping dingshan 4--Yun county 5--Xiangfan 6--Hefei
             M--montmorillonite   I--illite   K--kaolinite   M-I & I-M: mixed layer mineral

CATION EXCHANGE

Surficial physico-chemical properties of expansive soil are very active. From table 2, cation exchange content at the grain surface is generally greater than 20me/100g. From table 2, it is found that the exchanged cation Mg^{2+} of some expansive soil is very high, such as location 3 and 6, and their is a good relationship between it and the free swelling ratio. The higher soil's exchanged cation Mg^{2+}, the stronger soil's expansion. It is because the diameter of Mg^{2+} is smaller than Ca^{2+} and the potential of water affinity of Mg^{2+} is high than Ca^{2+}.

Table 2. Physico-chemical properities of soil samples

locations	exchange capacity (me/100g)	composition of exchange cation				TEBC (me/100g)	percentage of mg²⁺ in TEBC	degree of saturation (%)
		Ca^{2+}	Mg^{2+}	Na^+	K^+			
1	24.62	2.430	7.85	0.87	0.88	12.03	65	49
2	24.14	15.59	6.73	0.13	0.13	22.58	29	93
3	54.51	27.73	29.20	1.69	1.38	55.00	53	100
4	29.70	21.90	5.20	0.70	0.80	28.60	18	–
5	38.18	20.50	14.58	0.28	0.20	35.56	41	93.14
6	48.40	2.04	34.34	1.92	1.28	41.00	83	84

NOTES   1--Ninmin 2--Nanning 3--Ping dingshan 4--Yun county 5--Xiangfan 6--Hefei
             TEBC--Total exchange base capacity

DISPERSIBILITY OF CLAY GRAINS

Table 3 and Fig. 2 show that comparison between clay grain content of specimen adding dispersing agent and not doing. It is found that most expansive soils in China are a natural, highly dispersive one, because the results of grain analyses by above two methods are very close. This characteristic is important factor controlling soils expansion-shrinkage.

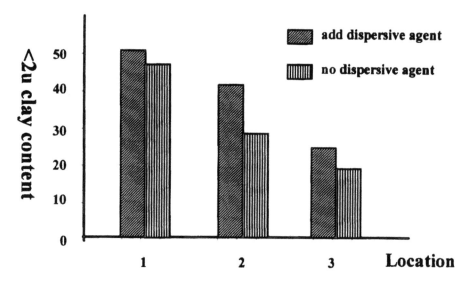

1–Niming, expansive soil   2–Liuzhou, laterite   3–Yun county, expansive soil

Figure 2. Comparison of clay content (<2u) between samples adding and not dispersion agent

MICROSTRUCTURE

Microstructure has a great influence on the degree of expansion and shrinkage of soils. For example, according to mineral composition of soils, the expansion and shrinkage of the soil from location 4 should be stronger than that of the soil from location 1. However, the study result is opposite. The particles of the latter soil are oriented to form a laminar-flow structure with weak cohesive force and have well-connected interlayer fissures. These factors are favourable to swelling and have a intergranular cohesive force greater than that of the latter soil. Their aggregates are arranged in disorder. The pores in the soil are more but isolated. These factors are not favourable to swelling and shrinkage. Therefore, although the soil from location 4 has a higher content of montmorillonite, it is weaker in expansion and the soil from location 1.

Table 3. Grain size of expansive soils in china

Area		soil grain group percentage (%)			
Province	city	>0.05mm	0.05–0.005mm	<0.005mm	<0.002mm'
Yunnan	Jijie			57.00	43.50
	Qujing	9.79	33.56	56.31	42.14
Guizhou	Guiyang	8.54	30.23	63.13	51.18
	Zunyi	8.10	29.52	62.38	54.46
	Chengdu	5.88	44.52	49.60	44.53
Sichuan	Nanchong	14.70	34.30	47.80	36.35
	Xichan	10.50	38.50	51.00	35.00
	Ningming	3.69	25.76	70.98	48.47
Guangxi	Guixian	8.10	19.00	73.38	58.38
	Wuming	11.75	33.25	55.00	42.00
Shanxi	An'kang	5.00	48.80	46.20	
	Jingmen	8.81	42.81	48.02	32.14
Hubei	Xiangfan	13.56	44.36	42.30	29.81
	Yun county 1	7.00	50.00	45.00	30.00
	Yun county 2	4.00	50.00	46.00	26.00
Guangxi	Liuzhou	6.80	47.00	46.00	32.00

CONCLUSIONS

(1) Expansive soil is widely distributed in China. It is completely estimated that economic loss caused by expansive soil exceeds 5 billion dollars annually in the world and that the area of houses destroyed by expansive soil is up to 10 million m^2 per year in China.

(2) Main clay minerals in expansive soil in China are illite, montmorillonite, kaolinite, and their mixed-layer minerals.

(3) Cation exchange content (CFC) is generally greater than 200 ueq/g. The expansive soils mainly with the exchange cation Na^+ and Mg^{2+} are stronger in expansion and shrinkage.

(4) Most expansive soils in China are natural highly dispersive soil with a clay content greater than 30%.

(5) Laminar flow microstructure is quite advantage to expansion and shrinkage of expansive soil, next is turbostratic, matrix microstructure.

REFERENCES

1. Li Shenglin et al . Studies on the Engineering Geology of Expansive Soils in China. Jiangsu Science and Technology Publishing House (1992).
2. Shi Bin & Li Shenglin . Relationships between the microstructures and Engineering Characteristics of compacted Expansive clays.   Chinese Journal of Geotech Eng. **10:6**, 80-87 (1988).
3. Shi Bin. Quantitative analysis of the microstructural changes of remoulded clayey soil in the process of compaction. Proc. of the 1st .Int. Conf. on Unsaturated soils, Paris (1995).
4. Mitchell J.K. . Foudamentals of Soil Behavior, John Wiley & Sons, Inc (1976).
5. Li Shenglin, Qin Sujuan and Shi Bin, Engineering Geological Characteristics of Expansive Soils in China, Processing in Geosciences of China, papers to 28th IGC, 1988,P205-208 (1988).
6. Shi Bin et al, The engineering properties of expansive soil in Yun County, Hubei Province, Chinese Journal of Nanjing University, No.3, **24:3**,406-420 (1988).
7. Shi Bin and Li Shenglin, Engineering properties of compacted expansive soils in Yun county, Chinese Journal of Nanjing University, **25:2**, 90-95 (1988).

Proc. 30th Int'l. Geol. Congr., Vol. 23, pp. 253-259
Wang Sijing and P. Marinos (Eds)
© VSP 1997

On the Swelling-Shrinkage Properties and Mechanisms of Compacted Expansive Soil

LI SHENGLIN, & DU YANJUN
Department of the Earth Sciences, Nanjing University, China

Abstract

In this paper, the swelling   shrinkage properties of compacted expansive soil in Huaiyin Section of Ning-Lian High-grade Road are introduced, and its swelling   shrinkage mechanisms are discussed based on the change of soil water content, dry density, material composition and microstructure. Finally, it's pointed out that in engineering practice this kind of compacted expansive soil called as "artificial expansive soil" should be paid enough attention in case great damage be caused.

Key words: compacted expansive soil, artificial expansive soil, swelling   shrinkage indexes

INTRODUCTION

Under the natural occurrence conditions of some clayey soils have certain swelling-shrinkage, but they are not expansive soil. When being disturbed, compacted and then used as packing filters for the embankment and roadbed, their natural structures are destroyed, water content decrease, dry density become high and swelling-shrinkage indexes increase, and thus they may change into expansive soil. We call this kind of expansive soil as "artificial expansive soil". Researches have shown most cracks of embankments and and roadbeds are due to the ignorance of this problem. So in our engineering works, this kind of expansive soil should be paid enough attention.

SWELLING-SHRINKAGE PROPERTIES

From Table 1 it can be seen that under natural occurrence conditions, swelling-shrinkage indexes of these soils are low except for the yellow silty soil. But they become high when being remolded and compacted, especially for the swelling force index. For example, swelling force of grey soil increases by 27 times varying from 22Kpa to 587Kpa, but of black soil increases by 38 times ranging from 10.1Kpa to 385 Kpa, while that of grey yellow soil increase by about 7 times changing from 40 Kpa to 276 Kpa. Only yellow silty soil, of which the swelling force increases by 6Kpa to 7.5 Kpa, still is not expansive soil after being remolded and compacted . Another example is that the swelling rate under 50 Kpa of in-situ soil is below zero and it increases above zero(except for the yellow silty soil)after being compacted. The increasing order of it is the same as that of the swelling force. Total swelling rate(δ_e) is a complex index which well describes the swelling-shrinkage properties of soil. δ_e of the in-situ soils become relatively higher after being compacted. δ_e of the grey soil is the largest one.It increases by 4.3% to 11.8%, nearly 3 times, and that of yellow silty soil is the lowest, only about -0.4%. Distinguished by δ_e these compacted soils are expansive soils excluded yellow silty one.

Table 1. Comparison of swelling-shrinkage indexes between compacted soil and in-situ soil

Index	Soil Type	In-situ Soil	Compacted Soil	I/C
swelling force	grey soil	22	587	27
	black soil	10.1	385	38
	grey yellow soil	40	276	7
	yellow silty soil	6	7.5	1
δ_{ef}	grey soil	-0.7	11.8	
	black soil	-1.9	7.3	
	grey yellow soil	-1.2	5.5	
	yellow silty soil	-3.7	-0.5	
δ_e	grey soil	4.3	11.8	3.0
	black soil	7.3	7.5	1.0
	grey yellow soil	2.9	6.2	20
	yellow silty soil	-0.7	-0.4	1.0
classification	grey soil	moderate	high	
	black soil	high	high	
	grey yellow soil	low	high	
	yellow silty soil	non	non	

δ_{ef} (%) stands for swelling rate under 50Kpa and δ_e(%) stands for total swelling rate.

MECHANISMS

Change of dry density and water content

In table 2, it can be seen that dry density of in-situ soil increases to the largest one while water content lows down to the optimum one. The change of dry density and water content are the direct reasons for different swelling-shrinkage properties between compacted soil and in-situ soil.

Table 2. Comparison of dry density and water content between compacted soil and in-situ soil

Soil Type	Dry Density(g/cm³)		Water Content(%)	
	In-situ Soil	Compacted Soil	In-situ Soil	Compacted Soil
grey soil	1.33	1.74	34.2	19.5
black soil	1.30	1.60	43.9	24.5
grey yellow soil	1.54	1.71	27.4	20.0
yellow silty soil	1.55	1.63	25.8	14.2

Based on the soil behavior study theory, the swelling of soil is due to the moisture film forming around the particles' outside layer result from the reaction between clay particles and water. As the thickness of moisture film increases, the volume of soil become larger and larger. Under the different dry density and water content conditions, reaction will be different. For example, as for the specimen of the same type soil with low water content the reaction is easy to take, and the moisture film may come up to the thickest one in a short while. Samely, when the dry density is high, clay particles per volume of soil is relatively higher, and thus the reactions between particles and water are more serious. This is benefit to the swelling. Compacted soil in Table 2 is in favor of it. Its dry density is higher and its water content is lower. However, in-situ soil has not these performances. So it is clear that dry density and water content play important roles in soil swelling.

Material composition
It is not the case that all of soils may become "artificial expansive soil" after being remolded and compacted. Only those which contain high affinity clay minerals, clay particles and compose closely to expansive soils may change into "artificial expansive soil" after compact. Take as an example, all the soil show in table 3 have high illite percentage and illite-montmorillonite mixed-layer mineral percentages are also high in grey soil and grey yellow soil. Meanwhile, montmorillonite and illite percentage are both affinity clay minerals, so the swelling of compacted grey soil and yellow soil and grey yellow soil is high. Besides mineral composition, clay particle content is anther important influence on compacted soil properties. Though yellow silty soil composes mainly of illite, yet it contents only 2.8% of clay particles. Therefore, its properties vary little pre- and post-lime improvement. Tahir's study also indicates that clay particle content influents not only on the change of soil's largest dry density but also directly on the shape of soil microstructure, therefore influences on soil swelling properties.

Table 3. Material composition of in-situ soil

Soil Type	Clay Mineral Composition %			Clay Particle Content %
	I	I-M	Ch	
grey soil	56	19	9	51.9
black soil	58	little	19	56.1
grey yellow soil	39	35	20	50.3
yellow silty soil	62	little	20	2.8

I stands for illite, I-M stands fore illite and montomorillite mixed-layer mineral and Ch stands for chlorite

Change of Microstructure
Table 4. Comparison of microstructures between compacted soil and in-situ soil

Soil Type	In-situ Soil	Compacted Soil
grey soil	turbulent structure	turbulent-orientation structure
black soil	aggregate structure	turbulent-orientation structure
grey yellow soil	matrix structure	turbulent structure
yellow silty soil	aggregate-skeletal structure	aggregate-skeletal structure

After long complicated natural functions, in-situ soil mass develops certain structure strength and strong connection force forms among soil particles. This can partly suppress the swelling. On the contrary, compacted soil has been ground, remolded and compacted, its original structure is destroyed, natural structure strength lows down and thus factors suppressing swelling are removed and makes it easy to swell. Table 4 which well describe the microstructure differences between compacted soil and in-situ soil can explain the phenomena rather clearly.

From Table 4 and Fig. 1 to Fig. 6, it can be seen, except for yellow silty soil other in-situ soils' microstructures have changed obviously after being compacted. The microstructures of compacted soils are mainly of turbulent shape or turbulent-orientation shape, their pores distribute constantly and form rows of channels(Fig. 2). When soil mass immerses into water, water can penetrate easily along these channels which makes aggregated particles react with water sufficiently. But microstructures of in-situ soil are mainly of matrix ones, and most of the pores are micro ones and they don't distribute constantly. Water penetrates into soil difficultly and it is harmful to swelling. The comparison of the microstructure between compacted soil and in-situ soil well explains this point. Microstructure of yellow silty soil does not vary obviously pre-and post-compact, the size of structure element does not change much either, the dry density and structure connection force of in-situ soils are close to that of compacted ones. Therefore, the swelling-shrinkage properties doesn't differ much between them.

From above, it is easily concluded the change of microstructure is the key influence on the swelling-shrinkage properties of compacted soil.

CONCLUSION

Having known the properties of compacted soil, in our engineering works we should concern on the study of it. Recent researches have indicated some clay soil with index plasticity varying from 13 to 17 and liquid limit from 35 to 40 is the most possible one to become artificial expansive soil after being remolded and compacted. To this kind of soil special improvement is necessary as it is used as embankment or roadbed packing filter, otherwise swelling damage may be induced.

REFERENCES

1. LI Shenglin et al. Study on the Engineering Geology of Expansive Soils in China, *Jiangsu Science and Technology Publishing House*, 129-140 (1992).
2. Tahir Ahmet Alwail. Collapse Mechanisms of Low Cohesion Compacted Soils. *Bulletin of the Association Engineering Geologists*, **4**, 345-353 (1992).

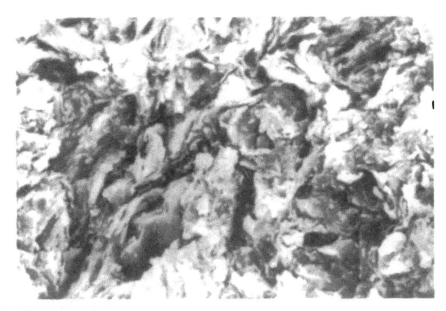

gure 1. The SEM photograph of grey in-situ soil

igure 2. The SEM photograph of grey compacted soil

Figure 3. The SEM photograph of black in-situ soil

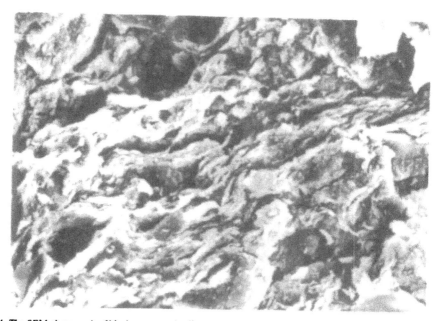

Figure 4. The SEM photograph of black compacted soil

Figure 5. The SEM photograph of grey yellow compacted silty soil

Figure 6. The SEM photograph of grey yellow compacted soil

Proc. 30th Int'l. Geol. Congr., Vol. 23, pp. 261-266
Wang Sijing and P. Marinos (Eds)
© VSP 1997

Research on Moistening and Demoistening Behaviours of Collapsible Loess

ZHANG SUMIN, ZHANG WEI & ZHENG JIANGUO
Institute of Geotechnical Investigation and Surve   Ministry of Machinery Industries   Xi'an   China

Abstract

In the past, collapsibility of loess, while soaked to saturation, was studied as a key subject. However in recent years, we made comprehensive researches on strength and deformation behaviours of collapsible loess during moistening and demoistening. This paper deals with the results.

Keywords   Collapsible Loess,Moistening,Demoistening, Collapsibility,deformation, strength

COLLAPSIBILITY OF LOESS DURING MOISTENING AND DEMOISTENING

The collapsible coefficient characters

To moisten and demoisten the loess from natural water content to various degrees in double   line method compression tests will result in a set of curve lines of collapsible coefficient (δs)   pressure (p)   as in Figure 1   . It shows that with the increase of water content, the collapsible coefficient decreases obviously, which indicates that the collapsibility of loess decreases with moistening and increases with demoistening.

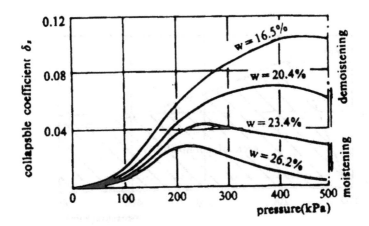

Fig.1 The collapsible coefficient curves in different water contents

It shows also that the collapsible coefficient curves have a peak value under different applied pressures. We call the maximum collapsible coefficient the peak collapsible coefficient and the corresponding pressure the peak collapsible pressure. They all decrease during moistening and increase during demoistening.

TERMINAL COLLAPSIBLE PRESSURE AND COLLAPSIBLE PRESSURE INTERVAL

We know from Figure 1 that there is a peak value in the collapsible coefficient curve, which means that when the pressure goes beyond a certain value, the collapsible coefficient decreases rather than increases progressively and when the pressure increases high enough, the collapsible coefficient will decrease to less than 0.015 approaching zero. That is to say, when soak to saturation under a certain high pressure, the loess will not deform noticeably, And this pressure can be called the terminal collapsible pressure, only when the pressure goes beyond the initial collapsible pressure and still less than the terminal collapsible pressure, and soaked to saturation does the loess appear in deformation corresponding to $\delta s \geq 0.015$. This pressure interval can be called the collapsible pressure interval, whose extent narrows with the increase of the original water content. Figure 2 is the curve based on the tests. The shadow in the Figure indicates the area of collapsibility. The front part ab of the curve is the so called initial collapsible pressure, and the back part bc of the curve is what we call the terminal collapsible pressure. When the applied pressure goes beyond the terminal collapsible pressure, the loess will be condensed by the spherical stress, so that the saturated soaking will not lead to obvious corresponding deformation with $\delta s \geq 0.015$. On the other hand, the Figure shows also that when the original water content goes beyond the corresponding ultimate water content of collapsibility w_{ul} of the curve peak b, under however high pressure when soaked to saturation, no collapsible deformation appears with $\delta s \geq 0.015$.

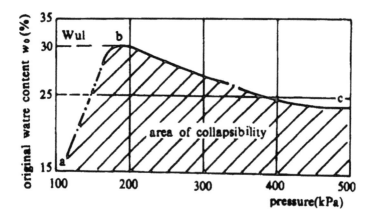

Fig.2 Initial collapsible pressure (ab)and terminal collapsible pressure (bc)

DEFORMATION CHARACTERS OF LOESS DURING MOISTENING

Moistening deformation model

Moistening deformation is an additional deformation to loess when water content increases, after deformation of loess under pressure reaches stable. This deformation mechanism can be expressed by a special structural model, called a moistening softening model: ZSM body, which is a parallel of non-linear Hooke bady (H) and a moistening softening body ZS , as shown in Figure 3 a . The structure of ZS body is two connected flakelets, During moistening by immersion, the rigid connection will be completely lost. The behaviours of ZS body can be expressed as:

during no moistening $\Delta\varepsilon = 0$

during moistening by immersion $\Delta\varepsilon = \Delta\sigma/E$

$$\left.\begin{matrix} \\ \\ \end{matrix}\right\} \quad (1)$$

Deformation model of collapsible loess can be expressed as series of a non linear Hooke body and n ZS bodies, as shown in Figure 3 b . During no moistening, the deformation of collapsible loess, ie loading deformation under the initial moisture state, is mainly controlled by the non linear Hooke body H_0 During moistening by immersion, with the increase of water content, the number of softened ZS bodies will increase, and every softened ZSM body will bring about a certain moistening softening deformation. The behaviours about the deformation model of collapsible loess can expressed as:

during no moistening $\Delta\varepsilon = \sigma/E_0$

during moistening by immersion $\Delta\varepsilon = \sigma/E_0 + \sum_{i=1}^{m}(\quad /Ei)$

$$\left.\begin{matrix} \\ \\ \end{matrix}\right\} \quad (2)$$

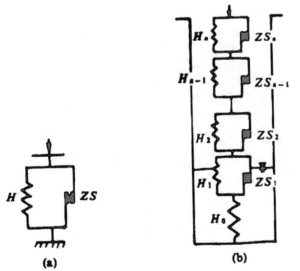

(a)   (b)

Fig.3 Moistening deformation model of collapsible loess

Where, E_0 is the modulus of the non-linear Hooke body under a given stress; E_i is the modulus of the ith non-linear Hooke body in the ith ZSM at a certain stress level; m is the number of the softened ZSM bodies during moistening by immersion and it is related to the degree of moistening.

Stress moistening path and curved surface of stress moistening deformation
Deformation of collapsible loess can be expressed as a function of stress level and water content. The space curves which describe the actions of force and water and corresponding process of deformation are called stress moistening path(as shown is Figure 4).The different pathes express the different cases.
i).   The path ABD means that the loess is soaked to saturation in natural state (line AB) and then loaded   line BD   .
ii).   The path ACD means that the loess is loaded in natural water content   line AC and then soaked to saturation under the stress   line CD   .
iii)   The stress moistening path can also express more complex processes, For example the path AB_1C_1D means that the loess is moistened nonloadedly   line AB_1,   -loaded with deformation under constant water content (line B_1C_1,)-soaked to saturation under the stress   line C_1D

The various levels of water and force actions will bring about various stress moistening path, which make up a curved surface in the space of σ-ε-w (stress-strain-water content). We called it the curved surface of stress moistening deformation. These two concepts above help us more comprehensively understand the theoretical meaning   and practical sense of initial collapse pressure, initial collapse water content and initial moistening

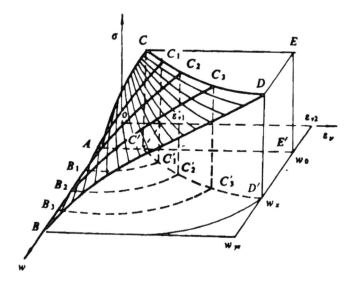

Fig.4 Stress moistening pathes and a curved surface of stress moistening deformation

STRENGTH CHARACTERS OF LOESS DURING MOISTENING

The results of triaxial shear test on collapsible loess with various water contents under different confining pressures show that, when water contents are the same, the test results are basically on a line. It is shown in Figure 5. The line is actually another expression of Mohr-Coulomb's strength criterion, So the studied collapsible loess, within the scope of water content variation in the test, basically follow Mohr-Coulomb's strength criterion, When water contents increase, the slopes of these lines decrease, and also their intercepts

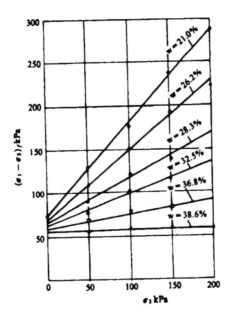

Fig.5 The relation between limit deviator stress(σ_1 $\sigma_3)_f$ and confining pressure σ_3

variate. It shows that, the ultmate strength of collapsible loess, expressed by limit deviator stress of triaxial test, is a function of confining pressure σ_3 and water content w , and also can be expressed as a curved surface in 3-dimensional space of σ_3-w- σ_1-σ_3 It is shown in Figure 6.

Index of strength, c,ϕ, can be expressed as a function of water content w, With water content being increased to a given number, the line of strength decreases and reaches limit equilibrium. This water content is called critical water content of collapsible loess, and it is one to one transformation to stress level on curved surface of strength.

The research on strength characters of collapsible loess during moistening has the following significance

i).The strength of collapsible loess continuously decrease with the increase of its water content. In fact, we don't think that, only when water content reaches a given value (for example, approaches or reaches saturation water content), will its strength have a abruptly drop.

ii). The strength characters of collapsible loess during moistening can be expressed by the curved surface of strength. For a given layer of collapsible loess, when its strength coefficient is obtained, according to the existing information, it can predict that, when water content increase, its strength would decrease, So we say this is a guidance to engineering practices in collapsible loess, subjected to moistening because of water rise and water infiltration etc.

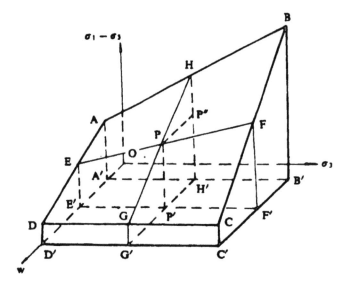

Fig.6 A curved surface of strength of collapsible during moistening

REFERENCES

Zhang   S   M. and Zhang   W.Collapsibility of loess during demoistening and moistening process. Proceeding 7th IAEG Congress   1847   1852   1994   .

Zhang   S   M. and Zheng   J   G. The deformation characteristics of collapsible loess during moistening process. Chinese Journal of Geotechnical Engineer : 21   31   1990   .

Zheng   J   G. and Zhang   S.M. The strength characteristics of collapsible loess during moistening process   Hydrogeology and Engineering Geology   6   10   1989   .

Proc. 30th Int'l. Geol. Congr., Vol. 23, pp. 267-275
Wang Sijing and P. Marinos (Eds)
© VSP 1997

Microstructure Effect on Loess Collapsibility

HU RUILIN, LI XIANGQUAN, GUAN GUOLIN, ZHANG LIZHONG
Institute of Hydrogeology and Engineering Geology, CAGS, P.R. China

Abstract

Microstructure of soil is a kind of physical state, which could be identified in some structure factors. From the angle of image processing, the technological course of quantitative study at microstructure and quantification methods of non-determinated structure factors have been put forwards. With the fractal structure analysis, the authors discuss the influence of natural structure and pressure on loess collapse, interaction law between water and specimen and its collapse effect, and then explain the mechanism of loess collapse with the fractal structure.

Key words: loess, collapsibility, microstructure effect

INTRODUCTION

Loess collapsibility is a long-standing problem that many engineering geologists have been following with interest. A lot of fruitful research work on it have been completed. People have received an intense knowledge that the loss collapsibility is controlled in nature by microstructure. However, since the shape of microstructure possesses non determination and the factor of shape is difficult to be quantified, the quantitative relationship between microstructure and collapsibility has not been found so far. Almost all of the research results on microstructure could not been applied to practical analysis and assessment on engineering property so that a lot of troubles have arisen in engineering construction. Therefore, making a thorough work on the mechanism of loss collapsibility, in the presupposition of quantification of microstructure and in the way of analysis on microstructure effect, has important significance to solve the nonlinear problem of engineering behavior of soil and to improve the quality of engineering investigation.

PRINCIPLE AND PROCESS OF STRUCTURE ANALYSIS

Structure state and its conceptive model
Soil microstructure is a kind of physical state, called as Structure State(S). This state could be identified anyway in some structure factors. The soil structure state is the function of a series of structure factors($X1,X2,...,Xn$). That is:

$$S = f(X_1, X_2, ..., X_n)$$

However, the relationship between structure state and structure factor is too complicated to be described in simple linear way. Despite this situation, the structure shape is fairly clear in concept. As common knowledge, the microstructure mainly consists of the shape and the pattern of particles(minerals or their aggregates), porosity and contact form(non-chemical linkage). These four parties are usually used as the structure factors of soil. According to engineering practice, these factors cannot respectively express the structure shape of soil. The simple overlay of their properties and function cannot flect the complete structure state. Therefore, we not only hope to analyse the characters of these factors, but fail to mechanically receive the relation and action between these factors. Systematicly grasping the behavior as a whole is very important.

In fact, structure factor only possesses qualitative significance here. Only realizing the quantification of structure factors first, can we identify the structure state in quanity. We call the parameters, quantitatively describing or constituting structure factor, as Structure Parameters. One structure factor may need one or more structure parameters to portray. Whole morphological system of soil microstructure is a complicatied, multilevel, intersect system. Its conceptive model is shown in Fig.1. This system is also dynamic, so its characters will remarkably change in different environments (e.g. different pressures).

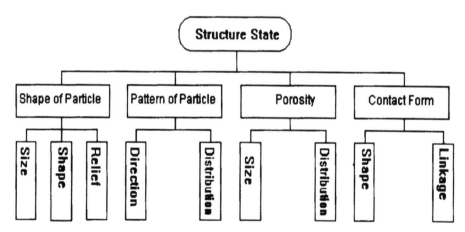

Fig.1 Conceptive model of microstructure

Process of structure analysis

Based on above train of thought, we adopt following technological process as Fig.2 to bring about the quantification of soil microstructure.

As the first step, according to the geometric features of image from soil specimen and image processing technique, we found the means to obtain quantitative information and sequently developed the Microstructure Image Processing System, called as MIPS for short. Meanwhile, in the light of Fractal Theory and starting with the identification of fractal feature of structure factor, we achieved the methods to quantify structure factors, which laies a foundation for quantitative structure analysis.

Afterwards, countering different engineering purposes, we had taken further steps to carry out the field investigation and indoor mechanical experiments for property analysis. With the help of MIPS, we can extract the quantitative parameters of specimen, and then do structure analysis corresponding to these parameters. Combining these results from structure analysis and making some of comprehensive judgement, we can obtain the results of evaluation on engineering behavior as a finality.

Algorithm on Fractal Parameters of Microstructure
The engineering property of soil is actually the synthetical expression of that of structure unit, and the property of structure unit depends on that of soil aggregats or even smaller single mineral to some extent. In view of this situation, soil structure behaves in remarkable administrative level and self-similarity. Serpinski Pad could be used to simulate the feature of granulated structure well, so soil microstructure has statistical self-similar fractal structure. We have enough reason to apply the theories and methods of Fractal Geometry for analysing and solving some of undeterminated problems in soil structure shape and then grasp the inherent law of engineering behavior of soil in quantity.

Facing the analysis of structure image, we, according to the fractal theory, get the algorithm on fractal parameters of microstructure shown on Tab.1. All of these parameters can be obtained directly or indirectly from the data or image by MIPS.

Tab.1 The algorithm on fractal parameters of microstructure

Name of Parameter (Fractal Dimension	Math's Model	ε	$N(\varepsilon)$
Particle Size(Dps)		Size of particle	Nember of particle
Particle Distribution (Dpd)		Border length of dividing net	Number of nets containing par-ticle
Surface Relief(Dpr)	$D = -\lim\limits_{\varepsilon \to 0} \dfrac{\ln N(\varepsilon)}{\ln \varepsilon}$	Length of scale	Number of step for measure-ment
Pore Size(Dbs)		Size of pore	Number of pore
Pore Distributio (Dbd)		Border length of dividing net	Number of nets containing pore
Contact-belt Distr. (Dco)		Border length of dividing net	Number of nets containing contact-belt
Particle Direction (Ddi)	$D = \lim\limits_{\varepsilon \to 0} \dfrac{\sum Pi \cdot \ln Pi}{\ln \varepsilon}$	Increment of directin angle	$N(\varepsilon)=\sum Pi \bullet \ln(1/pi)$.Here, Pi is the probality at which particle has the direction to dividing area.

ANALYSIS ON THE COLLAPSIBILITY OF LOESS

Loess collapse is a phenomenon that under certain pressure and soaked in water the structure of loess specimen will be destroied quickly and remarkable additional subsidence could takes place. With above technique and meessures, we have made some of quantitative research work on the mechanism of loess collapse due to microstructure effect. The samples for structure analysis were collected from Loess Plateau of China. They are terrace loess and plateau loess. The items of indoor mechanical experimeent include collapse-deformation test, water-effect test, common compression test, cousolidation test, shear test, granulometric test, sinkage test and so on. 355 pieces of thin sections have been prepared. About 30MB of data have been produced from image processing.

Natural structure state Influence on Loess Collapsibility
Fig.3-Fig.8 show the results of structure analysis on natural loess. In these figures, all the ordinates express the collapse coefficient(δs) displaying the intensity of collapse, but abscissas express respective structure parameters. It is clear that almost all of the parameters, except the fractal dimension of pore size and pore diamenter, are statistically related to collapse coefficient. Therefore, we think that although the loess collapse is controlled by lots of natural structure factors, the natural porosity shows little influence on it. The hypothesis that large pore leads to collapse is untenable.

Pressure Influence on Loess Collapse
As we know, the loess collapsibility is related to effective pressure exerted on specimen. It is common that the collapse coefficient of sample will increase with the pressure and then decrease. When the horizon of sample is deepen, this phenomena appears late. In other words, not only the pressure at which remarkable collapse takes place is high, but the maximum of collapse coefficient arises later. However, our research results(Fig. 9) indicates that the collapse coefficient dosn't increase in line with the pressure. In the middle pressures(0.2-0.4MPa) the increment will decrease or has little change so that the phenomenon called as Collapse Platform takes place.The distribution regularity of collapse platform is that:
   --The location of the platform for terrace loess is lower than those for plateau;
   --New loess often has lower position than the old.

In order to look into the case of generation of the collapse platform, we have done some of analysis under different pressures, which result is shown in Fig.10. We could find that on the platform the structure factors also behave in sudden change. Both the favorable elements and the unfavorable have reached certain balance so as to cause such a complex structure state bringing about this kind of steady macroscopic deformation. This is the reason why collapse platform exist.

Water Effect on Loess Collapse
Soaking in water is the essential requirement for loess collapse. With different water contents, loess will have different collapsibility. Some of domestic scholars(Zhang Sumin, Zheng Jianguo;1989, 1990,1991) deeply studied the water influence on mechanical

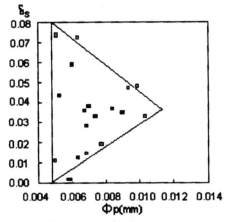

Fig.3 Relationship between pore
diameter and collapse coefficient

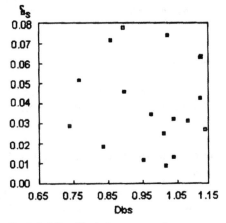

Fig.4 Relationship between particle
size dimension and collapse coefficient

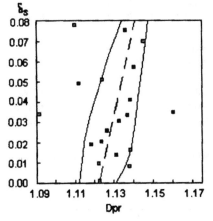

Fig.5 Relationship between particle relief
dimension and collapse coefficient

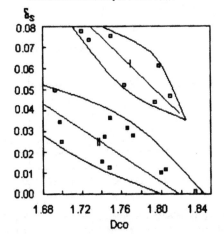

Fig.6 Relationship between pore size
dimension and collapse coefficient

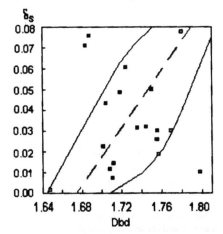

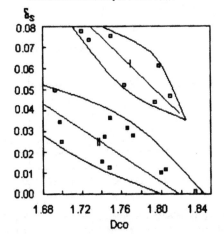

Fig.7 Relationship between pore direction
     dimension and collapse coefficient

Fig.8 Relationship between contact-belt
     distr. dimension and collapse coefficient

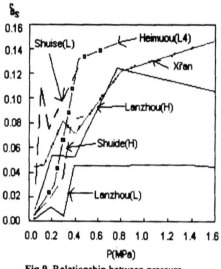

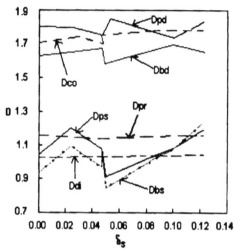

Fig.9 Relationship between pressure
     and collapse coefficient

Fig.10 Relationship between collapse
      coefficient and fractal structure factors

property of collapsible loess. They called the additional deformation due to different water
content after reaching a stable state as Water-increase Deformation. They think that this
deformation is briefly attributed to water influence on   special substance and
microstructure. Based on this knowledge, we intend to do some of initial studies at the
interaction between water and loess and its collapse effect in view of water influence on
structure state.

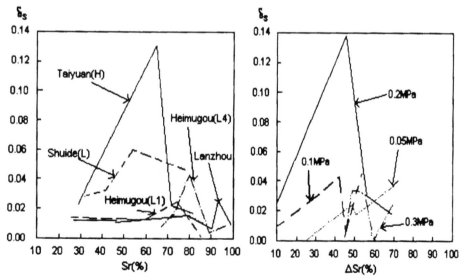

Fig.11 Relationship between degree of
      saturation and collapse coefficient

Fig.12 Relationship between increment of
      saturation degree and collapse coefficient

Fig.11 and Fig.12 demonstrate the results of water-effect test. From them, we can find that :

--Along with a raise in water content expressing in the Degree of Saturation (Sr) of specimen, the subsidence, expressing in collapse coefficient, also will increase. However, such a raise cannot be sustained. After reaching the maximum, it will decrease. The collapsibility curve shows a convex line with one peak.

--The maximum of collapsibility coefficient isn't located at the largest saturation. In other words, the collapse coefficient from the collapse test on the condition of complete soaking isn't the highest collapse coefficient, but slightly smaller. It is the reason that pore water increases in this situation with the result of effective stress and deformation going down.

--Under different initial state, the highest collapse coefficients are corresponding to different areas of degree of saturation, called as Peak Saturation. Usually, the peak saturation of plateau loess is larger than that of terrace loess. The peak saturation of the former has little difference from the saturation at complete soaking. In these figures, the peak saturation of plateau loess(Heimugou) is usually between 80% and 90% and the increment of saturation(ΔSr) is usually 50%-70%. The peak saturation of terrace loess (Shuide,Taiyuan or Lanzhou) is only 50%-70%.

--Under low pressures the peak saturation at 0.2MPa is the highest. The pressure divisions at different pressures are often stable. For example(Fig.7), the division of Taiyuan terrace loess is between 60% to 70%.

Taking Heimugou loess as a sample, we take further steps to study the law of structure change and the relation to collapse deformation. Based on structure parameters got before and after soaking, we calculated the Structure State Change Indexes(J_{sc}) at different water-increase levels. The formula of Jsc is:

$$J_{sc} = \frac{1}{n} \sum_{i=1}^{n} \frac{|J_{i0} - J_{i1}|}{J_{id}}$$

Where J_{i0}=value of structure parameter (i) before soaking;

J_{i1}=value of structure parameter (i) after soaking;

J_{id}=value range of structure parameter (i)

The larger Jsc, the greater the structure adjusts.

Fig.13 is the result of structure analysis. We can find that the structure state line (Jsc) is similar to   collapse line (δ_s). Structure state adjustment synchronizes with collapse deformation. Therefore, we think that the phenomenon of convex line directly originates in remarkable structure adjustment.

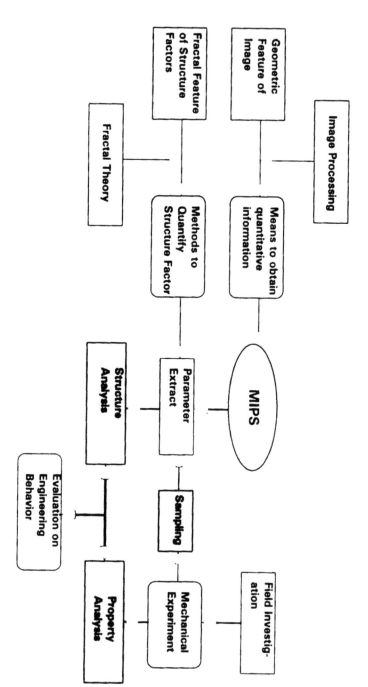

Fig.2 Technological process of quantitative analysis on microstructure

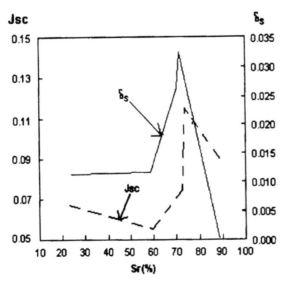

Fig.13 Relationship between degree of saturation and Jsc/ δ_s

CONCLUSIONS

Loess collapse results from the action of muti-factors. Besides the structure linkage and property of loess, the high state of fractal dimension is also the important aspect, which provides the space for collapse. Drived by the pressure and water, the loess at high dimension will change into the low-dimension, which is the micro-mechanism of macro-deformation of loess collapse.

REFERENCES

1.HU Ruilin et al, Quantitative microstructure models of clayed soils and their engineering behaviors, Geological Publishing House, 1995
2.HU Ruilin et al, Shape Factors of clayey soils and their processing technique of quantitative information, ACTA GEOSCIENTIA SINICA, Vol.17, Sep.1996
3.HU Ruilin et al, Fractal characters of loess Microstructure in Deformation Process and its Eng. Significance, Proceedings of 7th Congress of the IAEG, 1994, lisboa, Portugal
4.N. K.Tovey, Quantitative Analysis of Electron Micrographe of structure, Inter. Congress of Soil structure, 1973, Gethenbury

Proc. 30th Int'l. Geol. Congr., Vol. 23, pp. 277-284
Wang Sijing and P. Marinos (Eds)
© VSP 1997

Quantification Of Microstructure Of Fine Grained Soil

SHI BIN, WANG BAOJUN, & LI QI
Department of Earth Sciences, Nanjing Univ., China

JIANG HONGTAO
Department of Geography, Jiangsu Eduaction College, China

Abstract

Quantification of microstructure of fine-grained soil has been a difficult nut for long time. In this paper, authors introduce the computer image processing system and show some quantitative analyzing results of fine-grained soil's microstructure, and put forward three quantitative indexes for assessing soil microstructure, and finally make the quantitative appraisal for the microstructure of the compacted clay soil.

Keywords: quantification, fine-grained soil, computer image processing system, anisotropy, probability entropy frequency distribution function

INTRODUCTION

The microstructure of clay soils is an important quality index of soil mass, which reflects the origin of soil mass, on the other hand is an important factor that determines the physical, mechanical and other engineering properties of soil mass.

The microstructure of clay soils has been studied for several decades using a variety of techniques including optical microscopy, X-ray diffraction., transmission electron microscopy, and scanning electron microscopy, of which scanning electron microscopy, called SEM, is the most effective means. However, the researchers all face two difficult nuts. The first is how to obtain quantitative information of soil microstructure, the second is what quantitative indexes are taken to assess soil microstructure. Because of the complication of the microstructure of clay soils and the limitation of technique, relatively few studies have been made in quantitative analysis. Unitt (1975, 1976), Unitt and Smith (1976) developed first the Intensity Grandient Technique, a more simple way to examine orientation patterns than the use of optical diffraction of Fourier transform technique; Gillott (1980) has had some success in examining the outline shape of sand grains, obtaining the Foueier series describing this shape, and characterizing grains on the basis of the distribution and harmonics in the Fourier distribution; Tang Lurong (1980) put formula for assessing orientation of mineral grains; Tovey (1980), Tovey and Sikolov (1981), Smart and Tovey (1982), Tovey and Smart (1986), Osipov et al. (1989), Tovey and Krinsly (1992) have extended Unitt's method in the study of orientation patterns in electron micrography and applied computed image processing technique to the study of soil; Wu Yixaing (1991) studied the structural images of optical chips of the clay soils using the computed image processing system; Shi Bin and Sui Ding (1991) used a new technique-D/MAXIII A Auto Fabric Geniometer to determine the orientation of the mineral grains in soils. The above studies have made great contributions to the quantitative analysis of the mocrostructure of clay soils, however, there are many shortages in sampling, equipment, computer hardwares and softwares, quantitative method and image processing speed etc.

In this paper, authors introduce a Computer Image Processing System, called CIPS for short, and put forward three quantitative index to assess soil microstruture.   Finaly, using this quantitative technique and index, the microstrutures of some compacted clay soils are appraised.

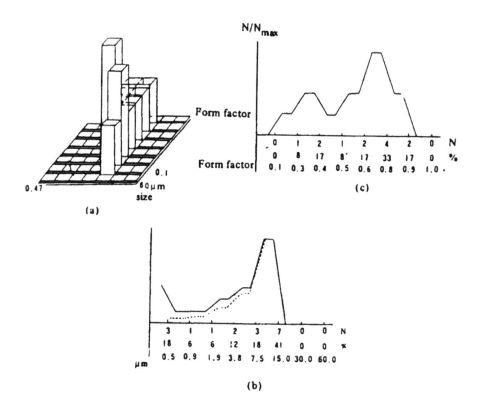

Figure 1.   Results of morphological analysis of SEM image of soft clay in Pacific Ocean.   (a) Size/form distribution; (b) pore size distribution (-), total area in intervals (......); (c) form factor distribution.   Results of morphological analysis; rel. Area 14.7%, total area 520um² , total perimeter 410um, quanty of pores 17, average area 31um², average perimeter 24um², average size 6.2um, average form factor 0.54[from 12]

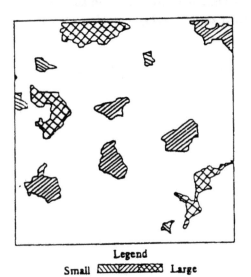 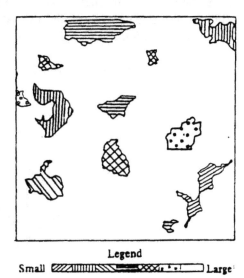

Legend

Small ⬚⬚⬚⬚⬚ Large

Legend

Small ⬚⬚⬚⬚⬚ Large

Figure 2. Distribution of void size after treatment of SEM image of soil clay in Pacific Ocean

Figure 3. Classification of factors of voids after treatment of SEM image of soft caly in Pacific Ocean

BRIEF INTRODUCTION OF CIPS

CIPS, which softwares for soil microstructure are made in Russion, has had great success in acquiring quantitative information of the microstruture of clay soils, especially in treating the orientation of the microstrutural units. CIPS connects directly with SEM and TEM, thus CIPS can directly acquire the image in SEM into the store. As a result, some man-made differences from photographing, edge enhansion, filling hole, separating particles. Etc. can be decreased. This is an outstanding advantage of CIPS. Of course, if you have the photos od SEM image, CIPS can also treat them by means of a good camera.

The two parts of soil quantitative information can be obtained after the SEM images are treated by CIPS. The first part includes the sizr, form, surface feature and quantity relation of the microstrutural units. These indices are the total area, total perimeter, average area, average perimeter, average formfactor and relations between them. Here formfactor can be defined as: $F=C/S$, where F is the formfactor, C is the circle perimeter that encircling the area should be equal to the actual area of the grain or void, S is the actual perimeter of the grain or void. F value is defined over(0.1), when F=1, the form is a circle; when F becomes smaller, the form does also narrower; when F trends to zero, the form trends to a line.

The second part is the spatial arrangement of the microstrutural units, which includes the distribution of oriention angles, main orientation angle and anisotropy. In addition, the VIPS can also give one- or two-dimensional Fourier analysis, X-ray map, etc. The detail illustration about CIPS is in Ref.[1]

Fig. 1-4 show that the results of quantitative analysis on soft clay in Pacific Ocean.

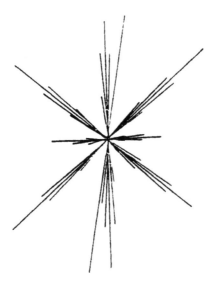

Figure 4. Orientation and isotropy analysis of soft clay in Pacific Ocean. Main orientation 85°, anisotropy 6.7%.
Interval (%): 0-10, 12.9; 10-20, 0.0; 20-30, 0.0;30-40, 0.0; 40-50, 30.1; 50-60, 0.0; 60-70, 0.0; 70-80, 0.0; 80-90, 0.0; 90-100, 29.0; 100-110, 0.0; 110-120, 0.0; 120-130, 0.0; 130-140, 0.0; 140-150, 28.0; 150-160, 0.0; 160-170, 0.0; 170-180, 0.0.

QUANTITATIVE INDEXES

Using above the quantitative techniques, the much information about orientation of clayey soil can be obtained.However,it is necessary to transfer the information into the quantitative structural parameter. Herein, author puts forward to three parameters.

Orientation index-Anisotropy
The analysing results of SEM image by the means of CIPS indicate that the distribution of orientation angles form a mirror and elliptical image from 0°~360°. In order to underst and the orientation characteristic in a whole, Tovey (1992) advanced such an index,which may be difined as :

$$Ia = \frac{R-r}{R} \times 100\%$$

where I_a is the index of anisotropy;R is the maximum axial length of the best fitting ellipes;r is the minimum axial length.I_a rangs from 0%,for random distribution,fully isotropy,to 100% for perfectly aligned distribution.

Order index-probability entropy

The analysing results show that anisotropy(I_a),which is statistically computed by the CIPS automatically,can fully describe the orientation characteristic of microstructural image,but cannot reflect the order extent of microstructural units.Herein,authors introduce a quantitative index-probability entropy from modern information systematic theory. Probability entropy(H) is defined by C.E.Shannon,a founder of modern systematic theory,as follow:

$$H = \sum_{i=1}^{n} P_i L_n P_i$$

Where P_i is probability that the signal is received, n is the number of signal types. Probability entropy(H) reflects the information content of each signal from information source.Herein,author recommends this index to appraise the order extent of arrangement of microstructural units,which is defined:

$$H_m = -\sum_{i=1}^{n} P_i Log_n P_i$$

as probability entropy of arrangement of microstructural units.Where P_i is probability of the microstructural units located in a certain orientation region,which equals to orientation intensity of the units at each orientation region; n is the number of orientation regions,so that n=18.It is very clear that H_m value is defined over [0,1].When H_m=0,all the microstructural units align in a certain orientation region,perfect order;when H_m=1,the microstructural unit arrangement is fully disordered.The larger H_m becomes , the more disordered the arrangement of the units does.

Frequency distribution function of clay aggregates

Барчен thought that the anisotropy of soil strength occurs when three exists orientation of clay aggregates.The destruction of soil mass results from the decrease of attraction between clay aggregates, not clay platelets.If the large clay aggregates are random, the soil strength may be isotropy,even though the microstructure inside the clay aggregates is anisotropy.This phenomenon can also be found in the test of soil creep.Therefore,frequency distribution function of orientation of clay aggregates is very important for analysing the stress-strain relationship of soil.From Figure 4, the distribution of orientation angles of clay aggregates is also axial symmetry,and similarily can be represented by the two-dimensional plot of Q(α) between 0°~90°. Where α is the plane angle of the clay aggregates from horizon orientation.Q(α) equals to a relative intensity of the orientation of clay aggregates at α direction,not the number of clay aggregates,which is given by computer image processing system such as CIPS.Clearly the $Q(\alpha)$ plot for a perfectly random sample is represented by a straight line that is parallel to the α axis.

All above three quantitative indexes can reflect the orientation of clayey soil,however there are obvious differences among them. I_a and H_m can indicate the alingment of microstructural units,but I_a shows the orientation of a whole soil microstructure,whereas H_m does the order extent of microstructural units.In many cases,the higher I_a is,the lower H_m is,but there are also exceptions. $Q(\alpha)$ can continuously reflect the orientation of clay aggregates ,which is the foundation for establishing micromechanical model of anisotropical clayey soil.

APPLIED CASES

Using above CIPS and quantitative indexes, authors make a quantitative assessment to five types of microstructures of a compacetd clay soil in Yun County, China.

The soil samples, located at point 1,2,3,4,5 on the ρ_d~W curve, are taken as microstructural samples (see Figure 5), which are observed under Scan Electronic Microscopy(SEM). Five microstructural types are obtained,based on the observation of SEM images of 88 compacted samples, which represent respectively the microstructure features of 5 samples located at point 1, 2, 3, 4, 5 on ρ_d~W curve. These microstructural types include loose collected-grained microstructure, densercollected-grained microstructure, clayclothed-grained microstructure, turbostratic microstructure, laminar microstructure.The engineering property of soil samples, and scanning electron photomicrographes and the detail description of five microstructural types are shown in the paper (3).   The quantitative analysis results are shown in Figure 5 and Table 1.

Table 1. Quantitative analysis results of 5 typical microstructures and expansion shrinkage properties of the soil

point	Relative area	Av.pore size	pore	size			Av. forming factor	Oriention index		Expansion=shrinkage		
No.	(pore)	/um	<0.5 mm	0.5~5 mm	5~50 mm	>50 mm		I_a (%)	H_m	swelling force /KPa	swelling ratio under 50 Kpa (%)	linear shrink-age ration (%)
1	9.2	10.5	16.7	24.7	50.2	8.4	0.57	5.1	0.98	150	2.3	2.5
2	7.4	5.2	21.6	38.2	40.2	0.0	0.53	4.5	0.98	505	15.3	4.2
3	6.5	4.2	24.4	45.7	28.1	0.0	0.49	3.6	0.97	450	10.2	8.0
4	6.7	4.8	23.1	51.4	25.5	0.0	0.32	2.9	0.93	18	3.5	13.1
5	7.2	5.7	18.8	58.4	22.8	0.0	0.20	51.9	0.85	0.0	-0.5	15.2

The above analysis results can quantitatively reflect the characteristics of microstructural changes of a fine-grained soil in the process of compaction, meanwhile, these applied cases also show that quantitative techniques introduced above are very useful for studying soil behavior.

CONCLUSIONS

The following conclusions may be drawn from the above results and discussion.

1. Computer image processing system is very useful to acquirement of the quantitative information of soil microstructure.
2. The three quantitative indices advanced in this note can fully reflect the orientation features of soil microstructure in difficult aspects.   Clearly, Ia represents the orientation feature of soil microstructure as a whole; Hm reflects the ordering extent of microstructural units and $Q(\alpha)$ can incessantly show the orientation distribution in the α direction from $0°$ to $90°$.

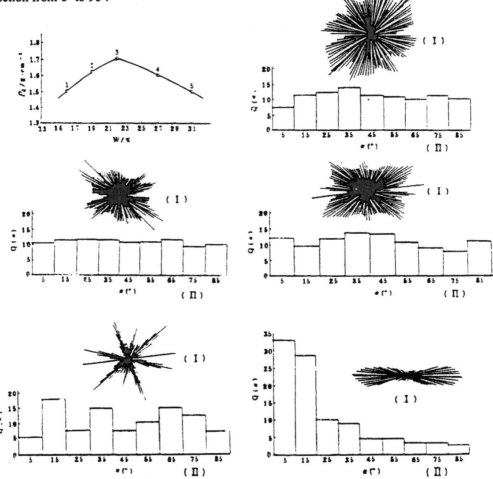

Figure 5. Orientation analyses of microstructures of remoulded clayey soil in the process of compacton. (I) Rose diagram of orientation angle distribution;(II) histogram of frequency distribution of orientation.

3. When remolded clayey soil is compacted under different water content, its microstructure will change. Five typical microstructures are found and analyzed quantitatively by computer image processing system.

Acknowledgments

The authors are grateful to Dr. Tolkachev, M. and   Prof. Li Shenglin for their help.

REFERENCES

1. Shi Bin, Li Shenglin, Tolkachev, M., quantitative approach on SEM images of microstructure of clay soil, Science in China, Ser.B, **38**:6,741-748 (1994).
2. Tovey, N.K., Krinsley, D.H., Mapping of the orientation of fine-grained minerals in soils and sediments, Bulletin of   IAEG, **46**,93-101 (1992).
3. Shi Bin, Li Shenglin, Relationships between the microstructures and engineering characteristics of compacted clayey soil, Chinese Journal of Geotech. Eng., **10** :6,80-88 (1988).

Proc. 30th Int'l. Geol. Congr., Vol. 23, pp. 285-294
Wang Sijing and P. Marinos (Eds)
© VSP 1997

Compression Settlement of Soft Soil and the Soil Improvement

WU XUYI   LUO CHENGPING
Institute of Investigation & Design PRWRC Ministry of Water Resources, Guangzhou 510611,P.R.China

Abstract

In the river mouth in delta area, bay or inland lake, there is a sedimentary layer of soft organic soil which includes mud, fibrous soil or swamp peat.The geotechnical indexes of these kinds of soil are very poor. One or two of them incline towards close to the viscous fluid. In project practice, it seems very difficult to analyze and treat the soil according to the classical Teazaghi theory. The settlement prediction and the soil improvement now is one of the important and difficult problems in engineering. The author has been working in this field for many years and read a lot of publications, and tries to put forward his ideas on the compression settlement of soft soil and the soil improvement.

Keywords:Soft soil, Settlement, Consolidation, Improvement.

INTRODUCTION

Soft soil deposited in slowly flow or static water environment has high water content, low density, great void ratio, high compressibility and high colloidal particle content. They are of very poor geotechnical characteristics (Table 1).

Table 1. Characteristics of some soft soil samples

sampling site	water content W (%)	liquid limit W_L (%)	dry density d (KN/m³)	void ratio e	liquidity index I_L	coefficient of compressibility $a_{v100-200}$ MPa^{-1}	nature shear strength C kPa ( )		Colloid content (0.002) (%)
Coast, Japan	713			9.900					
Zhuhai1, China	81.1	53.4	8.40	2.230	2.135	2.15	2	0.6	
Zhuhai2, China	5.2	52.7	8.20	2.305	2.820	3.56	1.4	0	
Haneda, Japan	170			4.200					
Zhuhai3, China	77.9	50.5	10.00	2.138	1.780	2.00	2	1.5	32.0

In Table 1, the maximum natural water content is $W_{max}=713\%$, and the maximum natural void ratio is $e_{max}=9.90$, and the maximum coefficient of compressibility is $a_{v100-200}$ =3.56MPa^{-1} ,the soil with so high water content is much different from the general mineral cohesive soil. Particles in the soft soil are disorderly   floating   in fluid. The normal component, N, and the tangential component, T, of the acting stress are all small[1]. Organic soil deposited in a reducing environment (most of its components are peat and humus soil) is really the colloidal matter mass form under gelatinizing. Under a electron microscope we can see that it is a small colloidal particle wrapped by two layers of water of combination, one being weak and another being strong (Fig. 1). The volume of the weak layer is much larger than that of the strong one[2]. Therefore, most of the water held by soil body is stored in the large pores among the particles, and the rest is stored in the small pores of the organic soil particles. In this paper, the large pores are called as   macro-pores   , and the small pores are called as   micro-pores   . Obviously, in the drainage consolidation process, after an additional stress σ'_o is added, water in the macro-pores is drained out first, and the soil body begins to be compressed. With this process, soil particles should be collapsed and a new void ratio appears[3]. If the additional stress is increased continuously, water in the micro-pores then will be drained out continuously until the consolidation is finished.

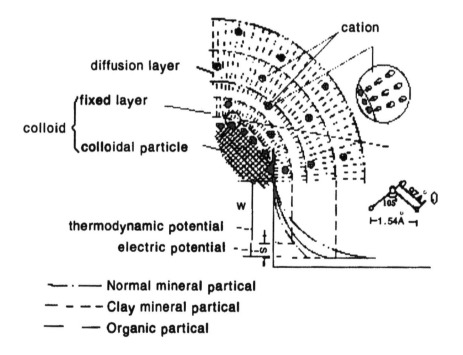

Figure 1. Diffusion double layer sketch of micelle.

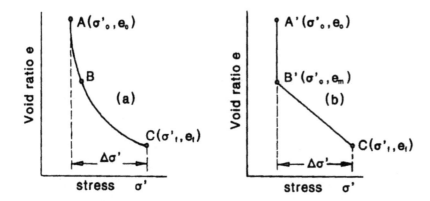

Figure 2. Sketch of stress σ' and void ratio e relation.
(a)Classical σ'-e curve, (b) Simplification of σ'-e curve

SIMPLIFICATION OF THE COMPRESSION CONSOLIDATION MODEL

In Fig. 2a there is a classical compression consolidation σ'-e curve. It is widely used in the project field. As for the soft soil with very high water content and void ratio, if it is acted by a small and constant active vertical force σ'_0, in it the one-dimensional drainage consolidation process can be continued. The soil's void ratio can become smaller and smaller under the constant force until a new balance appears. Therefore, in the consolidation, state of the soil body will change from the initial σ'_0 and e_0 into the σ'_0 and e_m. Based on this, Scott et al. simplified the σ'-e curve in Fig. 2a into the σ'-e broken line in Fig.2b[4]. In Fig. 2b,the initial state A' (σ'_0,e),intermediate state B'$(\sigma'_0,\ e_m)$ and final state C'$(\sigma_f'$, e_f)correspond to the states A, B, C in Fig.2a, respectively. This means that in the soft soil's one-dimensional consolidation, when on it a small additional constant active stress, σ'_0, is added, water will be drained out from the macro-pores with the regular rearrangement and collapse of the soil particles in the soil body. In this paper this is called as the compression settlement of macro-pores. When σ' is increased and then larger than the micro-pore pressure, water in the micro-pores will be drained out until the end of the consolidation process. In this paper it is called as the consolidation settlement of micro-pores.

Obviously, the one-dimensional compression and consolidation settlement of the soft soil is composed of the compression settlement of macro-pores and the consolidation settlement of micro-pores.

In practice project, when riprap of filling is put on the soft soil layer in a beach (this means that a small pressure is put on the layer), it can be found that the filling or stone settles continuously. If an undisturbed soft soil sample is sent to a laboratory, it can be seen, after the sample box is opened, that some water has gone out of the sample and the volume of the sample becomes smaller. These things indicate that drainage and consolidation of the soft soil body can happened with a small additional stress or gravity, making a strength proof of the simplification of the σ'-e curve described above.

COMPRESSION SETTLEMENT OF MACRO-PORES

In order to study the process of the compression settlement of macro-pores, first the compression mechanism of the soft soil should be understood. This means to understand the distribution of the pore pressure and the active stress inside the soil body.

The geotechnical centrifugal machine is an effective equipment to be used to solve the above problems. Several scholars[5] had done much work on this aspect. A soil sample, with 3 soil pressure cells and 3 pore pressure cells inside it and a sand bag packed of geotextile in its center to favour the drainage during the test (see Fig. 3a), can be put into the centrifugal machine to be tested, and some information can be collected. According to the general consideration, when a load is put on the soft soil, the pore pressure may suddenly increase, but the coefficient of the pore pressure can hardly equal to 1 (Fig. 4)[6].

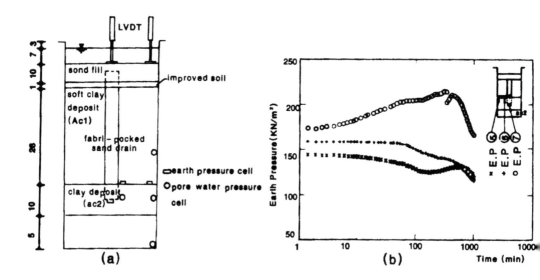

Figure 3. Distribution of earth pressure.
(a)Setup of sample in the centrifuge model, (b)Earth pressure curve with time. In Fig.3(b),earth pressure cell ⑥, show that earth pressure doesn't increase, but ⑦ increases beaus of centrifugal force.

Date got from the centrifugal machine show that when a load is put on the soft organic soil with very high water content, almost all of the additional stress changes into the pore pressure , and the active stress of the soil particles does not increase (Fig. 3b). Therefore, here the coefficient of the pore water pressure can be considered as B=1 (the coefficient of the pore air pressure is very small and can be omitted). In Fig.3b, the soil pressure cell No. 7 shows that the soil pressure increases slightly. This is a result of a force acting directly on the cell, which is a component of the force from the sand bag acting downwards caused by the centrifugal force(this role will be described hereinafter).

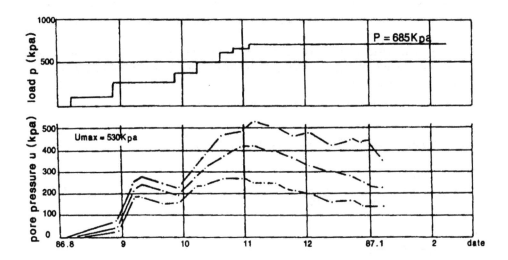

Figure 4. sketch of load and pore water pressure relation

Now, to compute the compression settlement of macro-pores becomes to compute the pore water pressure and the water amount drained out from the soil body.

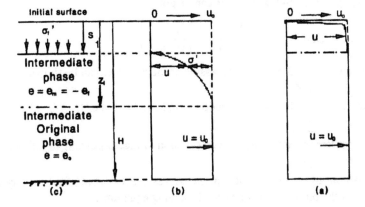

Figure 5.   Pore pressure and soil profile.
(a)Initial state, (b) Pore pressure, (c) Soil profile.

Fig.5 shows the pore pressure distribution in the process of the compression settlement of macro-pores. Fig.5a shows that almost all of the additional stress, σ'_0, change into pore pressure, that is, $u_0=\sigma'_0$. The coefficient of pore pressure is close to 1, and the increment of the active stress of the soil particles is zero. Fig.5b shows that the distribution of pore pressure at time t. In the intermediate phase there are two void ratios, e_m at the bottom and e_f on the top.

If the soft soil layer has a limited thickness of H and under it there is an impervious layer with high strength, under the active stress σ'_0 , the settlement stage can be divided into two parts. One is the compression settlement of macro-pores, and another is the consolidation settlement of micro-pores. The latter appears behind the former and is a small one.

According to the water drained out in the intermediate phase, the developing speed of the phase's limits can be determined. As per the hydraulic theory, the hydraulic gradient in the soil body surface can be shown as:

$$i = \frac{2U_0}{\gamma_w(Z_i - S_1)}$$ (1)

and the developing speed of the limits can be shown as:

$$Z_i\frac{dZ_i}{dt} = \frac{2KU_0}{\gamma_w(1-\Delta V)\Delta V}$$ (2)

where:
t-time;
K-coefficient of permeability;
U_0-pore(water) pressure;
γ_w-unit weight of water;
Δv-water volume drained out from an unit volume of undisturbed soil.

If the stress is quickly loaded and kept it constant, that is, Uo=constant, by integrating from equation (2), the following relations can be got:

$$Z_i = 2Bt^{1/2}, \quad B = \left[\frac{KU_0}{\gamma_w(1-\Delta V)\Delta V}\right]^{1/2}$$ (3)

The settlement caused by the compression of macro-pores can be shown as:

$$S_1 = \Delta VZ_i$$ (4)

therefore, S_1 can be shown as a function of the square root of t:

$$S_1 = 2B\Delta Vt^{1/2}$$ (5)

The final settlement caused by the compression of macro-pores is as follows:

$$S_{1f} = \Delta VH$$ (6)

When the limit of the intermediate phase arrives at the bottom of the soil layer ($Z_i=H$), the thickness of the phase is shown by the following equations:

$$H' = H - \Delta V H, \text{ or, } H' = H(1 - \Delta V) \qquad (7)$$

The total time, t, is shown as follows:

$$t = \frac{H^2}{4B^2} \qquad (8)$$

When t equals to $t_f(t = t_f)$, the whole process of the compression settlement of macro-pores has finished, and following it there will be the consolidation settlement of micro-pores.

CONSOLIDATION SETTLEMENT OF MICRO-PORES

After $Z_i = H$, the consolidation settlement of micro-pores becomes the main movement in the soil body which is in the intermediate phase. This is not much different from the consolidation settlement of the general mineral cohesive soil. But what should pointed out is that when t is smaller than $t_f(t < t_f)$, the consolidation settlement of micro-pores has begun and when t equals to $t_f(t = t_f$, that is ,U changes from $U = Uo$ into $U < Uo$), besides the finish of the whole process of the compression settlement of macro-pores, the finish of the first stage(S_{2m}) of the process of the consolidation settlement of micro-pores has also been obtained and then the degree of consolidation of the soil body is $U = 1/3$. There are a lot of ways in the classical theory to calculate the normal consolidation settlement of soil body, and the author does not describe them here.

COMPARISON OF THE RESULTS

In Fig.6, there are the observation data of a sea dike in Japan and the t-s curve made as per the settlement amount calculated by the equations described above.

The dike is on the marine soft organic soil. The thickness of the soft soil layer is 2.60m. On it there is a filled-earth dike with a height of 4m. The calculation results are t_f=426 days, the amount of the compression settlement of macro-pores S_{1f}=0.692m, the amount of the consolidation settlement of micro-pores S_{2m}=0.231m, and S=0.923m. The observation settlement amount is S'=0.926m, being very close to the calculated one, S.

In Table 2, there are the calculated and observed results of settlement amount of several sea dikes of filled-earth bodies on sea marine soft organic soil[7]. The relative settlement values are between 31.13-56.29%. It can be seen that the calculated results are rather close to their relative observation data.

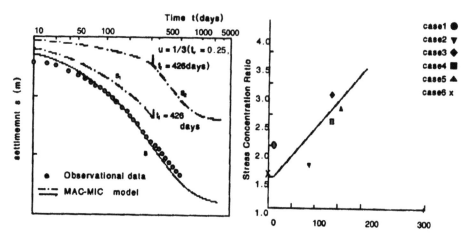

Figure 6. Settlement s-t curve   **Figure 7. Materials strength and stress**

Table 2. Some settlement results of calculation and observation data

Site	calc. value (mm)	obse. data (mm)	height of construction (mm)	relative settlement (%)	note
Coast, Japan	1315	1250	4000	31.13	
Zhuhai A, China	982	931	2650	35.09	
Zhuhai B, China		1790	3180	56.29	some lateral strain, and add settlement
Taniwabara, Japan		3484	8000	3.55	obse. 10 years and 5 months
Zhuhai C, China		2590	5775	44.85	some lateral strain

Note: 1.Settlement response to degree of consolidation U<=0.9.
           2.Relative settlement=settlement value/height of construction

THE SOFT SOIL IMPROVEMENT

The drainage consolidation is always an important mean for soft soil improvement, and there are many different methods to do it[8]. The sand pile or crushed-stone pile is one of them. But in the past time people paid little attention to the rate of stress concentration. The theory of the mechanics of materials shows that in non-homogeneous soil there may be the concentration of stress in some place. a soft ground improved with sand pile is in fact a large non-homogeneous soil body, and there can be the concentration of stress in it. The rate of stress concentration, n, is the ratio of the active vertical stress acting on the sand pile or on the crushed-stone pile to the mean active vertical stress acting on the soil body. Generally, the rate of stress concentration of the sand pile used to improve the soft ground is about 1.5. If the geotextile bags are used in the sand pile, the rate may increases with the increase of the materials strength of the geotextile(Fig. 7).

The sand pile or crushed-stone pile is a fine drainage path, and also they can bear the main of the structure loads on the ground. Therefore, it is an effective way for the soft soil improvement.

The lime or cement powder-sprayed pile is used in our country in recent years. It is also one of the effective ways for quick improvement of soft ground. listed in Table 3 are the physical features and the bearing capacities of a large power plant's soft ground in Guangdong province, obtained both in the natural state and in the state after the improvement with lime powder-sprayed pile. In fact, the proper ways of soil improvement and construction should be chosen according to the project conditions.

Table 3. Indexes comparison of nature soft soil and the soil between lime powder-sprayed piles

kind of soil	water content W (%)	void ratio e	liquidity index I_L	cohesion c (kPa)	internal friction angle(°)	safe bearing capacity (kPa)
nature soft soil	81.1	2,122	2.935	0.35	5.0	12
soil between piles	71.4	1.900	2.890	2.78	5.3	90*

Note: * safe bearing capacity mean the safe bearing capacity of composite ground resulted from large scale static loading test.

CONCLUSIONS

In this paper, the compression curve of the soft organic soil with high water content is simplified. And the phase-change model in the compression process of the soft soil is used. The compression settlement of macro-pores and the consolidation settlement of micro-pores are used to explain the compression and consolidation process of the soft soil. Also, the test results of the centrifugal machine is used to test and verify the mechanism of the compression settlement of macro-pores of the soft soil. It is pointed out in this paper that the compression settlement of soft soil is composed of the compression settlement of macro-pores and the consolidation settlement of micro-pores. This has been proved by the measured data. And it is indicated that, based on the consideration of the drainage path and the rate of stress concentration, the sand pile or crushed-stone pile is one of the effective ways of the treatment of soft ground. The lime powder-sprayed pile is introduced and the feasibility of the use of the sprayed pile in the soft soil improvement is also verified by using the test data and the static load test data.

Acknowledgments

The author likes to express his sincere appreciation to several experts, who are working in the rock and earth mechanics field, for their kindly advice and help in his writing the paper.

REFERENCES

[1] T. William lambe: Soil mechanics, SI version", Massachusetts Institute of Technology, 1979.

[2] Luo Guoyu, Li ShengLin: Engineering geology   ,Geology Press, Oct.1982.

[3] K. koyure and M. aoyama: Approximative prediction of the settlement of peat deposit", Proceedings of international conference on soft soil engineering, Science press. 1993.

[4] Scott, R. F.   Consolidation of sensitive clay as phase change process", Journal of Geotechnical Engineering, ASCE, 115, No.10, 1989.

[5] M.kitazume et.al: Centrifuge model tests on the consolidation behavior of soft clay improved by fabric-packed sand drain, Proceeding of the international conference on soft soil engineering, Science Press. 1993.

[6] Feng Guangyu: Geotechnical centrifugal model test on soft ground of Dai-huang exp. way", Journal of Ya ngtze-River Science Academy Vol.4 No.2.

[7]
                              No.468,1993-6.

[8] Lin Zongyuan: Geotechnical processing manual, Liao-ning Science & Technology Press. 1993.

Proc. 30th Int'l. Geol. Congr., Vol. 23, pp. 295-301
Wang Sijing and P. Marinos (Eds)
© VSP 1997

Engineering Geological Characteristics and Evaluation of the Soft Soil in the Lower Yellow River

WANG XIYAN, MA GUOYAN, & LI HONGXUN
Department of Geology, Reconnaissance, Planning, Design and Research Institute of Yellow River Conservancy Commission, No. 109 Jinshui Road, Zhengzhou, China 450003

Abstract

A large quantity of analyses and studies are made regarding the distribution, engineering geological characteristics, liquefaction, seepage deformation, settlement deformation and sliding deformation of the soft soil in the Yellow River lower reaches from the point of view of theory combined with practice. So far the most comprehensive engineering geological characteristics and evaluation of the soft soil in the foundation of dikes along the lower Yellow River are obtained.

Keyword: distribution pattern, engineering geological characteristics, evaluation, lower reaches of the Yellow River

INTRODUCTION

The Yellow River, flowing towards east through a canyon in Mengjin county of Henan Province, runs into its downstream course. After crossing Zhengzhou in Henan and Jinan in Shangdong, it empties itself into the Bohai Sea in Kenli county of Shandong province (Fig.1). The total length of the lower river course is about 800 km.

Due to sedimentation year by year, the river bed raised 3--5 m, even 10 m at maximum, above ground. So the Yellow River becomes a "suspended river" and the water divide between Hai River and Huai River.

Loose sand and sludgy soft soil are widely dispersed in the lower reaches plain. The followings are introduction on distribution, characteristics of soft soil and problems encountered in engineering practice. '

DISTRIBUTION OF SOFT SOIL

The lower Yellow River plain, based on the genesis of the deposits, can be classified into 3 zones, i.e. alluvial, alluvial-lacustrine and alluvial-marine plains as shown in Figure 2.

The suspending river topography of the alluvial plain can be roughly classified into 3 units, namely meandering unit, natural levee and flood depression, as shown on Figure 3. The meandering unit consists mainly of medium sand and silty fine sand. The natural levee varies from silty fine sand, sandy loam to loam in lithology. The flood depression consists of clay and loam.

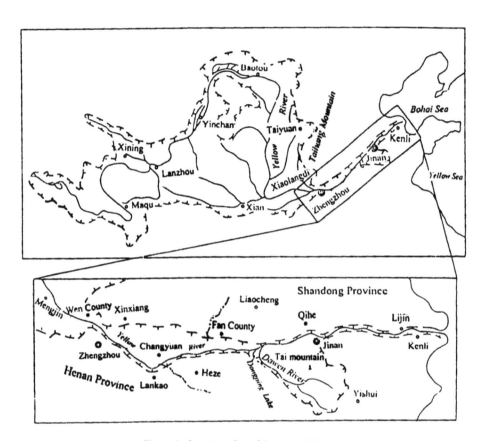

Figure 1   Location plan of the lower Yellow River

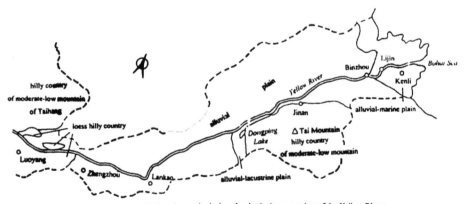

Figure 2 Engineering geological zoning in the lower reaches of the Yellow River

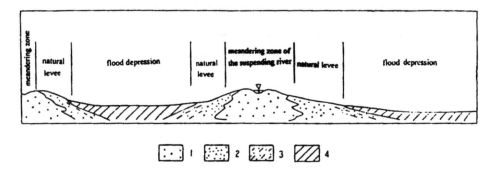

Figure 3 Diagrammatic section of the suspending river landform

1.fine and medium sand; 2.silty fine sand; 3.sandy loam 4.clay and loam

Because of frequent migrating, dyke-breaching, flooding and wandering of the Yellow River, different topography and lithology are superimposed, so the lithology exhibits very complicated changes. Therefore, the foundation here is mostly in double-layer or multiple-layer. Thickess of the sandy soil in the meandering unit relates to duration of the Yellow River flowing. Clayey soil in bankside depression, or sludgy soil in water-logged swamp are all soft soil. Recharged by lateral seepage of the Yellow River, the soft soil with high water content is distributed along the Yellow River.

5 - 8 m below the ground in the Yellow River mouth delta is marine sludgy soft soil. 2 - 6 m thick lacustrine sludgy soft soil adjacent to the Dongping Lake at the west foot of Taishan Mountain is buried at a depth of 2 - 4 m, with its bedding plane slightly dipping to the lake centre. Furthermore, in the inrush mouth of the Yellow River dike, rotten stalks or sludgy soft soil are found and scouring depth of the breaching reaches 36 m below the ground.

ENGINEERING GEOLOGICAL CHARACTERISTICS OF THE SOFT SOIL

The grain size of sandy soil becomes finer and lithology becomes silty fine sand, sand and sandy loam gradually from west to east, and grain size becomes coarser from surface to depth, reflecting the Yellow River sorting along its course and laws of both gentle river course gradient and finer grain size of the deposits due to sedimentation.

Because of the high silt content, the permeability coefficient of the Yellow River alluvial sand layer is generally less than that of Wenhe River developed at the western foot of Taishan Mountain (Table 1).

The Yellow River alluvium is characterized in engineering geology by greatly changed facies, antisotropy, developed micro bedding plane, and nonhomogeneity. Because of great changes of the Yellow River in flow quantity and velocity, grain size of the deposits varies considerably, different alluviums are superimposed and microbedding is developed, which directly affects the permeability and mechanical properties of the alluvium.

Table 1. Comparison of permeability coefficient in different river alluvium

Alluvium in	Yellow River	Wenhe River
Lithology	Permeability coefficient (m/d)	
Silt	0.5 - 2	
Fine sand	3 - 4	4.5 - 20
Medium sand	16 - 18	20 - 40
Coarse sand	30 - 35	50 - 150

Silt content of the clayey soil is generally in the range between 58% and 69%, which has direct impact on property of the soil. When soaked up, the strength of dry sandy loam is rapidly reduced, and the bearing capacity is only 70 - 90 kpa; when dry, its bearing capacity is as high as 110 - 130 kpa.

There are some sludgy soil balls in alluvium, which greatly affects the shear strength. Some data indicated that 10% increase of sludgy soil balls in fine sand would lower 4 to 6 degrees of inner friction angle of the fine sand.

Marine, lacustrine and swamp sludge or sludgy soil are the weakest soil due to their low shear strength and high compressibility.

ENGINEERING GEOLOGICAL EVALUATION OF THE SOFT SOIL

The main engineering geological problems in dyke building, hydraulic structure and bridge construction in the lower reaches are: liquefaction, seepage deformation, settlement and sliding.

Liquefaction of foundation
Saturated loose silty fine sand, sandy loam and light-medium loam extensively distributed in the lower Yellow River are prone to liquefaction. When strong earthquake occurs, increased pore water pressure in the soil will lead to water spraying and sand boiling, thus shear strength and bearing capacity is reduced or even faded away, or shifting deformation occurs.

Various methods are used for liquefaction analysis including comprehensive geology analysis, S.P.T test, shearing wave velocity method, H.B. Seed, standard blasting test and dynamic triaxial test, as well as finite element analysis on the critical dikes. Shallow and loose sand down to depth of 10 - 16 m may liquefy, and the dyke toe may liquefy locally under the impact of earthquake intensity VII. Overall stability of the dyke will be affected by VIII intensity earthquake.

Investigation to historical earthquake damage has drawn the same conclusion as mentioned above. For example, Heza Earthquake of M=6.9 in 1937 spreading to the Yellow River dyke between Dongming county and Zhencheng is VII in intensity, causing serious sand boiling and water spraying in the ground, and a 100 m long, 30 cm wide longitudinal fissure in the dyke toe. The seismic intensity is VII east of Lijin, when Bohai Earthquake took place in 1969, resulting in many fissures in the dyke, sand boiling and ground subsidence, new dyke avalanching in the river mouth area, but the old dyke is not damaged at all.

The most serious liquefaction areas lie in: bank beach in the river mouth, top of the delta, fossil channel and dyke breached mouth. This is because that extensively distributed loose silt or fine sand or sludgy soft soil in the areas, especially newly accumulated loose and soft silty fine sand, silty soil or sandy loam in the delta close to the sea, and marine sludgy soft soil in the depth, are prone to liquefaction by earthquake vibration, and damage resulted from earthquake is often serious.

Seepage stability of foundation
● Seepage deformation
Dyke (dam) embankment in the foundation consisting of either loose sandy earth or double-layer of sand and clayey soil will result in great water head difference in two sides of the dyke (dam), and increase of seepage pressure. If seepage gradient exceed impermeability gradient of soil, the composition and structure of the soil in the dyke (dam) may result in change or damage, namely seepage deformation or damage.

The water head difference between upstream and downstream of the Dongping Lake is 4.5 m, creating a great quantity of piping and sand boiling in the box dam toe and earth borrow pit.

Since 1949, seepage deformation has occurred in 114 localities at the toe of the dyke back. Dyke breaching occurred many times due to seepage deformation.

● Characteristics of seepage deformation
Silty fine sand in the lower reaches of the Yellow River is uniform in grain size, non-uniform coefficient is generally less than 5. Therefore, seepage deformation principally takes form of shifting. When the shifting occurs in a sudden, all the soil grain moves simultaneously, which is more dangerous than piping of corse sand.

Seepage deformation can be categorized into 4 types, i.e. piping, sand boiling, surface earth building and frost boiling.

● Distribution law of seepage deformation
Seepage deformation takes place mostly in double-layer soil structure of thin clay at top (usually less than 2 to 3 m thick) and sand layer below, or in the formation with the surface layer being silty fine sand or light sandy loam.

Seepage deformation is mostly distributed in seepage concentrated convex bank of the Yellow River or the low land close to the area where tributaries are merged with the Yellow River.

Seepage deformation often occurs in water gully, paddy field, pond and earth borrow area, generally 10 - 20 m from the dyke back toe. The farthest piping was in Kangtun, Shandong Province, which is 190 m from the dyke toe, where the piping with a out fall of 25 cm in diameter took place 4 times in the flood season of 1982.

Serious seepage or seepage deformation usually occurs in the dyke where fossil channel or dyke breached mouth are located. Based on statistics, 37% of the seepage deformation occurred in dyke-breached mouth.

Of the 4 types of seepage deformation, piping and sand boiling are predominate.

- Antiseepage gradient of soft soil

Antiseepage gradient of the soil is determined by comprehensive analyses based on laboratory test, formula calculation, empirical value and analogy methods. Measured gradients for various soil in the completed projects where seepage deformation occurred in the lower reaches are: 0.62 - 1.03 for loam, 0.73 - 0.75 for sandy loam, 0.5 - 0.6 for silt, and permissible seepage gradients are 0.3 - 0.6, 0.35 - 0.52 and 0.25 - 0.34 respectively.

Settlement and sliding deformation of the foundation

Roughly, 3 forms of settlement and sliding deformation of the foundation can be categorized based on the engineering geological characteristics and genesis of the deformation.

- Settlement of sludgy soft soil

Lacustrine, marine and swamply sludge or sludgy soft soil, mostly in soft plastic or flowing plastic state, are unfavourable soil as foundation of structure, and are prone to uneven settlement and sliding because of low shear strength and high compressibility. When a water gate founded in such soft soil, concrete piles or replacement of the soil are employed to improve the foundation.

- Uneven settlement of the foundation of new and old alluvium

The Yellow River dike has been built over 130 years and the foundation has been consolidated and compacted. When the old dyke is heightened and widened, the new foundation is usually in the bank beach consisting of newly accumulated loose and highly
compressible sludge. Therefore, uneven settlement often occurs along the contact between the old and newly raised dike, and then longitudinal fissures appear along the dike.

- Avalanche in the complicated soil foundation of the old dyke breached mouth

Rotten stalks and sludge in dyke breached mouth are prone to uneven settlement and sliding. For example, a 81.6 m long sliding occurred in the dyke breached mouth in Wangjialihang in Licheng county of Shandong province, where the dyke crest moved outward about 1 to 3 m, avalanching height is 6.6 m at maximum, and the centre of the sliding is just located in the dyke breached closure, where the soil is black plastic sandy loam and loam sandwiched by rotten stalks or soft plastic clay.

Proc. 30th Int'l. Geol. Congr., Vol. 23, pp. 303-313
Wang Sijing and P. Marinos (Eds)
© VSP 1997

Distinctive Particle Types in Ground Materials: Implications for Engineering Geology

IAN FRANKLIN JEFFERSON & IAN JAMES SMALLEY
Collapsing Soils Research Group, Department of Civil and Structural Engineering, The Nottingham Trent University, Nottingham, NG1 4BU, U.K.

Abstract

Soils are particulate materials. Their properties are determined by the nature of particle interactions. These interactions may be affected or modified by the water in the system, however, it is the particle matrix that controls the overall material behaviour, Surprisingly, little effort has been made to examine and define these particles. As an initial basis for the study of engineering soils, five basic particle types and two interaction models are proposed. These are: type A, the active clays, type B, the inactive clays, type C, fine primary mineral particles, type D, silt and type E, sand; with either long range or short range contact bonds acting between particles. Much of soil behaviour then becomes comprehensible when the basic materials types are examined. For example, the complex collapsing behaviour of post glacial quickclays, previously thought to primarily consist of primary clay, could be explained.

Keywords: collapse, bonding, quickclay, loess, particle nature, soil classification.

INTRODUCTION

Some large scale engineering geology failures have occurred as a result of a lack of appreciation of the nature and properties of ground materials used in construction. Unfortunately, traditional classifications are crudely based on nominal particle diameter, placing soils into almost artificially assigned size fractions. Good examples of poor material appreciation leading to engineering failure include the Susuma and Teton Dam collapses [21, 18]

More attention therefore, needs to be paid to the nature and properties of definable ground materials. These definitions should be considered from a material perspective, so improving judgements made on the engineering behaviour of soils. Such an approach would have to account for the geological controls experienced by the different soil types, and so utilise key aspects of fundamental science. This is a stand point keenly advocated by Mitchell [12]. The importance of this view point has been further highlighted by the significant increase in the discussion of soil fundamental behaviour given in the third edition of the famous monograph 'Soil Mechanics in Engineering Practice' [21].

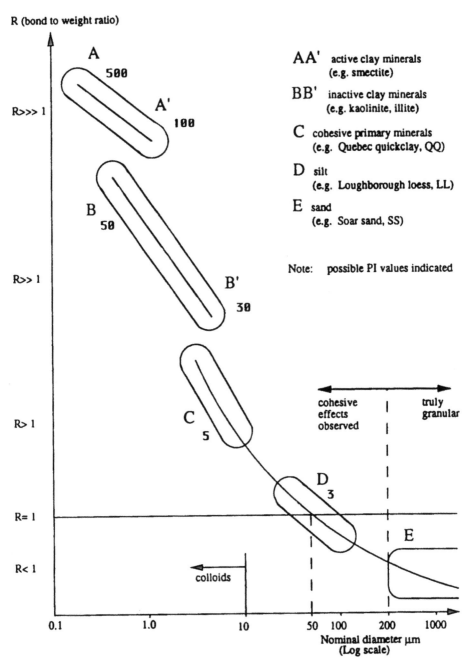

Figure 1. The R-size diagram [6], where R is the bond to weight ratio plotted against nominal particle size (μm) on a log scale. This shows the five particle types arranged according to the relative significance of their surface bond activity. The greater R, the greater the significance of surface forces on the overall behaviour of soil materials.

This paper by adopting this philosophy, aims to propose and then clearly set out five basic particle types. This will allow a more scientific approach to soil classification to be adopted, by overcoming the disadvantages of the more traditional 'black box' approach typically used by soil engineers. The proposals examined in this paper are extensions of the ideas presented and discussed at the 1994 N.A.T.O. Collapsing Soils conference [4]. There is unavoidably an element of oversimplification in this approach, but it is felt that this is necessary when dealing with such a complex material as soil undoubtedly is.

A useful differentiation of soil materials can be introduced via the 'R-size' diagram (see Fig. 1), where R is the inter-particle bond to weight ratio [6]. For sands R is low, i.e., weight forces are appreciable compared to the surface bond force. However, R is very high for a smectite clay where the bond forces are considerably greater than the corresponding weight force. At R equals 1 is the point at which the cohesive/ cohesionless transition occurs, a point fundamental to collapsible soils (see Jefferson and Smalley [6] for further details).

To further aid this method of particle differentiation, two simple dichotomous bond models will be used: long and short range bonds. This is essentially the approached pioneered by Cabrera and Smalley [2] in their Quaternary and Geomorphological studies of very sensitive soils. The long range bond is in essence a 'clayey' bond exhibited by high plasticity (as measured by the Plasticity Index, PI) soil systems. By contrast the short range bond is a contact bond resulting from the direct interaction between two grains. The latter systems have low to zero PI's and no strength when disturbed.

This paper's ultimate goal is to provoke discussion on these important issues, particularly now, with the growing importance of environmental considerations in engineering geology. This necessitates consideration of both mechanical and physico-chemical aspects of soil behaviour.

FIVE PARTICLE TYPES

Jefferson and Smalley [6] attempted to define six particle types, but this may well be too complex an approach, and five seems to be a more satisfactory number. These can be clearly seen in the R-size diagram presented by Jefferson and Smalley [6], see Fig. 1. It is proposed in this paper that these five particle types are further defined, such that they incorporate an understanding of the geological controls that operate on their formation (see Table 1). These are not a natural supply of a whole range of particle sizes as some classification systems seem to suggest. However, the five particle types defined in Table 1 take account of the modal sizes and distinctive particle types that occur naturally. It is therefore, necessary to appreciate these geological factors when the engineering behaviour of different soils are being discussed.

Table 1. The five particle types encountered in engineering soils.

Name, Symbol & R-state	Formation Controls	Shapes & Sizes	Interaction	Engineering Problems	References
A-A' Active Clay (Smectite) R>>1	Intense weathering of (feldspars) volcanic ash	Very small platy particles (> 10 A × up to 10 µm) Water envelope nm thick	Classic long range bonding; high PI (PI > 200)	Expansive Soils	Jefferson & Smalley [5]; Mitchell [12]
B-B' Inactive Clay (Kaolinite, illite, etc.) R>1	Weathering of feldspars	Small usually platy particles (0.003 × up to 10 µm) Water envelope nm thick	Mixture of long & short range bonding; low PI (PI ~ 30 - 50)		Smalley et al. [20]: Mitchell [12]
C Small Inactive Primary Minerals R>1	Crushing of larger particles; fragment formation, glacial grinding, impact chipping	Fine silt platy particles (2 - 5 µm)	Short range bonding; system exhibits cohesive effects	Very sensitive clay flowslides	Cabrera & Smalley [2]
D Silt (medium or coarse silt) R = 1	Crushing of larger particles; fragment formation. Size distribution is defect controlled.	Flat particles in 8:5:2 proportions, usually quartz (20 - 60 µm)	Essentially short range bonding; can be modified by clay at bond points	Hydroconsolidation subsidence and collapse	Rogers et al. [14]
E Sand (quartz sand) R<1	Eutectic reaction in source granite. Defects in quartz crystals.	Angular to round, mode size 200 to 600 µm. (Initial mode ~ 600 µm)	Short range bonding	Liquefaction of saturated sands	Krinsley & Smalley [9]

Note: PI = Plasticity Index as determined by the Atterberg Limits [22].

Type A: The Active Clays
This group represents material typically in the sub-micron size range and corresponds to a high R. The term 'active' is similar to the term 'activity' introduced into geotechnical nomenclature by Skempton [15], and is an effort to come to terms with the wide range of properties associated with materials called 'clay' by engineers. Skempton's actual parameter indicated that the same amount of 'clayeyness' could be produced in a soil by varying amounts of different clay minerals, e.g. a small amount of smectite compared to a large amount of kaolinite. This term indicates the highly significant nature of physico-chemical or surface forces on the behaviour of these clay soils. The active clays are the smectites, e.g. montmorillonite and its relations. These have high PI's and epitomises the concept of long range bonding, discussed below. Particle types within this group have marked double layers, allowing cohesive plastic soil systems to develop, hence strength loss on disturbance is minimal (see Fig. 2). For soils within this group inter-particle contacts are likely to occur through their respective double layers.

Type B: The Inactive Clays
This group represents clay materials whose inter-particle interactions are less 'active' and exhibit a more significant weight force component, i.e. contacts are more solid-to-solid in their nature. It is felt that it is essential to distinguish materials from group A and B, due to their intrinsic differences. In essence group B materials are significantly larger and less 'clayey' than group A materials. Making this separation allows clay materials to be classified taking into account their respective activities. Clays within this group include kaolinite and illites, although this is largely a northern hemisphere perspective. For example minerals like halloysite, allophane and imogolite would have be added to this group, to make it truly global. Halloysite is an important addition to this group, due to its distinctive particle shape and remarkable inactive aspects. A coastal flowslide at Tauranga in New Zealand was caused in part by the predominance of spheroidal halloysite in the soil system [20]. Other examples of the influence of halloysite include the Susuma Dam failure [21].

Overall, it may be that a three micron kaolinite particle has more in common with a three micron quartz fragment, than with a montmorillonite particle.

Type C: Fine Inactive Primary Minerals
In many traditional soil classification systems it is difficult to distinguish the particles that are truly fine silt and not just silt sized clay minerals. This is an important size range (2 - 5 μm) and needs to be carefully separated from not only the fine silt sized clay minerals, but also from the modal silts at 20 - 60 μm (an order of magnitude larger). The main reason why the group C particles need to be defined is because they comprise the quickclays: the extremely sensitive post glacial clay soils that are prone to flowslide failure [11, 19]. Cabrera and Smalley [2] produced a model of quickclays, that required them to consist of type C particles with short range inter-particle bonds. This seriously challenged the idea that quickclays were clays. Cabrera and Smalley [2] examined the geology, distribution,

mineralogy and bonding to produce a seminal view of quickclay nature that is now widely accepted. This depends on the recognition of the key role played by type C particles.

Type D: Silt

Materials in this group lie either side of the line R equals one, which corresponds to the divide between cohesive effects and cohesionless effects dominating [6]. Group D materials typically have a 20 to 60 μm mode size, and correspond to the silt of loess and alluvium deposits. This group is the basic structural material for metastable and collapsible soils. These soils are particularly problematic because they are neither fully cohesive nor fully cohesionless. It is with the silty soils that the new materials understanding will be of most benefit. A classic example where this approach would have been of benefit is the collapse of the Teton Dam [13, 18]. This appears to have been largely due to a failure to appreciate the true nature of the Idaho loess, which was unfortunately used to construct the large core. The loess material used was found to have a PI of 3. It is therefore, a classic short range bonded silt, and was used where a long range bonded clay would have been better suited [13].

Type E: Sand

The nature and formation of quartz sand is not well understood [17, 9]. Sand is made inside granite rocks and is the result of eutectic type reactions, whose product is a fine mixture of feldspar and quartz. The quartz develops tensile stresses as it goes through the high to low transition. These geochemical factors continue to shape the sand grains: it is not produced by the breaking down of large materials. Materials in this group have R values less than 1, although two sub-groups may exist: bond affected and truly cohesionless sand (see Jefferson and Smalley [6] for further details).

THE TWO BOND TYPES

Long Range Bonding

The most remarkable property of a clay mineral soil is its plasticity. A clay soil can be deformed and still maintain its strength. An ideally plastic soil could be deformed indefinitely without loss of its mechanical properties. This plasticity is possible because the particles are held together by long range bonds; there is an overall 'cohesiveness' in the system. The negatively charged clay mineral particles are kept together by being in an environment that is rich in positively charged solute ions (see Fig. 2). Cabrera and Smalley [2] considered this was analogous to the Drude-Lorenz model of metallic bonding. In the ideal metal, the positive metal ions sit in an environment of negative electrons, giving the metal an overall cohesive force and hence its strength. This bond nature also allows metals to have plasticity. The classic example material that exhibits these properties is copper, which can deform in a very remarkable way. This is very similar to the behaviour of sodium montmorillonite, which due to its intrinsic long range bonding exhibits a high cohesive ability and a high plasticity. If the soil system is plastic enough, i.e., has enough internal long range bonding, it will not suffer brittle failures and thus will tend not to be metastable.

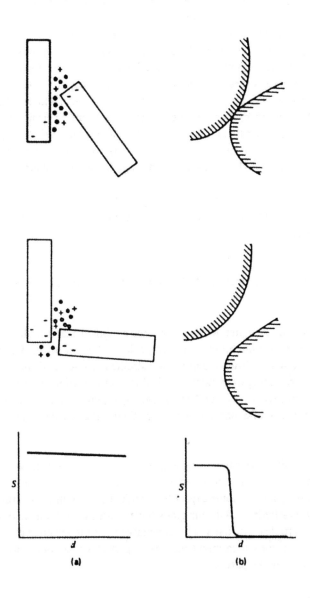

2. Long range and short range bonding [3, 2], showing how strength (s) is related to displacement (d). ng range bond can be deformed without loss of strength, i.e. active bond (a); the short bond loses strength rbance, i.e. inactive bond (b).

Short Range Bonding
The basic alternative to a long range bond is the short range bond, in which there is no 'action-at-a-distance'. This is a brittle bond, essentially a contact bond, that loses its strength when disturbed. This is the bond to be expected between uncharged particles, and unlike the long range bonds, it is a short directed particle to particle bond (see Fig. 2). Invoking the short range bond allows a form of explanation to be advanced for quickclay behaviour. If the quickclays, despite their name, consist of type C primary mineral particle, the inter-particle bonding will be of the short range variety. This allows the open single particle structure to disintegrate when disturbed and accounts for the catastrophic loss of strength that characterises these materials. This strength loss is measured as sensitivity; a very sensitive clay, such as the material involved in the St. Jean Vianney landslide of 1971 [19]. To establish that short range bonds predominate in the St Jean Vianney material, it was necessary to show that it contained little clay mineral material, and ideally no type A material. Termogravimetric methods indicated a type B content of about 9% and a high primary mineral content [19], suggesting short range bonds control the material properties.

This two bond type model is probably too abrupt and extreme to explain all soil behaviour, but serves to provide an initial bases, form which various soil types can be defined using a material view point.

IMPLICATIONS FOR ENGINEERING GEOLOGY: PROBLEM SOILS

Expansive Clays
These are classic type A materials, that have very high PI's and high smectite contents. They present problems from an engineering geological stand point, due directly to their high type A soil contents. They are prone to extremes of swell and shrinkage as moisture content changes, presenting tremendous difficulties to foundation engineers [22]. Actually type A clays do not give a very good analytical signal, but methods are available that can show the predominance of expansive clays, e.g., thermogravimetric analysis [5].

Sensitive Clays
Terzaghi et al. [22] listed four types of metastable soil: (1) extrasensitive clays, quickclays, (2) loose saturated sands susceptible to liquefaction, (3) unsaturated primarily granular soils, typically loess, and (4) some saprolites, residual soils with open structures that have been formed by weathering. All four types have short range bonds and are made up of primary mineral particles, the first and most remarkable is the sensitive clay group [10, 11]. A working definition for which might be, type C particles in an open structure with short range bonding between the particles.

Collapsing Loess
This is another metastable system, probably the classic collapsing soil [14]. It can be considered classic because it is the most widespread collapsing deposit in the world, causing the most persistent geotechnical problems. Essentially a type D material in an open

structure with modified short range bonding. The inter-particle bonds are subtle and complex, forming a structure made up of primary mineral particles in which the bond points are modified by clay minerals. The inactive bond contacts are modified by small amounts of type B materials. Wetting the type B material allows the main contact to be mobilised, resulting subsequently in collapse [14].

Residual Soils

Metastbility is induced in a residual soil by weathering. In the case of a granite, the feldspars weather away and the eutectic quartz is left forming a very open network. It is a brittle system prone to collapse, but due to its special nature cannot be simply discussed in five particle terms. These represent fringe materials, and as such will not be discussed further.

DISCUSSION

Obviously the true clay minerals, the type A and B materials, have charged particles. This yields their special relationship with water, which results in the long range bonding (see Fig. 2). This produces the strong cohesiveness and plasticity associated with these materials. The separation of the two groups is necessary due to their distinctive intrinsic differences. The engineering behaviour of group A materials is physico-chemically dominated compared to the more mechanical nature of group B material's behaviour. Although, in both cases bond forces are still very much greater than the corresponding weight force (see Table 1). In essence the difference is their activity, as discussed above (see the Five Particle Types Section).

Quartz sand, the type E material is a blocky primary mineral particle and as a result has no significant relationship with water, with short range bonds dominating. The difference in particle nature and system properties from type A and B to type E materials is great; greater than tends to be truly appreciated. Using the material approach mooted in this paper, these differences become considerably more obvious. Hence, making material behaviour that much easier to appreciate. In the transition zones, C and D, problems can arise and it is with these materials that mistakes can be made due the misinterpretation of the soil material's intrinsic behaviour. These are important groups due to their metastability and warrant further discussion.

Type C materials (2 to 5 μm, see Table 1) correspond to a size where primary mineral particles exhibit surface activity, e.g. follow a Schulze-Hardy sequence [8]. Clearly, then these materials will show a certain amount of cohesiveness and hence is why R is greater than 1. However, due to the lack of long range bonding, resulting from a lack of strong double layer associations [1], the surface bond force is less than that of the type B and in turn type A materials.

The important zone, where R equals 1, corresponds to the type D materials. These materials are the most important group of metastable soils (see above). It has been estimated that R equals 1 occurs at 50 µm, strong evidence for which has been provided experimentally [7, 16]. This was further indicated by work presented by Kitchener [8]. Smalley [16] and, Jones and Pilpel [7] examined the flow characteristics of dry, clean particulate materials flowing through a range of different sized orifices. Their results showed that as the particle size reduces, so initially the flow rate increased, followed by a gradual reduction with further reductions in particle sizes. Eventually, both researcher found that the flow ceased at 50 µm. This was attributed to surface forces, which imparted a cohesive force equal to that of the weight of the particles. Moreover, Kitchener [8] cited work that showed that quartz systems up to 30 µm exhibited electrolytic effects. However, beyond 50 µm no such effect was observed. This strongly supports the suggestion that R equal 1 at 50 µm. This could go some way to explain the unique behaviour of soil in this size area; the type D materials which form the metastable collapsible structure that cause engineering problems throughout the world [14].

CONCLUSIONS

In order to avoid potential misunderstandings, that in the past have been associated with soils, a more materially based approach, utilising fundamental science should be used. This approach will greatly help with the appreciation of the true behaviour of complex materials such as soils. Using this philosophy, this paper has set out five basic soil types. These are: active clays, inactive clays, fine inactive primary minerals, silt and sand. These have been distinguished according to the relative significance of surface (bond) to weight forces, and geological controls acting on the particles themselves. This allows the 'R-state' (bond to weight ratio) to be defined, thus allowing the relative importance of 'activity', i.e., physico-chemical aspects of soil behaviour, to be used. These concepts are clarified using two simple dichotomous bond models: long and short range bonds. In this model, the long range soil bonds are analogous to metallic bonds typically exhibited by metals such as copper. It is these bonds that yield the 'cohesiveness' and high plasticity observed by active clay soils. By comparison short range bonds are brittle contact bonds typical of primary minerals such as quartz. Overall, by using these five basic types it is possible to truly appreciate the nature of soil behaviour. From definitions similar to those proposed in this paper, it has been possible to better explain the behaviour of quickclays.

Overall, although the proposals discussed in this paper are an oversimplification of soil behaviour, it is deemed necessary in order to elucidate this complex problem. It is hoped that the proposals discussed in this paper will help to focus attention toward a more materially based approach to solving engineering problems.

REFERENCES

1.   R.H. Bennett and M.H. Hulbert. *Clay Microstructure*, D.Reidel Publishing Co., Holland (1986).
2.   J.G. Cabrera and I.J. Smalley. Quickclays as products of glacial action: a new approach to their nature, geology, distribution and geotechnical properties. *Eng. Geol.* 7, 115-133 (1973).

3.   J.E. Costa and V.R. Baker. *Surficial geology building with the earth*, John Wiley & Sons, Inc., New York (1981).
4.   E. Derbyshire, T.A. Dijkstra and I.J. Smalley (Eds). *Genesis and Properties of Collapsible Soils*. NATO ASI Series C:Mathematical and Physical Sciences, **468**, (1995).
5.   I. Jefferson and I.J. Smalley. Classification of expansive soils in arid regions: thermogravimetric investigation of smectite content. In: *Engineering Characteristics of Arid Soils*. P.G. Fookes and R.H.G. Parry (Eds). pp. 95-98. A.A. Balkema, Rotterdam (1994).
6.   I. Jefferson and I.J. Smalley. Six definable particle types in engineering soils and their participation in collapse events: proposals and discussions. In: *Genesis and Properties of Collapsible Soils*. pp. 16-31. NATO ASI Series C:Mathematical and Physical Sciences, **468**, (1995).
7.   T.M. Jones and N. Pilpel. The flow properties of granular magnesia. *J. Pharm. Pharmac.* **18**, 81-93 (1966).
8.   J.A. Kitchener. Flocculation in mineral processing. In: *The Scientific Basis of Flocculation*. K.J. Ives (Eds). pp. 283-328. Sijthoff and Noordhoff, The Netherlands (1978).
9.   D.H. Krinsley and I.J. Smalley. Sand. *American Scientist*, **60**, 286-291 (1972)
10.   J. Locat. On the development of microstructure in collapsible soils. In: *Genesis and Properties of Collapsible Soils*. pp. 93-128. NATO ASI Series C:Mathematical and Physical Sciences, **468**, (1995).
11.   N.H. Maerz and I.J. Smalley. *The nature and properties of very sensitive clays: a descriptive bibliography*. University of Waterloo Library, Waterloo, Canada (1985).
12.   J.K. Mitchell. *Fundamentals of soil behavior*, 2nd Ed. John Wiley & Sons, Inc., New York (1993).
13.   C.D.F. Rogers, T.A. Dijkstra and I.J. Smalley. Discussion: Human factors in Civil and Geotechnical engineering failures, by G.F. Sowers. *J. Geotech. Eng., ASCE*, **120** (8), 1446-1449 (1994a).
14.   C.D.F. Rogers, T.A. Dijkstra and I.J. Smalley. Hydroconsolidation and subsidence of loess: Studies from China, Russia, North America and Europe. *Eng. Geol.*, **37**, 83-113 (1994b).
15.   A.W. Skempton. The Colloidal activity of clays. In: *Proc. 3rd Int. Conf. Soil Mech. & Found. Eng.* pp. 57-61. 1. Zurich (1953).
16.   I.J. Smalley. Flow-stick transition in powders. *Nature*. **201**, 173-174 (1964).
17.   I.J. Smalley. Formation of quartz sand. *Nature*. **211**, 476-479 (1966).
18.   I.J. Smalley and T.A. Dijkstra. The Teton Dam (Idaho USA) failure: problems with the use of loess material in earth dam structures. *Eng. Geol.*, **31**, 197-203 (1991).
19.   I.J. Smalley, S.P. Bentley and C.F. Moon. The St. Jean Vianney quickclay. *Canadian Mineralogist*. **13**, 364-369 (1975).
20.   I.J. Smalley, C.W. Ross and J.S. Whitton. Clays from New Zealand support the inactive particle theory of soil sensitivity. *Nature*. **288**, 576-577 (1980).
21.   K. Terzaghi. Design and performance of the Susuma Dam. *Proc of ICE*, **9**, 369-394 (1958).
22.   K. Terzaghi. R.B. Peck and G. Mesri. *Soil mechanics in engineering practice*, 3rd Ed. John Wiley & Sons, Inc., New York (1996).

Proc. 30th Int'l. Geol. Congr., Vol. 23, pp. 315-322
Wang Sijing and P. Marinos (Eds)
© VSP 1997

Estimation of the Residual Shear Strength of Cohesive Soils Using a Neural Network Model*

ZHANG DEZHENG
Anshan Engineering & Research Incorporation of Metallurgical Industry, 114002, China

GAO QIAN, & SU JING
University of Science and Technology Beijing, 100083, China

Abstract

The residual shear strength of cohesive soils is an important index in study and designing of geotechnic engineering projects, a neural network model that has been developed to estimate the residual shear strength with the physical property parameters of cohesive soils. It has been studied with the model that relation of the physical property parameters, which present the properties and conditions of cohesive soils, and its residual shear strength, based on the further analysis of residual shear strength mechanism and its influence factors. The results of calculation and analysis show that the model is more precise than the empirical formulas are. The better self-learning, and calculation capacity and mistake tolerance make the neural network models more exact and useful.

Keywords: Cohesive soils, Physical property parameters, Residual shear strength, Neural network model

INTRODUCTION

It is indicated in the engineering practice that cohesive soils in slope of fissured hard clay, the shear zones of old slides, intercalated soft layers or under the condition of large shear strain are loose and texture bonds weak because of action of long-term stresses loads, the influence of water and other factors, and thus the shear strength of the cohesive soils is residual shear strength. Therefore, the residual shear strength, instead of the peak strength that are measured in the ordinary test, should be taken as the calculating parameters in the design and research of a corresponding engineering project.

There are some methods that can be used to measure the residual shear strength of cohesive soils such as direct shear test of reiteration, circling shear test, triaxial shear test, etc. The characteristics of all these test methods are complex testing procedure, affected by many of factors, and time and money consuming. Therefore, it is useful in practice of test and   research as well as engineering design to estimate   the residual

* This paper is partly financial suppoted by the Engineering Geology Mechanics Open lab of Chinese Academy of Scinces.

shear strength of cohesive soils.

RESIDUAL SHEAR STRENGTH AND ITS INFLUENCE FACTORS

In the course of a shear test cohesive soil sample under a normal effective pressure, if the sample is sheared continuously after the peak shear stress, as the shear displacement increases the shear stress will decrease obviously and reach a relatively stable value at last. This value can be used to determine the shear strength so called residual shear strength of cohesive soils. Cohesive soils is a multistate system, which consists of solids, liquids and gases, and has complex texture and structure. In different geotechnical engineering, it behaves variably. The characteristics of cohesive soils, such as mineral and granulometric composition, texture, structure, etc., are reflected by the physical and mechanical properties of them in different aspects. The results of the test show that the shear strength of cohesive soils is dependent on its bond and compactness. The essential factors affecting the shear strengths are cohesive soil's mineral and granulometric composition, compactness, texture and structure feature, stress conditions, etc. In the process of a sample sheared (or in the condition of large strain), the original structures and cementing bond are destroyed and the clay mineral grains are adjusted to a large extent and arranged in a certain direction, because of shear stress. Consequently, the negative potential on the surface of clay particles controls the shear strength, the residual strength, which is effected by the mineral and granulometric composition of clay particles and the chemical properties of the bond water on the surface of clay particles.

The clay particles mainly consist of clay mineral, sesquioxide, secondary dioxide, and some particles of primary mineral such as feldspar, quartz, mica, etc. The clay particles have properties of colloid with large specific surface area because of its micro size. The clay mineral and granulometric compositions of cohesive soils can be revealed by the physical property parameters, such as content of clay particles, specific surface area, plasticity index, plastic and liquid limit. The content of clay particles shows the proportion of clay particles in soils; the specifically surface area indicates the granulometric and mineral composition of clay particles; and the plasticity index implies the mineral and granulometric composition of cohesive soils.

EMPIRICAL ESTIMATION OF THE RESIDUAL SHEAR STRENGTH

There are a lot of factors affecting the residual shear strength of cohesive soils. The physical property parameters, which present the behavior of cohesive soil, reveal the close relationship between the physical property parameters and the residual shear strength in different degrees. As the statistic analysis of the test data proved, each of the physical property parameters is well related to the residual shear strength. The relationship is shown by following empirical equations with one unknown, which are proposed by the former:

$$\Phi_r = 453.1/W_L^{0.85}$$
$$\Phi_r = 46.4/0.446I_p$$
$$\Phi_r = tg^{-1}[1.0/(0.513-162.04/B)]$$

where, Φ_r presents residual shear strength, degree; W_L is liquid limit water content, %; I_p is plasticity index, %; B is specific surface area, m^2/g.

From the empirical formulas mentioned above, the residual shear strength of cohesive soil can be expressed with different physical property parameters. It is obvious that there are complex nonlinear relationships between them. The formulas can be used to estimate the residual shear strength with the physical property parameters.

NEURAL NETWORK MODEL

Because of the complex nonlinear relationship between the residual shear strength and the physical property parameters of cohesive soils, it is difficult to develop an accurate equation for describing the relationship between the shear strength and the physical property parameters in consideration of all the influence factors by a statistic analysis method. As a result, the nonlinear mapping capacity of the neural network may be used to express the relationships and correlativity and to estimate the residual shear strength with the physical property parameters.

The neural networks have strong nonlinear dynamic characteristics, and are suitable for dealing with the fuzzy, random and low precision data without knowing their distribution patterns and the relationship of variables.

A neural network trained by a backpropagation algorithm is a nonlinear network that can complete the mapping from a subset A in n dimensions to a subset of F(a) in m dimensions in Euclidean space, that is $A \in R_N \rightarrow R_M$, through the compound mapping of nonlinear handling neurons for the training samples. The training algorithm of the backpropagation network consists of a "forward pass" calculation and the calculated error propagating backward through a "reverse pass" procedure. In the forward pass, the input propagates from the network input layer to its output layer through the hidden layers one by one and finally the output is got.

Suppose a sample set has S vectors, the input and desired output of the neural network can be described respectively:

$$X = [X_1, X_2, ..., X_S]^T$$
$$B = [b_1, b_2, ..., b_S]^T$$

where, $b_k (k = 1,2,...S)$ is the test result related to the sample k and, also the expected output of the neural network.

Based on the relationship between the physical property parameters and the residual shear strength of cohesive soils, it can be considered that there is a nonlinear mapping relationship F between X and B:

$$b_k = F(X_k), k = 1,2,...,S$$

The nonlinear mapping can be developed with a neural network model, and F can be approached gradually by adjusting the network's weights and thresholds during training of the neural network. The algorithms for training the neural network consist the following steps:

(1) Analyzing the test results and determining the input samples, and changing the testing data into the input vectors for the network model;
(2) Initializing the weights and thresholds of the neural network;
(3) Developing the configuration and choosing the momentum coefficient of the model;
(4) Providing property matrix X and desired vector B of the samples are provided for the network;
(5) Calculating the neuron's output layer by layer through the forward pass calculation;

$$O_{kj} = f(NET_{kj}) j = 1,2,...,N_L$$

$$NET_{kj} = \sum_{i=1}^{N_{L-1}} W_{ij}O_{kj} + \theta_j, \quad j = 1,2,...N_L$$

(6) Calculating the output error of the model;

$$E_r = \frac{1}{S}\sum(T_{kj} - O_{kj})^2$$

(7) If the error E is less than the given small positive value for convergence, go to (9);
(8) Adjusting the weights and thresholds of the network layer by layer through the revere pass calculation and go to (5)

$$W_{ij}(t+1) = W_{ij}(t) + \eta\delta_k O_{kj} + \alpha[W_{ij}(t) - W_{ij}(t-1)]$$

$$\theta_j(t+1) = \theta_j(t)\eta\delta_{kj} + \alpha[\theta_j(t) - \theta_j(t-1)]$$

where, t is the number of iteration of the neural network, $\eta (\eta \in (0,1))$ is the training rate coefficient, α is the momentum coefficient (and $\alpha \in (0,1)$);

(9) The output of the network is changed into the matrix which represent the actual meaning of the sample.

Based on the experience in using the neural network and the results of calculation and analysis, the network model, estimating the residual shear strength of cohesive soils, has been developed. After training, the model has acquired the knowledge of the relationship between the network input and its output, and then stored it into the matrixes of weights and thresholds and the structure parameters of the model. Hence, the trained model is suitable to estimate the residual shear strength of cohesive soils by its physical property parameters.

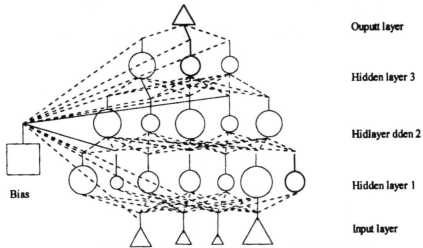

Fig. 1 The Network Model for Estimating the Residual Shear Strength of Cohesive Soils

APPLICATION

In the paper the developed neural network model, for estimating the residual shear strength is comprised an input layer, three hidden layers and output layer(Fig. 1). The input layer has four neurons that correspond to the clay content, specific surface area, plastic limit and plasticity index, respectively. In the output layer, there is a neuron that represents the residual friction angle estimated by the model. The number of neurons in the three hidden layers are 7,5 and 3 respectively. The model is used for the calculation and analysis of the test data of samples collected from the different geotechnical engineering projects in the Northeastern of China. Part of the fifty samples provided for the network training are listed in Table 1.

The data is served the neural network for training after preparation. The network's error reach to 0.001 after iteration of 18540 times, and the network gets convergence.

The finishing of the training means that the model has acquired knowledge from the samples and then, it can be used to estimating and calculating. In order to assess the estimation precision and the utility of the model in the estimation of residual shear strength of cohesive soils, some of the training and examining samples were calculated with the model, and compared with the empirical formulas' results(Table 2).

As shown in Table 2, model's estimating results for the training and examining samples are very close to the actual values. The maximum and minimum value of the relative error for the training samples is 0.147% and 0.003% respectively, and the maximum value of the relative error for the examining samples is 3.267%. The results of the empirical formulas have larger error than the network output. For the estimated results of the residual friction angle, the model has   highest precision, the equation 2.3 and 2.1 are the moderate among the methods, and the equation 2.2 has larger error relatively.

Zhang Dezheng et al.

Table 1 Physical Property Parameters for Part of the Training and Examining Samples

No.	Clay particle content $P(\%)$	Liquid limit $W(\%)$	Liquidity index $I_P(\%)$	Special surface area $B(m^2/g)$	Remarks
1	66	42	21	159.82	Samples
2	63	58	29	290.55	for
3	56	98	50	619.27	training
4	68	27	12	64.65	
5	67	35	16	108.12	
6	36	65	37	169	
7	51	62	28	681	
8	58	40	17	146	
9	55	67	27	688	
10	41	69	37	238.65	
11	31	60	28	66	
12	61	74	36	398.1	
13	51	92	46	551.3	
1	46	78	42	379.4	Samples
2	54	54	27	203	for
3	51	86	45	496.5	examining
4	35	26	10	34	

Table 2 The Estimated Residual Friction Angle(degree) Comparing between the Model and the Empirical Formulas

No.	Testing values	Calculating values of the model	Relative errors of the model,%	Results of formula 1.1	Results of formula 1.2	Results of formula 1.3	Remarks
1	12.02	12.0298	0.082	11.97	18.87	13.12	sample for
2	12.02	12.0092	0.090	10.37	14.57	11.86	training
3	10.98	10.9687	0.103	8.14	9.09	11.20	
4	18.00	18.0041	0.023	15.7	27.47	19.60	
5	13.01	13.0051	0.038	9.31	12.95	12.95	
6	13.01	13.0072	0.022	9.31	12.95	12.95	
7	11.31	11.3266	0.147	10.76	13.50	11.31	
8	14.04	14.396	0.003	12.95	19.80	13.50	
9	11.86	11.8516	0.071	10.76	12.95	11.31	
10	12.02	12.0227	0.022	8.31	12.41	12.19	
11	21.01	21.0086	0.007	10.54	14.04	19.29	
12	12.02	12.0367	0.139	9.43	11.86	11.53	
13	12.02	12.0269	0.057	8.47	9.65	11.31	
1	11.53	11.5270	0.026	8.92	11.31	11.59	examining
2	13.50	13.2844	1.597	10.76	15.11	12.41	samples
3	10.98	11.3387	3.267	8.53	10.20	11.37	
4	24.70	25.0936	1.594	16.70	28.34	24.23	

When the clay content is less than 30% or the plasticity index is smaller, the errors are larger for the calculation of the residual shear strength with the empirical formulas, while the network still has higher precision. Consequently, it can be seen that the precision of the network model is better than the formulas because of its taking accounting of influence factors, such as the material and granulometric component of cohesive soils,etc, sufficiently.

The values from the empirical formulas show that the estimation errors are different with each of the relevant physical property parameters, and which explain that the residual shear strength correlated with physical property parameters differently. In general, the correlativity is only known through the systematically analyzing and comparing. On the basis of the sensitivity analysis of the model for the inputs, it can be studied and assessed that the physical property parameters affect on the results of estimation. The variety of the model's output can be observed by changing its input values. When the input vector is changed one percent, the percent changing of the output for the training samples are listed in Table 3. When the network input varying, the output is more sensitive to the changing of input neuron that represent the special surface area, the secondary is the clay content, and the water content and plasticity index are not so sensitive. The block size of the input neuron in the trained model, shown in Fig.1, demonstrates the relative characteristics too. Thus, it can be recognized that the special surface area of cohesive soils correlate close with the residual shear strength, and the correlativity is secondary for the clay content. The specific surface area not only indicates the grain size of clay, but also reveals its component, in turn, it controls the residual shear strength behavior of cohesive soils. It is oberviously shown in Table 2, the formula 2.3 , which represent the relationship between the specific surface area and the residual shear strength among the estimation equations, is more precise. But the application of the formula is limited because the specific surface is measured difficultly. The sensitivity analysis of the model reveals that the clay content is more relative than water content and plasticity index with the residual shear strength, and the clay content is easily tested in a laboratory with a specific gravity meter. For a common project, the residual shear friction angle can be acquired from the formula of the lay content related to residual friction angle, which is developed with the ordinary nonlinear regression analysis method.

CONCLUSIONS

On the basis of computing and analysis, it has been shown that the neural network is suitable for describing the complex nonlinear relationship between the physical property and the residual shear strength. The trained model can be used to estimate the residual shear strength with higher precision by the physical property parameters such as specific surface area, clay content, liquid limit, and plasticity index, which indicate the material and granulometric component. Among the physical property parameters, the clay specific surface area relates with the residual shear strength very well, and the clay content is secondary. Comparing with the estimation of the empirical formulas, the model has better precision. For the reason of the self-learning ability, the precision and the estimation capability of the model will be improved as new sample is added to the training sample set.

Table 3   The Percent Change of the Output for the Model as the Input Changed 1%

No.	Input related to clay particle content	Input related to liquid water content	Input related to plasticity index	Input related to special surface area
1	-5.002	-0.884	-4.699	-8.812
2	-5.819	6.362	-9.517	-8.457
3	-5.343	0.141	0.188	-0.021
4	-0.011	-0.002	-0.001	-0.000
5	-0.838	-0.655	-0.279	-0.061
6	0.006	0.612	-0.177	-0.746
7	0.486	7.011	-6.629	-3.769
8	4.641	0.173	-0.023	-1.053
9	5.902	6.636	-6.631	-6.575
10	-14.554	-9.002	-11.659	-11.375
11	0.002	0.127	-0.611	-0.007
12	2.719	6.585	6.631	-5.894
13	-5.345	-0.666	-0.419	-5.206

REFERENCES

1. Cheng Xiangjun, The Principle and Application of Neural Network, National Defense Industry Publishing House, 1995
2. Tang Daxing, Geotechnical Engineering, Geological Publishing House, 1987
3. PD Wasserman, Neural Computing Theory and Practice, Van Nostrand Reinhold-New York, 1989

GEOENVIRONMENT AND GEOHAZARDS

Proc. 30th Int'l. Geol. Congr., Vol. 23, pp. 325-341
Wang Sijing and P. Marinos (Eds)
© VSP 1997

The Increasing Role of Urban Geoscience at the End of the Twentieth Century

JOSEPH McCALL
Honorary Fellow, Liverpool University, England
Chairman, International Working Group on Urban Geology

Abstract

Urban geoscience is assuming an ever increasing importance. Unfortunately there is a tendency to spurn the potential of scientific solutions to the present problems of escalating urbanisation and to rather concentrate on sociological problems. The International Working Group on Urban Geology (sponsored by COGEOENVIRONMENT, a commission of IUGS, IAH and IAEG) has recently prepared and had published a multiauthor definitive volume treating the subject in a broad manner. This contribution to the Symposium B4 on Urban Geology at the 30th IGC in Beijing expresses the philosophy behind this volume. The topic has been treated from a realistic viewpoint, acknowledging the equal importance of resource aspects and environmental constraints. Some consideration has been given to the reasons for the increasing incidence of urban geohazards: among them are the increasing size of urban developments and conurbations, the extension of urban development onto unsuitable ground, the lack of urban planning with early input of geoscience data and evaluation, the decay of old cities and towns and the too rapid and uncontrolled growth of new cities and towns. Much of the increase is real, though it may in part be an artifact of increased public and media awareness. The global effects on the environment of urbanisation are briefly considered and the global escalation of population.

Keywords: Urban, Land-use planning. Environmental geology, Geohazards

INTRODUCTION

Geology is generally considered to be an open-air pursuit, whether followed professionally or as an amateur interest, and, as Robinson and McCall [1] have remarked : "All geologists rightly exult in field-work, which takes them out of town into the countryside".

Yet, more and more, with the escalating increase in global population and urbanisation extending to the growth of megacities in the "Third World", existing cities and towns sprawling outwards more or less unchecked, and older cities decaying and changing their industrial base, are geoscientists being called on to tackle problems in the urban setting and also to consider problems raised by urbanisation but affecting areas outside the urban limits and even the environmental health of the globe in general [2].

Nearly 50% of the global population now live in towns and cities. Table 1 shows some statistics indicating the recent increase in urbanisation. All the signs are that urbanisation will continue to increase globally after 2000AD.

Table 1. The change in urban populations by percentage, regionally and global: 1960 to 1990

Region	Urban Population as % of total: 1960	Urban Population as % of total: 1990
Africa	18.3	33.9
North & Central America	63.2	71.4
South America	51.7	75.1
Asia	21.5	34.4
Europe	61.1	73.4
Former USSR	48.8	65.8
Oceania	66.3	70.6
World	34.2	45.2

Source: World Resources Institute, 1992, quoted by Pickering and Owen [11].

The increasing demand for geoscientific input to planning of urban land use, rectifying problems of decay and poor prior procedures, rehabilitating land after the closure of extractive and other industries, designing new constructions, and environmental assessment and monitoring, may well take up the employment-slack caused by the decline of traditional areas of geoscience employment.

The germ of an idea for a book on Urban Geoscience was initially developed by Joe McCall and Ed de Mulder, now Chairman and Vice-Chairman of the International Working Group on Urban geology (sponsored by COGEOENVIRONMENT, IAH and IAEG), and was further developed by a group of geoscientists active in AGID, COGEOENVIRONMENT, the British Geological Survey and Department of the Environment, and the Engineering Group of the Geological Society. Outline schemes were submitted by members of this ad hoc group and it was agreed to prepare the book as a definitive account, carefully designed and structured, and not a random set of articles: a book targetted at the geoscientist, the engineer and the land-use planner, and scientists of other disciplines - the problems of urbanisation can only be solved by a multidisciplinary approach. The synopsis of the book which emerged is given below:

Contents of "Urban Geoscience"

Chapter 1.   Urban Geoscience - Introduction.   *E.F.J de Mulder*
Chapter 2.   Mineral Resources.   *G.J.H.McCall and B.R.Marker*
Chapter 3.   Urban Soils.   *T. Simpson*
Chapter 4.   Urban Groundwater.   *A.R.Lawrence and C.Cheney*

The various chapters were prepared by a number of authorities, each with expertise in the particular topic. We took the headings Minerals, Soils, Groundwater, Natural Hazards and Man-made Hazards as the topics for the five Chapters which follow the overview introduction: these chapters describe the urban resources and constraints. The next chapter ties up some loose ends remaining from these, defines the role of the Earth Sciences in assessing urban resources and constraints, and emphasises the relationships between the urban setting and the area outside its limits. The succeeding three Chapters cover positive aspects of utilizing the resources, Urban Development: Identifying Opportunities and Dealing with Ground Problems; The Integrity of Building and Construction Materials, and The Problems of Geoscience Education in the Urban Setting. At the heart of the problems facing geoscience in the urban sector lies the land-use planner and the final chapter represents a tieing together of the previous text from the viewpoint of a senior land-use planner, Robin Mabey, who stresses The Need for Earth Science Information in Urban Planning. A certain amount of overlap between the Chapters has been retained editorially, for this emphasises the interrelationship between the Chapter topics and also the difficulty in drawing hard and fast boundaries between natural and human influences on the processes described.

It is useful here to consider the subdivision into topics and the thinking behind this subdivision.

THE SUBDIVISION INTO TOPICS

Resources and Constraints
Mineral Resources   "All materials used to create cities are derived from the Earth" [3].

As the above quote emphasises, mineral resources, including energy resources, are fundamental to the existence of cities and towns (Table 2). It is essential to approach mineral resources in the light of the demands of the real world, and not from an all too common simplistic viewpoint of mineral resources being 'environmentally bad'. Many cities and towns were originally sited on account of nearby mineral resources and/or energy resources to support their industries. Many of these have now exhausted their resources and this, together with and changing needs and markets, has led them to seek a new industrial base. Many other cities and towns have an indirect relation to minerals - ports for handling, financial centres of the minerals trade etc. The principle direct requirements of cities and towns are minerals for construction and infrastructure, energy minerals to support activity and raw materials to support local industry.

Table 2. The estimated annual consumption of mineral commodities by a city of 250,000 people (based on figures presented at various times by G.W. Luttig and F.C.Wolff).

Mineral commodity	Tonnes per annum
Sand & gravel	1,650,000
Hard Rock	500,000
Petroleum	600,000
Coal	500,000
Limestone	350,000
Steel	50,000
Cement	25,000
Clays	100,000
Industrial sands	80,000
Rock salt	50,000
Gypsum	20,000
Dolomite	2,500
Phosphate	2,000
Sulphur	7,000
Peat	6,500
Natural freestone	6,500
Potash salts	6,000
Aluminum	5,000
Kaolin	4,000
Copper	3,500

Minerals have been considered under SUPPLY (both of those that support construction and infrastructure and those that support industry, including raw materials and industrial minerals used in the processes, Table 3), also energy minerals (Table 4): under DEMAND, PRICE & TRANSPORT: EFFECTS ON THE ENVIRONMENT: and REHABILITATION OF LAND.

All proposals for mineral development, whether in urban areas or outside them, should be subject to a rigorous, independent environmental audit, before permission to extract is granted or refused.

Urban Soil   The public image of an urban area is a continuous spread of houses and other buildings separated by traffic-ridden roadways.   The amount of greenspace actually there is quite surprising, because when driving through a townscape one only sees the houses, not the back gardens and only the walls of parks, in most cases. The topic of urban soils has been virtually ignored until recently when it was made the subject of a book by Bullock and Gregory [4].   It has been covered from the aspect of LAND USE (including soil variability and main functions);   the EFFECTS OF URBANISATION ON THE SOIL RESOURCE (loss, damage to the soil structure, causes of deterioration,

Table 3. The more important industrial minerals (those used in industrial processes)

Usage category	Minerals
Abrasives	corundum, diamond, quartz sand, magnetite, spinel
Ceramics	clays, shales, potash, pyrophyllite
Chemical industry raw materials	coal, limestone, potash, salt, silica, sulphur
Filtrants	diatomite, pyrophyllite
Glass constituents	glass sand, borax, potash
Insulators	asbestos*, diatomite, graphite, perlite,
Lubricants	graphite, talc
Pigment bases	black titaniferous, sands - ilmenite, rutile
Plastic bases	coal, oil
Soap and dye constituents	potash, salt
Sorbers, fillers and coaters	bentonite, diatomite, fuller's earth, kaolin ("china clay")
Steel industry fluxes	dolomite, fluorite, limestone

* demand for asbestos has decreased significantly on account of its carcenogenic properties

Table 4. Sources of energy - minerals and others

Type of energy source	Source
Mineral	Oil (from wells, oil shale, tar sands), Gas (natural hydrocarbon and methane or "coal gas" derived from coal) Coal, Peat, Nuclear energy from radioactive minerals
Others	Hydroelectricity, Direct water power, Tidal & wave energy Wind power, Geothermal energy, Solar energy, Wood and waste burning (including methane generation from waste)

acidity, anthropogenic materials and contaminants, made ground): and finally URBAN SOIL MANAGEMENT. Soil loss is the most sigificant result of urbanisation, but also important is the reduction of the ability fo the soil to act as an effective medium for plant growth. There is a need world-wide to protect urban soils from further loss and degradation, and to improve the quality of endangered urban soils.

Groundwater Many cities are dependent on groundwater for part, or even all of, their water supply. Urbanisation results in important changes to the groundwater balance by replacing and modifying groundwater mechanisms and the effects of abstraction. Leakage and seepage from sewers and mains may affect shallow aquifers and eventually the groundwater system as a whole. Pollution of aquifer sources under or close to the city may cause deterioration in quality or depletion, even removal, of the source aquifer. Restoration of an aquifer to acceptable standards once polluted is more difficult than restoration of surface water. Protection must therefore always be the preferred policy. The IMPACT OF URBANISATION has been considered under impermeabilisation and drainage; the effect of introduction of piped systems, sanitation services, irrigation of amenity areas; land surface storage and disposal of industrial effluent; MODIFICATIONS due to on-site sanitation systems, sewered systems, pluvial drainage, industry and lagoons and solid waste disposal are also considered in detail. Two Tables, 5 & 6, show the degree of access of the population to water supply and sanitation in four

Table 5. Impact of urbanization on groundwater recharge for selected cities

City	Aquifer	Climate type	Sanitation arrangements	Increase in recharge (mm/a)	Main source of recharge	References
Liverpool (UK)	Triassic sandstone	humid temperate	Sewered	+55	Leaking water mains	price & Reed 1989
Birmingham (UK)	Triassic sandstone	humid temperate	Sewered	+55	Leaking water mains	Price & reed 1989
Wolverha-mpton (UK)	Triassic sandstone Permo-carboniferous Red Beds	humid temperate	Sewered	+10-50	Leaking water mains	?
Merida (Mexico)	karst limestone	tropical	unsewered	+500	(i) leaking water mains (ii) pluvial drainage (iii) on site sanitation	Morris et al 1994
Santa Cruz (Bolivia)	alluvial outwash plains deposits	tropical	unsewered	+150-170	(i) leaking water mains (ii) pluvial drainage (iii) on site sanitation	Morris et al 1994
Hat Yai (Thailand)	coastal alluvium	tropical humid	unsewered	+60	leaking water mains	?
Lima (Peru)	alluvial gravel	tropical arid	unsewered	+700	leaking water mains irrigation amenity areas	Geake et al 1986

Table 6. Urban population with access to water supply and sanitation

Indicator	City			
	Metro Manilla	Jakarta	Calcutta	Madras
Total population(millions)	6.4	6	9.2	5
Area (km²)	646	550	800	1170
Urban density (cap/ha)	98	200	115	43
% population in substandard housing (slums)	45	40	33	60
% living in squatter-illegal settlements	30	na	na	25
% with access to water (house connections)	43	47	48	40
% garbage collected daily	70	25	55	78
% human access to human waste disposal system	60	42	45	58

Source: Lea and Courtney (1986)

third world cities and the impact of urbanization on groundwater charge in three western cities and four third world cities. Urbanization increases recharge but lowers groundwater quality. These diagrams indicate that the problems are greatest in 'third world' cities, where there is a common absence of sewerage.

Natural Hazards   Geohazards can be classified under those that are Intensive and rapid acting and those that are pervasive and of slow onset (for example, as classified by McCall [2]) (Tables 7 and 8), but, for the purposes of the book in question, they have been divided them into separate chapters on "Natural" and "Man-made" geohazards respectively (Table 9), though the distinction between the two can be blurred.

Table 7.   Intensive, rapid-onset geohazards (from McCall [13]).

- Earthquakes and active faults
- Volcanic eruption (lava flows, ash falls, nuees ardentes, lahars)
- Tsunamis
- Landslides, avalanches, debris flows: rapid subsidence
- Flooding, storms, cyclones, tornados
- Wildfire related to earthquakes, combustion related to hazardous gases such as methane
- Collapse of solid waste stacks

Table 8.   Pervasive slow-onset geohazards and other usually non-intensive processes (from McCall [13])

- Foundation problems: peatlands and other compressible soils, low areas subject to flooding
- Permafrost conditions requiring special and expensive building and construction techniques
- Subsidence related to excessive groundwater abstraction, mine workings, hydrocarbon extraction or karstic conditions
- Ground fissures, including those related to expansion and shrinkage
- Neotectonic uplift
- Groundwater pollution, by sewage, leachates from waste dumps, nitrates from agriculture or other sources, benzene, toxic metals and chlorinated   organic solvents, toxic metals.
- Surface water pollution
- Depletion of the groundwater resource, including loss of recharge due to urbanisation
- Rising groundwater levels due to cessation of or diminished abstraction, irrigation returns etc.
- Rising sea level (? possibly due to global warming) - coastal cities and towns
- Sea water intrusion in coastal situations
- Contaminated soils from urban industrial sources
- Saline soils
- Compressibility, shrinkage, heave of soils
- Radon emanation
- Erosion and deposition
- Soil loss due to deforestation and urban development
- River bank failure and silting
- Coastal changes due to urbanization

Table 9. Natural and man-made geohazards considered in the book "Urban Geohazards" - McCall et al. [1]

Natural geo-hazards	Man-made geohazards
Earthquake, Volcanic	Surface movement resulting from the abstraction of fluids
Tsunami, Extra-terrestrial	Mine-related subsidence, Contaminated land
Landislide, avalanche, Subsidence	Fluctuation of groundwater levels
Hazardous gases (methane , radon)	Modification of groundwater quality due to urbanization
Fire, Flooding, cyclones	Waste disposal
River bank failure, silting	
Coastal erosion, Expansive and collapsing soils	
Neotectonic fissures, Permafrost	

Though such natural hazards as earthquakes, eruptions, tsunamis, landslides and floods will always be with us, increasing urbanisation increases their effect, because it concentrates populations in "target" areas. Urban sprawl causes building and construction on unsuitable and threatened sites, and construction methods may be poor. There has been an evident increase in the incidence of geohazards world-wide, including urban geohazards, and much of this observed increase is real, though part may be due to increased public and media awareness. Reasons for the real increase include the increasing number of large urban developments world wide, the increasing size of individual cities, towns and conurbations, the extension of existing city suburbs onto unsuitable land, the decay of old cities and towns and the development of new cities without adequate land use planning involving the input of geoscience information and evaluation.

Earthquakes constitute the greatest single natural urban geohazard (Fig. 1). Prediction is difficult and pinpointing the actual site of the epicentre even more so: at present predictions can only be made of the likely occurrence of a major quake within a broad area. Evacuation of a city has been successfully carried out once in China (the Haicheng earthquake), following a prediction, but there are major sociological problems facing evacuation of a western city such as Los Angeles (even in the case of Haicheng there was an earlier abortive evacuation, and repeated abortive evacuations are likely to occur with decreasing public response). Mass panic effects are another major problem in the case of sudden evacuation. Volcanic hazards emanate from fixed sources (for example, Fig. 2) and it is much easier to define the zones in which the danger lies for this hazard.

The remote possibility of extraterrestrial impact (Fig. 3) has been considered, together with the more familiar natural hazards and also the pervasive, harmful effects of radon emanation,and the problems of permafrost. In some cases, all that can be done is to avoid unsuitable locations for new urban development and ensure the availability of emergency planning in the case of existing development: much can be achieved however by means of engineering, for instance in design of buildings to counter earthquake shocks and mitigation of landslide risks. Both the radon and permafrost hazard can be 'engineered out'. Warning systems are effective in the case of volcanic eruption, lahars and tsunamis but again tend to meet with sociological problems.

Figure 1. A toppled building, Santa Rosa California, after the San Francisco earthquake of 1906 (from a teaching collection, source unknown)

J. McCall

Figure 2. The spine of viscous lava extruded from Mt Pelee, Martinique in 1902-3. In mid-1903 it collapsed with a violent pyroclastic flow emission (photo. A. Lacroix)

Prior evacuation has however proved effective in the case of some volcanic eruptions - those which are not instantaneous, unlike that of Mt Pelee in Martinique in 1902 - and successful evacuations have been achieved in Rabaul and Montserrat recently.

Man-made hazards Though the distinction between man-made hazards and natural hazards is not clear-cut, five man-made urban hazards were selected for detailed coverage. The first is subsurface movements resulting from abstraction of fluids (Including water and hydrocarbons) and minerals. Both can destroy buildings and constructions, can cause flooding, personal injury and loss of life. The second hazard is that of contaminated land, generally the product of decayed industries; besides the health hazard, there is the need to restore waste land in urban settings and bring it back into use, whether for new industry, housing or recreational purposes. The third hazard is the fluctuation of groundwater levels, in particular rise in the level of the water table, Many cities in both the developed and developing countries are so affected and this causes flooding of underground structures and basements, damage to foundations, invasion of grossly contaminated water causing a health risk and increased flow in surface drainage networks, again of contaminated water. The fourth hazard is the modification of groundwater quality by urbanisation (mentioned above under groundwater). The main source in developed countries is areas of contaminated land, where plant used to manufacture chemicals, iron and steel, town gas and other commodities has been left derelict, inevitably leading to groundwater contamination. Corrosion of fuel storage tanks is another source and also the use of chlorinated solvents in the motor industry. In developing countries the main source is unsewered sanitation. The fifth hazard is waste disposal and arises from the need to dispose of vast amounts of household, domestic and industrial wastes generated in urban environments, most of which going to landfill and generating leachate and gas contaminants.

Addressing urban resouces and constraints: the role of geoscience
Investigations and responses A historical review of the patterns of development and growth of cities and towns, from the past to the present led into a consideration of INVESTIGATIONS AND RESPONSES, covering site investigation, proactive and reactive repsonses and stressing the need for balanced response. The problem of 'planning blight' is relevant here (an almost insoluble problem which has to be faced); and the interrelationship between the urban area and the surrouding rural area is very important. Lack of the introduction of Earth science information and expertise into planning and decision making leads to increased costs rectifying mistakes and on occasion to disaster. The problems raised need to be faced, not avoided. The geoscience input must be early in the planning and decision making process.

Urban development - identifying opportunities and dealing with problems The ENGINEERING PROPERTIES OF ROCKS AND SOILS were first considered and then the various procedures of thorough GROUND INVESTIGATION. CONSTRUCTION and other site procedures were discussed including tunneling. The various types of FOUNDATIONS were illustrated. The last section covered DEALING WITH PROBLEMS, during and after development commences; MONITORING and REMEDIAL ACTION.

The Integrity of Building Materials   This covers a most important but often overlooked topic.   The chapter concentrates on natural stone and concrete, but other materials briefly mentioned are bricks, tiles and reconstituted stone.   The modern use of stone for facings rather than as a structural component is an important consideration.   Durability and testing are considered in some detail.   Water absorption, porosity, frost resistance, acid dissolution, thermal stability and staining are important factors in the case of natural stone. The durability of concrete, the effect of various types of cement and agregate materials, and the water used in manufacture are considered, and the causes of external and internal attack, special consideration being given to physical, and chemical reacations.   reactions with the aggregate are particularly important (Table 10).

The Urban Planner's need for Earth Science Information   In the final chapter, the fact   was recognised that market mechanisms cannot cope with the need to provide urban infrastucture and employment growth, while at the same time maintaining environmental safeguards.   The U.K. Planning System at all levels was cited as an example of what is needed in this respect.   The essential need for Earth science input into the planning process is not fully appreciated at the present time and needs to be stressed.   For example, Pearce [5] reports that at the recent UN Habitat II conference in Istanbul on Urbanisation, negligible attention was given to this important need, the discussions being almost entirely about sociological issues.

Urban Geoscience Education   This chapter was added a something of an afterthought, but it is nevertheless thought provoking.   It deals with the problems of access to field study for urban students at schools and in tertiary education.   The possibilities of using substitutes for the field experience such as tours of interesting stone buildings within the town, gravestone studies, theme parks and museums, and even "visual reality" were considered.   With increasing financial and other constraints on 'in-the-field' teaching and learning, the need for such programmes is becoming more and more urgent if geoscience is to be maintained within national curricula.

Discussion

In hindsight,   one topic that was not covered in this project was the application of geoscience to urban archaeology and this was an unfortunate omission, for there is a growing application of geoscience techniques to urban archaeological research (for instance, the use of ground radar to locate old tombs and cavities) and archaeological data is very important in connection with planning urban development.

This project was a start, no more, and could well be followed up by more specific texts dealing will specific aspects of urban geoscience (for example hydrogeology, engineering geology etc,). The project was generously supported by AGID, and if such more specific volumes are to be produced, there will be a need for other sources of finance to be explored.

CONCLUSIONS

Table 10. Reactive forms of silica (from Geological Society Engineering Group 1993)

Variety of silica	Common Geological Occurrence
Opal	Vein material and vugh filling in a variety of rock types, a constituent of some types of chert a replacement for siliceous fossil material and a cementing material in some sedimentary rocks
Volcanic Glass	A constituent of some volcanic rocks ranging from acid to basic composition. Volcanic glass devitrifies over geological time and devitrified glass may be potentially reactive.
Tridimite and Cristobalite	High temperature metastable polymorphs of silica found as a minor constituent in some acid and intermediate volcanic rocks.
Microcrystalline and	The principal constituent of most cherts and flint. Vein material cryptocrystalline quartz and vugh fillings in a variety of rock types, groundmass mineral in some igneous and metamorphic rocks, cementing material in some sedimentary rocks.
Chalcedony	A fidrous variety of microcrystalline quartz found as a constituent of some cherts and flint. Vein material and vugh fillings in a variety of rock types, cementing material in some sedimentary rocks.
Strained quartz	Found especially in metamorphic rocks but also in some igneous rocks, subjected to high stresses. Also occurs as a detrital mineral in clastic sediments. Current opinion is that strained quartz itself is probably not reactive and reactivity may be associated with poorly ordered silica at the highly sutured grain boundaries commonly associated with strained quartz.

How important is urban geoscience?   It is easy to imagine the reader saying "This is something that has just been invented by geologists as a means of empire building", but the truth is that , at the threshhold of the 21st Century, this is a matter of immense importance.

The compilation of the book "Urban Geoscience"   has led to the conclusion   that:-

a)   supplies of Earth materials are essential for maintaining the fabric of cities and the industries within them.

b) it is essential to exploit resources in the most efficient but least environmentally damaging ways, and to safeguard these against sterilisation by other development or contamination (soils and water) in the interests of sustainable development.

c) it is important to recognise that action based on Earth-science information can significantly reduce the costs of development and the risks of losses of investment and productivity due to damage to properties, injuries and loss of life; and can contribute also to improvement of the urban environment through achieving conservation objectives.

This can only be achieved if there is a proper use of Earth science information at the outset - the site selection and planning stages, as well as the site investigation stage. This requires good sources of information (e.g. applied maps and data bases), easily accessible to be drawn upon.

Similarly, the assessment of potential problems for existing urban development can only be foreseen if there is a good understanding of the ground conditions. Given the large amount of investment bound up in urban areas, it makes good sense to identify the constraints on future urban development, the hazards and risks and the opportunities for cost-effective development at an early stage, and to institute monitoring and remedial actions where necessary.

All too often, Earth scientists are regarded as "bringers of problems" because disadvantages of sites are identified by them.   It should, however, be borne in mind, that it is better to discover a potential problem early so that it can be dealt with in a cost effective manner, rather than to discover it later on, leading to delays to development and even damage.   There may be concern that information on ground problems may affect local land or property prices, or may drive away buyers or investors. This may be so where there is insufficient information and lack of confidence.   However, a much greater lack of confidence is caused when an adverse event such as, for example, a collapse due to unsuspected mine workings under an urban housing development takes place and attendant publicity leads to public alarm.

It should be remembered that Earth science information is as much about identifying the opportunities at sites through lower costs and lesser risks where ground conditions are sound or preventative works can be undertaken for a realistic cost.   Earth scientists need to pay more attention to presenting their findings in terms of cost advantages and opportunities - to be taken seriously, Earth scientists must be more prepared to assess the costs and benefits of their work.   There is, at present, very little literature indeed on this subject, apart from sporadic collections of figures, on costs of specific problems. There is a need for much more research on this to provide a better basis for discussion with policy makers. economists and those who may fund the preparation of databases and applied Earth science maps.

There is also a need to better educate the public and decision makers about the value of using Earth science expertise and information. This requires action at all levels: schools, universities, adult education, liaison with politicians and decision makers.   Earth scientists should be prepared to communicate their ideas at all levels, including widely-read, non-technical media rather than only producing technical papers. Professional organisations could give a lead in securing such a more open approach to dissemination of information.   In addition, there would be great value in other professionals, such as land-use planners, having an Earth-science module incorporated in their training, in those countries where this is not already the established practice (Worth [6]). Conversely, there is a neeed for Earth-scientists to become better acquainted with the requirements of those who need to be aware of, and to utilise, the information which they can provide.

Looking at urban geoscience from a wider, global viewpoint, "the hazard of overpopulation and excessive human activity adverse to the environment.... is one that overrides all other geohazards" (McCall [7]).

The problems engendered by escalating urbanisation and industrialisation go hand in hand with those of escalating population. Prince Philip [8] observed "the concentration of large populations means that the pressure on the surrounding land to produce the food and food requirements virtually denies it any form of life... at the same time the cities create vast quantities of waste and noxious effluents which pollute the air, and everything downwind, the waters, and everything downstream". London had a population of 1,114,000 in 1801, by 1939 it was 8,700,000 and by 1991, 9092,024 (Clout [9]: Times World Atlas [10]). Those who take the trouble to plot the figures for London may find false comfort in the recent tailing off in the escalation at the end of the graph for London, but this of course partly reflects a shift of growth of population (and industry) outside the "green belt", designated to contain the urban area, was established in the 1960's..

The global population reached 4.4 billion in 1980, it is predicted to reach 6.1 billion by 2000 and 8.5 billion by 2025 There must surely be a limit to the population that this finite World can sustain and there must surely be a limit to the population growth and sprawl of urban areas taken up by megacities such as London, Mexico City (predicted 31 million inhabitants by 2000), New York, Sao Paulo, Cairo, Tehran, Shanghai and Tokyo, before such cities strangle themselves by their growth, congestion, pollution and ever increasing demand for resources from outside. Sir John Houghton, (Times, London, issue 65504, February 16th 1996], is reported, on the day that this review article was completed, to have stressed that global warming will cause immense scale food water and food shortages (and possibly wars related to these) - but why only worry in isolation about global warming (to which urbanisation is a major contributor and which remains uncertainly defined in its future scale and detailed effects) when we already have the same global deficiencies and adverse effects being produced by galloping urbanisation and population escalation?

The rates of increase suggest that something drastic must indeed be done to curb these expansions during the 21st Century, but Earth scientists cannot solve the attendant problem. They can only strive to ensure that from now on a fuller use is made of geoscience expertise in supporting land use planning, new urban design and construction, rehabilitation of decayed urban land, and controlling urban pollution, thus limiting the adverse environmental effects of urbanisation. Such measures may even then amount to no more than a "first-aid approach" to the real global problem, but they are surely better than nothing.

Acknowledgments

Figure 3. Trees felled by the Tunguska, Siberia, event in 1908, for a distance of 40 km from the site of impact (from McCall [15]).

The author has reproduced some textural and illustrative material from individual chapters in AGID Sp. Publ No. 20 "Urban Geoscience" in this paper and acknowledges the source of these in the Chapters by A.R.Lawrence and C.Cheney, B.R.Marker and B.F.Miglio.

REFERENCES

1. G.J.H.McCall, G.J.H., E.F.J. de Mulder and B.R. Marker. Urban Geoscience. AGID Sp. Publ No. 20, A.A. Balkema, Rotterdam, 273pp. (1996)
2. G.J.H.McCall. Geoindicators of rapid environmental change: The urban setting. In: Geoindicators: Assessing rapid environmental changes in Earth systems. A.R. Berger and W.J.Iams (Eds.), Balkema, Rotterdam, 311-318 (1996)
3. R.F. Legget. Cities and Geology. McGraw-Hill, New York (1973).
4. P. Bullock, and P.J. Gregory, P.J. Soils in the Urban Environment. Blackwell Scientific Publications, Oxford, 174 pp (1991).
5. F.Pearce. Megacities plan spurns science. New Scientist, 8 June. p.4. (1996)
6. D.H.Worth, D.H. Planning for the engineering geologist. In: Planning and Engineering Geology. Culshaw, M.G. Bell, F.G., Cripps, J.C. and O'Hara, M. (Eds.), Geological Society Engineering Geology Special Publication No. 4, 39-46 (1987)
7. G.J.H.McCall. Natural and Man-Made Hazards: their increasing importance in the end-20th Century World. In: Geohazards - Natural and Man-Made. G.J.H.McCall, D.J.C. Laming and S.C. Scott (Eds.), AGID Spec. Publ. No. 17, Chapman & Hall, London, 1-4 (1992)
8. Prince Philip, Duke of Edinburgh. Down to Earth: The Human Population Explosion. Collins, London, 116-147 (1988).
9. H. Clout (editor). The Times London History Atlas, Times Books, London, 191 pp. (1991).
10. The Times World Atlas, Times Books, London (1992)
11. K.T.Pickering, and L.A.Owen. 1994. An Introduction to Global Environmental Issues. Routledge, London and New York (1994)
12. J.P.Lea, and J.M.Courtney . Cities in Conflict. Studies in Planning and Management of Asian Cities, World Bank, Washington.
13. G.J.H.McCall,(Geohazards and the urban environment. In: Geohazards and Engineering Geology. J.G. Maund (Ed.). Special Publication of the Engineering Group of the Geological Society, Geological Society Publishing house, Bath, England (in the press).
14. Geological Society Engineering Group. Aggregates. Geological Society Engineering Group Sp. Publ. No. 9, Geological Society Publishing House, Bath, England, 339 pp (1993).
15. G.J.H. McCall, Meteorites and their Origins. David & Charles, Newton Abbot, 352 pp. (1973).

Proc. 30th Int'l. Geol. Congr., Vol. 23, pp. 343-355
Wang Sijing and P. Marinos (Eds)
© VSP 1997

Geological Processes and Stability of Urban Areas

OSIPOV V. I.

Institute of Environmental Geoscience, Russian Academy of Sciences, Moscow, Russia

Abstract

For the nearest decade, the Earth's urban population will first exceed the half total number of the people living on our planet. The concentration of people in towns, powerful technological impact on the geological environment, insufficient geotechnical preparation of areas, excessive extraction of natural resources-all these increase the vulnerability of humans and urban technosphere under the action of natural disasters. However, we can decrease their consequences and, in a number of cases, even prevent hazardous natural processes in case of realization of reasonable city designing policy, creation of a system of prediction and prevention of hazardous phenomena.

Keywords: urbanization, geological processes, megapolices, risk control

URBANIZATION PHENOMENON

From the ancient historical eras, coinciding in time with the Holocene beginning (8-10 thousand years ago), till the early XIX century, the Earth's population changed insignificantly. It varied from some tens to some hundreds million people depending upon disease epidemics and starvation waves. The Earth's population reached 1 million people only in 1830 (Fig. 1) [11]. But since the industrial revolution began, the situation changed, and the rapid growth of population started. It assumed a catastrophical scale in the second half of the XX th century: the Earth's population reached 2 milliard in 1930, 3 milliard-30 years later (in 1960), and 4 milliard -15 years later (in 1975). Afterwards, in a 12-year period (1987) population of our planet exceeded a 5-milliard boundary; and it is expected that more than 6 milliard inhabitants will be in the world in 1999, while in 2010 and 2025- more than 7 and 8 milliard people correspondingly [11, 17].

The urban population grows even quicker. In 1800 the absolute majority of people lived in rural areas (town dwellers made only 3% from the total population of the Earth), however in 1900 the urban population reached 13.6%, in 1950-28.8% and in 1990-42.9%. As expected, more than a half of the world inhabitants will live in towns in the early XXI century (Fig. 2), [18]. Presently urbanization assumed a new quality, it became a really global process with catastrophically growing scale and rate. Beginning with 1970, the total Earth population grew by 1.7% annually on average, while the number of urban dwellers increased annually by 4% in the same period [18].

It is interesting, that through the historical course (from 800 till 1800 A.D) population of the largest cities changed insignificantly with falls and gradual rises connected with frequently spreading diseases (Fig. 3), [7].

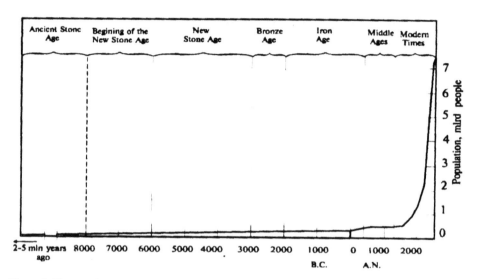

Figure 1. The growth of the Earth's population in Holocene.

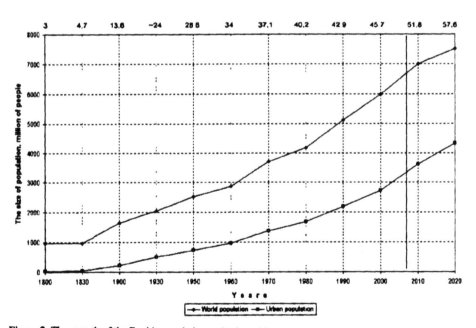

Figure 2. The growth of the Earth's population and urban citizens in 1830-2020 years.

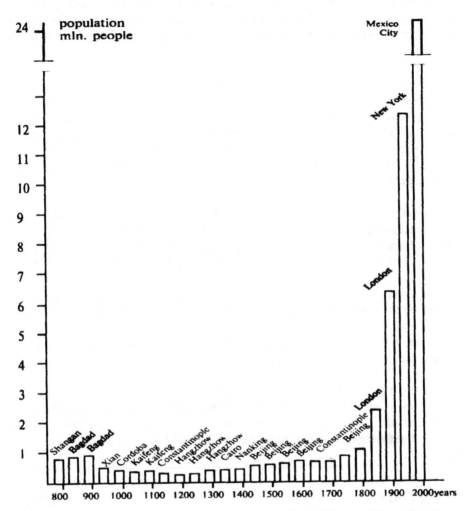

Figure 3. The population growth in the largest world cities from 800 till 2000 years. Urban citizens became more protected from mass diseases in the second half of XIX century and in the XX century, which resulted in the rapid urbanization growth.

Against the general background of urbanization, the number of megacities increases rapidly. According to UN data in 1800 there was only one city-Beijing -with the population exceeding 1 million, while in 1900 there were 16 such megacities already, in 1959-59.511 megacities are expected in 2010, and 639- in 2025, 486 (76%) of them being in developing countries [18]. If in 1900 only one city-London-had the population of more than 5 million people, in 1950 their quantity increased up to 8, and in the year of 2000 there will be 45 such cities. And, at last, the amount of gigantic megacities with population above 10millionincreases rapidly; if in 1950 there were just 3 of them (New York, London and Shanghai), in the year of 2000 there expected to be 24 such megacities.

The growth of large cities in the developing countries goes in advancing pace. In such cities as Mexico, Sao Paulo, Djakarta, Lagos, Karachi, Dakar, etc., The population increases twice every 15-20 years. Megacities in the developed countries grow slower (Fig. 4).

The present and future megacities have no precedents in the history. If the present-day rate of population growth is preserved, than in 2010 Mexico, for example, will amount 30 million people, which will exceed the total predicted population of Canada. Urbanization calls for greater urban areas. As predicted, the total urban square will increase by 2.6 million Km^2 up to 2020. The territory of Mexico-city, for example, enlarged from 130 to 250 Km^2 during 1940-1990; and the territory of Moscow-from 326 to 994 Km^2 through the same period.

Cities grow rapidly in the developing countries due to the intense migration of people from the rural regions. Settlers have to occupy places of city territory the worst for living, the most hazard-prone: hill slopes, flood plains, swampy and coastal territories. Besides, settling in these places goes on with poor construction codes and inadequate housing.

Even in developed countries, the rapid rise in population of megacities causes the local authorities to condense the urban building, thus increasing an impact upon the environment. The same reasons makes the local authorities to build in former "poor lands" within the precincts of a town, that is in the least favorable natural conditions for housing and living. People concentration in cities, poor construction codes and inadequate housing, intense anthropogenic impact on the environment, insufficient engineering preparation of urban territories, excessive extraction of mineral resources- all these factors increase the vulnerability of people and the urban anthropogenic sphere to the impact of natural hazards. Cities become centres of emergency with increasing frequency, where death and sufferings of people, huge destruction and losses assume the mass and ever rising scale.

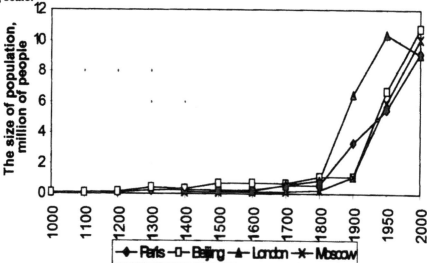

Figure 4. Population growth in some megacities.

NATURAL DISASTERS AND STABILITY OF MEGACITIES

Humankind was fatally afraid of natural disasters from the very early in its history. History knows a lot of examples, when towns once prospering, fell into decay due to natural disasters and then were completely razed to the ground. Despite the progress in science and technology, the protection of mankind from natural hazards decreases constantly. According to the date of World Conference on Natural Disaster Reduction (Yokohama, 1994), [12] the damage from natural disasters increases annually by 6%.

3.6 million people died all over the world, and 3,0 milliard suffered from natural disasters during the last 30 years (1962-1992), with total economic losses having constituted US$ 340 milliard. At present about 250000 people are lost annually from natural disasters, with the damage value being not less than US$ 40 milliard. The damage from natural disasters is forecasted to grow sharply in the years 1990-2000, and will average US$ 280 milliard. But it is evident already, that even this figure will be exceeded essentially, since just the Hunshin earthquake (Japan) on January 17, 1995 caused the economic damage to Kobe-city and its environs of about US$ 60-100 milliard. The next regularity is observed: the social risk (i.E. The risk of people death) rises rapidly in the developing countries, where the quick increase in population goes without adequate investments in construction codes and safe housing. Thus, the ratio of those suffered from natural disasters to the total population is twice higher in Asia than in Africa, 6 times higher, than in America, and 43 times higher than in Europe [12]. At the same time in developed countries with considerably higher protection of people the economical risk is greater due to superhigher concentration of material values accumulated in cities. For example, 4 major natural disasters in USA (namely, the earthquakes in Loma Prieta and Northridge, the Andren hurricane and the Midwest flood), having occurred in recent years (1989-1994) caused huge material damage of US$ 88 milliard with relatively small number of deaths.

The earthquakes in the XX th century are left to be one of the most dangerous disasters among the other natural hazards. The earthquake in Kanto (Japan) in September of 1923 destroyed the city, 142807people having been buried under wreckage and the total material loss averaged US$ 200 milliard. The earthquake in Ashkhabad in April, 1948, brought death to more than 100000 citizens of Turkmenistan capital. An overwhelming seismic catastrophe took place in the city of Tangshan (P. R. China) in July 1976. It killed 242000 people, that is 20% of the city population. The recent earthquakes in Armenia (Spitak, December 1988), in Iran (July, 1990), Japan (Hanshin, January 1995) resulted in death of 25; 40, and 5.5 thousand people correspondingly and caused severe material damage: about 8 million m² of housing were lost in Spitak and Leninokan, while in Kobe-city (Japan) and neighbour regions 74442 buildings were destroyed.

But not only earthquakes cause disaster to ancient and modern cities. Major calamities occur in connection with floods, fires, landslides, subsidence, collapses, etc. Moscow, for example, suffered from various natural disasters more than 130 times in the course of its 850-year long history. The most dangerous of them were the major fires that occurred approximately 70 times (Kremlin burnt 12 times, Kitai-gorod -11 times and Zamoskvorech'e-6 times), floods (25 times), hurricanes and storms (more than 30 times).

A lot of examples of people death and huge material damage due to landslides and mudflows are known. The largest catastrophe occurred in Gansu province of China in 1920, where Haiyan earthquake resulted in vast activization of landslides. More than 100 thousand people died under sliding masses [10].

616 people were killed and 130 thousand buildings were destroyed as a result of a major mudflow in Kobe (Japan), 1938. A catastrophical mudflow destroyed Alma-Ata in 1921, with about 500 people having been killed. The eruption of volcano El Ruis in Columbia (1985) resulted in large mudflows that buried 23 thousand people (from the total 30-thousand population) of the town of Armero. Buildings and trees in the city were cut as if by an enormous bulldozer [1].

The cities in Peru, Japan, USA, Italy suffer great losses annually from landslides and mudflows. In USA, for example, the annual damage from landslides constitutes $ 1-2 milliard, in Japan-$ 1.5 milliard, in Italy -$ 1.1 milliard. 725 Russian cities (that is 68% of 1064 cities with population exceeding 100 thousand) are subject to sliding phenomena.

NATURAL-ANTHROPOGENIC GEOLOGICAL HAZARDS

The problem of people safety and urban stability is complicated sufficiently nowadays since the new hazards, named natural-anthropogenic hazards, develop besides natural ones on urban territories with urbanization growth. Natural-anthropogenic hazards comprise processes and phenomena developing in geoenvironment under the human impact. On the territory of cities with the most intense human impact on the environment, this impact shows itself as development of new processes and activization of slowly-developing natural ones. It does overwhelming material damage to the cities, causing premature deformation of buildings and accelerated destruction of underground communications-lines. These phenomena aggrovate ecological situation and social-psychological tension among citizens, in some cases posing danger to their lives. We may cite as an example the territory of Moscow, where a number of natural-anthropogenic hazards of considerable geological risk for citizens and anthropogenic sphere develops. The danger of geological risk exists nowadays on 48% of the Moscow territory, 12% being the territories of potential risk. Only 40% of Moscow territory may be considered geodynamically stable (Fig. 5). A number of major industrial facilities (such as heat and electric power plants, bases of liquefied gas, refrigerating mills using ammonia gas, etc.) Are situated in sites of natural-anthropogenic geological hazards. In case of an accident due to geological reasons, an emergency situation may arise in the city.

Such processes as induced seismicity, land subsidence, ground-water level rise, karst, suffosion and induced physical fields- should be mentioned among the most widespread natural-anthropogenic hazards developing in megacities.

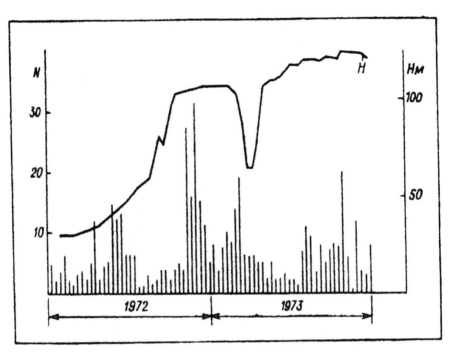

Figure 5. Induced seismicity during filling in the Nurec reservoir on the Vakhsh river. N-quantity of earthquakes, H-variation in water level in reservoir, m.

Induced seismicity

Large water reservoirs, injection of fluids into deep horizons of the Earth's crust, and huge underground explosions, are the main factors of induced seismicity. When the source of this effect lie near cities, they may pose a severe threat to the whole agglomeration. Cases of induced seismicity with filling in reservoirs were observed in many countries: France, Spain, Switzerland, USSR. The induced-seismic activity is proved to correlate with water level in the reservoir. The data obtained for the Vakhsh river in Tadshikistan are very visual in this respect (Fig. 6).

Induced seismicity may be caused not only by major-reservoir construction, but also by injection of fluids into deep geological horizons. This necessity occurs when disposing contaminated waters constructing the underground repositories for liquids and gases, perimeter flooding of hydrocarbon deposits for sustaining the stratum pressure, and in some other cases. For example, at Romashkinskoe oil deposit in Tatarstan (Russia) the essential rise in regional seismic activity with earthquakes up to 6 points of magnitude was observed as a result of perennial perimeter flooding [6].

The underground explosions, including nuclear ones, are the immense factor of seismic activity. Explosions themselves cause seismic effects and in combination with discharging of accumulated natural stresses may provoke rather dangerous induced earthquakes with magnitude up to 5-6 points [14].

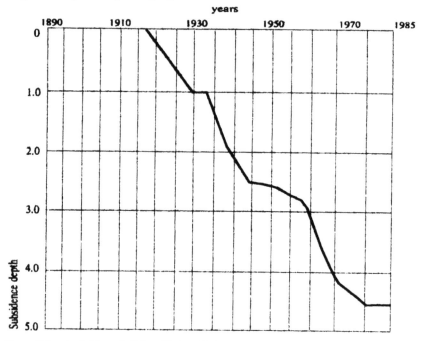

Figure 6. Land subsidence in Tokyo (Koto region).

The underground nuclear explosions are usually set off at remote testing sites far from large cities. But the fact that it is possible to discharge natural stresses beforehand with their help is regarded by scientists as one of the possibilities of planned mitigation of seismic hazard in urbanized areas.

Land subsidence

Land subsidence phenomena occur in urban territories due to human-induced factors. Ground-water extraction is one of the subsidence reasons. The Japanese experts first paid their attention to this effect in connection with land subsidence in Tokyo, Osaka, Niagata and several others cities. Fig. 6 displays the observation data on subsiding the north-eastern part of Tokyo (the Koto region), where in 1970-1975 years. The maximal value of surface level settling was reached (about 4.5m) [8].

The land subsiding in the Mexico-city, having began in the end of the XIX century due to the intense groundwater extraction, achieved afterwards a catastrophical scale. The rate of surface-level settling in some sites of the city reached 30 cm/year in 1948-1952. In late 70s the whole territory of the city subsided for more than 4m, with its north-eastern part-for 9 m. Up to now the subsiding rate is stabilized by reducing the volumes of groundwater pumping and by supplying the city with water from elsewhere.

The surface settling occur also with the extraction of oil and gas resources. Subsidence develops in vast areas including territories of neighbour cities. The most impressing example is the town of Long Beach near the city of Los Angeles. Oil and gas extraction in this region resulted in land subsidence in the city with ever-increasing rate, which achieved 30-70 cm/years up to 1952. The subsidence funnel had the form of an ellipse with axes 65 and 10 km long and square about 52 km2. Up to the early in the 50s, the maximal subsidence value reached 8.8 m with horizontal displacements - 3.7m [14]. Industrial facilities, houses, traffic routes, and the seaport were damaged badly.

Underflooding
Underflooding is the process of groundwater-table rise up to 3 m or less from the surface. It occurs due to disturbance of natural run-off of surface waters during engineering lay-out of the city territory (relief leveling, filling up of gullies and minor-river valleys), and also due to water leakage from underground pipe-lines. Underflooding causes the increase in seismicity of cities, decrease in bearing capacity of soils, and as a result, untimely deformations and damage of buildings and underground communication-lines.

Urban vulnerability to underflooding is extremely high in such countries as Russia, China, England, France, Germany. In Russia, for example, 800 thousand ha of urban areas suffer from ground-water level rise conditions. Underflooding is registered in 792 (74%) cities from 1064 with population of above 100 thousand . Practically all the major cities, such as Moscow, St.-Petersburg, Novosibirsk, Omsk, Rostov-on-Don, Tomsk, Khabarovsk, Novgorod, Yaroslavl, Kazan, etc., Are being affected by underflooding. Water-level rise is registered on 40% of Moscow territory, for instance, and the affected area is expected to constitute 50% in the year 2010.

The cities built on loessial subsoils suffer the most intensely form underflooding. As a result of moistening loesses, they are subject to deformations (compaction) or loess collapse. The collapse value may vary from the first centimeters to some meters, depending on the thickness of the loess mass. Collapses cause damages in buildings and constructions, underground communications, transport routes, etc.

Karst and suffosion collapses
The intense pumping of groundwater and disturbing hydrodynamic regime in areas affected by the ancient karst, may result in development of so-called karst-suffosion processes, leading to forming funnel sinks of natural-anthropogenic genesis. Karst and suffosion development creates serious difficulties for urban construction and often poses a threat to safety of megacity dwellers. In the city of Wuhan (China), for example, the karst collapse of 23 m in diameter and 9.93 m deep occurred in May, 1988. The collapse square amounted 415 m^2, and the volume- 1663 m^3. The collapse funnel enclosed 10 buildings, ruined underground communications, a highway and a electric-power line. High vulnerability to karst and suffosion is registered in other China cities also: thus, 34 sink holes were formed in the city of Guiyang during the period of 1964-1980, and in the city of Kunming 64 cases were marked during the same time [5].

From 1964 till 1970, 6 cases of karst collapse of 2-10 meter of diameter and up to 10 m deep were registered in the north-eastern district of Paris. 54 collapses occurred in Russia in the vicinity of Dzerzhinsk within the area of 283 km2 from 1935 till 1959, while around the city of Ufa more than 80 karst-suffosion collapses have been registered during the last 65 years [9].

During the recent 25 years, 42 karst collapses occurred in the north-western part of Moscow, not subject to such a phenomenon earlier. Those funnel sinks were of 40 m in diameter and from 1.5 to 5-8 m deep. As a result of this, 3 5-storeyed building were affected, with inhabitants having been resettled, and houses demolished.

Human-induced physical fields
The intense anthropogenic activity in cities results in forming there the human-induced physical fields, such as vibration, stray electric currents, and thermal fields.

Vibration fields impact dynamically the subsoil, causing decrease in their bearing capacity. It affects negatively both the maintenance of buildings and constructions and living conditions of people. The electric fields raise the corrosion activity of subsoil to the underground constructions and pipe-lines. The metal corrosion intensifies significantly under the influence of there fields, accelerating the still pipes decay in 5-10 times.

As estimated, about 30% of all pipe breakage in Moscow is caused by electrocorrosion from the stray electric currents. The research fulfilled permit to classify 24% of Moscow area as a territory of high corrosion hazard, where the stray electric fields exceed the natural background in hundreds of times [15].

Alteration in the thermal regime in the city affects biota and ground waters, increases the aggression of the subsoil and ground waters to underground constructions and lines, and in some cases makes unexpected difficulties in engineering works.

MANAGING THE STABILITY OF URBAN AREAS

City dwellers are not helpless, however, facing the natural and natural-anthropogenic hazards. Studying natural conditions and close cooperation between architects, designers, planners, geologists and engineers assists in solving many issues of rational urban land-use and urban stability providing.

The World Conference on Natural Disaster Reduction (held in the city Yokohama in 1994), elaborated a new concept based on forecast of the timely preparedness to natural disasters [16]. This new concept concerns the disaster response as a compelling limited measure, which does not solve the problem as a whole, but yields only temporary results at a very high cost. As regards urban areas, the main constituents of this new concept are as follows: engineering-geological zonation of urban areas; monitoring and prediction of hazardous phenomena; making timely decisions on management.

Engineering-geological zonation
The engineering-geological zonation of the city area is one of the most important measure in mitigating the risk from natural and natural-anthropogenic disasters. The zonation

should be executed by a number of geological factors: relief, rock composition and properties, hydrogeological conditions geodynamical-process development, etc. In the maps of engineering-geological zonation the urban area is divided into zones according to land-use fitness and stability to natural and natural-anphropogenic impacts. This zonation contributes to making appropriate planning decisions and to optimal investments in urban-land development meeting safety requirements.

The seismic-microzonation maps are complied for seismicity-prone regions besides the maps of engineering-geological zonation. The main objective of these maps is subdivision of the city area by the seismic-hazard level (earthquake magnitude) with account of all specific factors, influencing on spreading the elastic waves in geological environment.

Monitoring and forecast
Possible changes in urban geoenviroment should be taken into account during carrying out the planning and constructing activities. This requires the constant control over the urban geoenvironment by means of the monitoring system. The objects of monitoring are specific in every city, depending on geologic, geomorphologic, climatic and other conditions. The database and observation rows are compiled on the base of constantly replenished monitoring information, which are then processed in accordance with the tasks set. The alteration regularities (trends) in objects (processes) under monitoring are further find out to forecast the hazard development.

Making decisions
Observation and forecast data and recommendations provided by scientists form the base for making managing decisions by the city authorities. According to their objectives these decisions are divided into three types: a) adjusting urban planning and land-use; b) aimed at carrying out preventive measure; c) emergency decisions.

Decisions, that regulate urban planning and land-use, include the admitting the legislative standards and directives of urban engineering developments. These documents contain the specific requirements on design and planning solutions, types of foundations and constructions to be used, on development of karstified, landslide-prone or flooded urban areas. In addition to acting standards, the special decisions are often carried out concerning the preventive measures and extra investments in increasing stability of areas and safety of buildings. These measures include construction reinforcing, arrangement of drainage systems, erecting protection walls, increasing the bearing capacity of soils by cementation, strengthening, organization of protection from human-induced physical fields, etc.

Preventive measures include also imposing restrictions on ground-water extraction in the entire city, and in its separate districts; on sewage-waters disposal in deep horizons of geoenvironment; on major underground explosions in the city vicinity.

At last, the emergency decisions are made on the base of short-term forecasts and operational information concerning the hazard heralds when there is no time to undertake the preventive measures. They include the urgent notification of urban population about the forthcoming disaster, population removal (resettling), providing people with shelter, and mobilization of special subdivisions (including military) to response the disaster.

RISK REGULATION BY INSURANCE

The state and city government cannot bear the burden of all expenditures connected with prevention of, preparedness to and response on natural disasters. It is necessary to city dwellers, especially to those owning private capital and investing in construction and realty, to participate in providing safety for themselves and their economic assets at risk. The state and private insurance companies should play an essential role in this process. According to world-statistics data, the expenditures of insurance companies are constantly rising lately and comprises now about 40% of the total economic loss from natural disasters [2]. To set more favorable conditions for their financial activity, the national insurance companies may in their turn conclude contracts with international re-insurance companies, thus involving the latter into the common process of insurance activity. In case of major expenditures for their commitments, insurance companies may use funding of their international partners and thus escape bankruptcy.

Activity of insurance and re-insurance companies coincide in many ways with state and megacity interests, as finally it permits to provide the prompt aid to victims and contributes to safety of people. The latter is achieved by insurance regulation of risk, that is by including in contracts with insurants the requirements on meeting the established regulations on safe disposal of their property, awareness of and meeting the rules of behavior in emergency, etc. Refusal from insurance in case of the object location in hazard-prone area of city may be regarded as one of the extreme measures.

To estimate the amount of insurance payment, the insurance companies need data on the risk of the whole city and its separate districts from the specific natural hazard. To obtain such information the company should collaborate closely with experts assessing the risk, and evidently to invest partially in the risk assessment study.

Divergence in information on natural hazards and related economic, environmental and social losses permits insurance and re-insurance companies to develop probable scenarios of services provided by insurance companies for every megacity. To estimate the possible financial losses the ensurers set the common financial liability it by the losses coefficient, calculated from assessment of natural hazards and, forecasted risk. The sum resulted will be the insurance payment, which determines the amount of insurance allocations. Thus, the insurance companies can achieve the positive balance between incomes and expenses and to escape bankruptcy only on the base of scientific risk assessment.

REFERENCES

1. A Decade against Natural Disasters. *World Meteorological Organization, Geneva, 1994, no 799, 20p.*
2. Berts G. A. Global Warming and Insurance Operations. *Priroda i resursy, 1991 vol. 27, no. 3-4 (in Russian).*
3. Confronting Natural Disasters. An International Decade for Natural Hazard Reduction. National Research

Council *U.S. National Academy of Sciences. U.S. National Academy of Engineering. National Academy Press, Washington D. C., 1987, 60 p.*

4. Fukuoka M. Some Case Studies on Landslides in Japan. Landslides and Mudflows. *Report on Alma-Ata International Seminar, Alma-Ata, 1981. UNESCO, p. 333-352.*

5. Geological Hazards of China and their Prevention and Control. *Geological Publishing House, Beijing, China, 1991. 260 p.*

6. Induced Seismicity. *Moscow: Nauka, 1994. 219 p (in Russian).*

7. Jones, Barclay G and Kandel W. A. Population Growth, Urbanization and Disaster Risk and Vulnerability in Metropolitan Areas: A Conceptional Framework. *World Bank Discussion Paper 168, Washington D. C.: World Bank, 1992. Pp. 51-76.*

8. Karbonin L. The Surface Subsidence -- A disaster of a Global Scale. *Priroda i resursy, UNESCO, 1985 vol. XXI, no 1, pp.2-21 (in Russian).*

9. Kutepov V.W., Kozhevnikova V.N. Stability of the Karst Areas. *Moscow: Nauka, 1989, 150 p (in Russian).*

10. Landslides and Mudflows, *Vol. 1, The Center of International Projects of GKNT, Moscow: 1984, 350 p. (in Russian).*

11. Nabel B.J. Environmental Science. The way the Word Works. (Third Edition). *Prentice-Hall. Inc. Englewook Cliffs. 1990, Vol. 1, 424 p.*

12. Natural Disasters in The World. Statistical Trend on Natural Disasters. *Natural Land Agency: Japan, IDNDR, Promotion Office, 1994, 18 p.*

13. Nigel H. And Puente S. Environmental issues in the Cities of the Developing World: The Case of Mexico City. *Environmental Issues in the Cities of the Developing World, 1990, 2(4), pp. 500-532.*

14. Nikonov A. A. The Man Affects the Earth's Crust. *Moscow: Znanie, 1980, 47 p.*

15. Osipov V. I. Zones of Geoloigcal Risk on the Territory of Moscow. *Vestnik Ross. Akad. Nauk, 1994, vol.64, no.1, pp.32-45.*

16. Osipov V. I. Natural Disasters in the Limelight of Scientists. *Vestnik Ross. Akad. Nauk, 1995, no.6, pp.483-495.*

17. United Nations. Estimates and Projections of Urban, Rural and City Populations. 1950-2025: the 1982 Assessment. *New York, United Nations, 1985.*

19. United Nations. The prospects of World Urbanization. Revised as of 1984-85.(ST/ESA/SER.A/101) *New York: United Nations, 1987.*

19. World Resources Institute, United Nation Environment Programme and United Nation Development Program. *World Resources. 1990-91. New York, Oxford University Press, 1990.*

Proc. 30th Int'l. Geol. Congr., Vol. 23, pp. 357-365
Wang Sijing and P. Marinos (Eds)
© VSP 1997

Sustainable Development and Mineral Resources Accounting of China

DU DONGHAI
Institute of geology, Chinese Academy of Sciences, Beijing 10029, China

WANG ZHIXIONG
State Science .& Technology Commission, Beijing 100862, China

LONG YULIN
Department of Accounting, the Hong Kong Univ. of Science & Technology, Kowloon, HK

Abstract:

China is a developing country, and its economy grows fast in recent years. Mineral resource is one of the main support for the development of the socio-economy of China. With the great development in mineral resources exploitation and utilization, there are great damage on ecology and environment as well as the depletion and deterioration of natural resources. In order to put the socio-economic development into a sustainable model, we should take natural resources into the system of national economic accounting and calculate the value and cost of natural resources and environment. In this way, we can achieve the aim of sustainable development of the socio-economy under the limitation of natural resources of the earth.

Keywords: Sustainable Development   Mineral Resources Accounting

THE CONCEPTION OF SUSTAINABLE DEVELOPMENT

Modern science and technology have had great impetus on the development of human's productivity. The development of science and technology has enlarged the area that can be explored by us and has improved our ability to utilize nature for our own benefit. Human civilization has been brought up to such a high level that can't even be imagined by people in the past. However, the development and production of human society have caused many social problems in the meantime. During the past several decades, the dramatic increase of world's population, shortage of food supply, deterioration of ecological environment, depletion and excessive consumption of natural resources and energy have developed into such an uncontrollable state, that even the proper environment for human existence is seriously threatened by these problem. And it already became a worldly problem and the focus of concerning.

The world's population reached 5 billion on July 11th, 1987, increasing steady on a rate of about 70 million annually and now the world's population is 5.8 billion. To feed so many people on the earth, we need more and more natural resources such as land, water or mineral resources. The maximum production in pursuit of profit of human beings has caused depletion of resources and degradation of environment. Many resources are nearly exhausted because of excessive consumption. Some kinds of non-renewable resources which were once thought abundant and wouldn't be used up, i.e. mineral resources, are exhausting. The citizen of the earth village became more and more aware of the fact that we are faced with the problem of ruining our home by ourselves. The serious fact forced people to reexamine the relationship between human and nature.

The concern of world's resources crisis and environmental problems rose up in the 1960s. A future science research institute, named Rome Club, which mainly consists of European scholars, entrepreneurs and politicians, firstly put forward the conception of global problem. In June, 1972, the UN's Conference on Human Environment was held in Stockholm, and at this conference, *the Declaration of Human Environment* was passed. This showed people's concern about environmental problems. In the same year, *the Limit of Growth*, a book brought out by Rome Club also had caused a great shock. The conclusion of *the Limit of Growth* is that: The unlimited growth of socio-economy is unrealistic, and to wait for nature to stop the growth of socio-economy is not a way human being willing to accept. Where is the way for us? Human beings should constraint their growth or to develop in coordination with nature, this is the best way. Since 1970s, there came out a lot of theoretical research papers on sustainable development. Sustainable development defined as an scientific terminology appeared in *the World Natural Resources Protection Outline* at 1980. Though the document is mainly put forward to deal with natural resource protection, its content is far more than pure natural resources protection, instead, it regarded protection and development as two aspects of the same issue that cannot be separated and put nature protection into the frame work of social development. In this document the definition of sustainable development is: to improve the living standard of human beings without exceeding the capacity of natural ecological supporting system.

The work done by World Conference on Environment and Development (WCED), which was established on November, 1983, has an important impact on the formation and the development of the conception of sustainable development. The WCED was set up according to the UN's decision and get the UN's support in its work. Under the guidance of Mrs. Bulunterland, former prime minister of Norway, on the basis of research of important economical, social and environmental problems of many countries all over the world, they brought out the report of *Our Common Future*, and it was also called *Bulunterland's Report*. Here the definition of sustainable development is: while we satisfied the need of the current generation, we should not damage the ability of the later generation to satisfy their need. This definition is widely used in the research of sustainable development problems and accepted by most people.

Our Common Future put forward the principle of Fairness, Sustainability and Common, advocated fair distribution of resources, and took the demand of the later generations into consideration. It also tried to find a sustainable economic development model on the basis of protection of the earth's natural ecological system, to achieve the aim that human being and nature living in a harmonious way. *Bulunterland's report* was passed on the 42nd UN's conference and became an important theoretical document in the field of environmental protection and economic development.

The UN's conference on Environment and Development (UNCED) held in Ri De Janeiro, Brazil, in 1992 was a historical conference on social development and environmental problems of human beings, and it marked an important milestone for awakened the world to the need for a kind of development that doesn't jeopardize the existing condition of future generations. The leader of most of the countries had attended the conference, so the conference is also called the earth's summit. In the main theme of sustainable development, the UNCED pushed environmental and developmental problems up to a position of the greatest concern. Many decision and documents are made by the conference and the most important ones are *Rio Declaration* and *Agenda 21*, in which sustainable development will not only be a theory but also be carried out in practice. The Rio Conference secured a set of agreements between governments which marks a significant advance in international cooperation on development and environment issues (Ghali, 1992). And it marshaled political commitment to these arrangement at the highest level and placed to issues of sustainable development at the heart of the international agenda .

MINERAL RESOURCES AND SUSTAINABLE DEVELOPMENT

Environment and development are two major concerns of the international community. Yet, neglect of environment in the process of industrialization, particularly the irrational exploitation of mineral resources, has caused global environment pollution and ecological degradation, and has posed a real threaten to the survival and development of mankind. Economic development is essential to the very survival and progress of mankind. Furthermore, it provides a material guarantee for the protection and improvement of the global environment. Economic development should go hand in hand with environment protection for the sustainable development of socio-economic (Li Peng, 1992).

Mineral resources and sustainable development
Mineral resources are non-renewable resources which should be extraordinarily cherished, rationally allocated and efficiently utilized. China is rich in mineral resources in terms of their total amount, but the per capita figures are less than half of the world's average. Currently, 95% of the energy and 80% of industrial raw materials needed for economic construction depend upon the mineral resources supply. The proven reserves of minerals are evidently insufficient and will become acutely in short of supply during the 21st century and won't ensure the sustainable development of national economy. Meanwhile a lot of problems in the exploitation of minerals coupled with low level of integrated development and utilization of resources have aggravated the gap between demand and supply. Efforts should be directed towards both increasing production and reducing resources consumption. Apart from reinforcing the geological exploration of mineral, and boosting their proven reserves, it is necessary to stick to the basic rolling of protection, conservation and rational utilization of resources, and heighten the public awareness concerning the significance of rational exploitation and utilization of mineral resources as required by the sustainable development of economy and society.

The irrational exploitation of mineral resources leads to not only waste of mineral resources but also degradation of ecological environment. According to statistics, in China occupation of space and surface subsidence resulted from chaotic dumping of tailings and waste of large-scale mining have amounted to 2 million hectares and are still growing, which had been bringing about atmosphere and water pollution and triggering subsidence, landslide, mud-rock flow and other geological disasters. Hence, to effectively curb the irrational development model and to reduce the environmental cost incurred due to exploitation of mineral resources do represent an urgent task that has to be tackled during development and utilization of mineral resources in China to fit in with the requirements of sustainable development.

Mineral Resources are material foundation of the development of human society. Social economic development of a country is closely connected to the exploitation of its mineral resources, and the stock of mineral resources is an important part of national wealth. It is the basic raw material employed to improve people's living condition. With the economic development, the role played by mineral resources became more and more important. But the types and quantities of mineral resources are limited. On certain technological level, the ability and sphere of human beings to exploit and utilize mineral resources are also limited. The scarcity of mineral resources forced us to make the best use of it and allocate it in an optimal way in the social production.

The exhaustibly of mineral resources leads to the depletion and the crisis of resources (Li, 1991). The development of socio-economy has been restricted by the drastically growth of world population, ecological destruction and environmental deterioration. In China, mineral resources were once used freely. The maximization of output value without any care about the protection of the mineral resources' base in the process of production lead to depletion of mineral resources. The economic output value increment is at the expense of constant degradation of mineral resources. This kind of economic growth may exhaust natural resources, and cause unrecoverable destruction of environment.

Now Chinese are faced with the challenge of how to efficiently manage natural resources and how to exploit mineral resources rationally. In order to use mineral resources at a rate compatible with their reproductive capacity, we must realize the scarceness and the real value of the mineral resources.

Mineral Resource Condition of China
Mineral resources are crucial to human society. The production and consumption of mineral resources is directly connected with the degree of economic development. Since mineral resources are exhaustible, the shortage of mineral resources will be a very serious problem.

China's economy is developing at a rate of more than 10% per year. With such a high speed of economic growth, the demand for the raw materials and energies is also very large. China is at the beginning of middle stage of industrialization, and the resource consumption speed is increasing steadily. The comparison between 1985 and 1953 shows that the national income increased 6 times from 1953 to 1985. In the meantime the consumption of energy resources, iron ore and nonferrous metal of 1983 have increased respectively 14, 24 and 23 times as much as the consumption of 1953. However, mineral resources situation of China is not promising. China, with its large territory, has abundant resource reserve (Table 1.). However, because of its large population, per capita resource of China is very low. The potential reserve total value of China is listed in the third after the United States and former Soviet Union in the world, but the per capital value is listed at the 53rd and per unit area value is arranged in the 24th in the countries of the world.

Per capital mineral reserve of China is not optimistic, and some kind of mineral resources are in serious shortage (Table 2.). The main kind of mineral reserve except coal are far less than the average per capita amount of the world. For example, per capital iron of China is only 13% of that of the world average, and oil is only 34% of that of the world average. With the development of China's industry, the demand of the mineral resources has increased at a speed much faster than that of supply. It was estimated that in the year 2000, main mineral production except coal won't meet the demand of the socio-economic development (Table 3.) The production of oil is only about 82% of the demand and the production of iron and Cu are about 77% and 79% of the demand.

Table 1. The potential total value of mineral reserve of some country (1990)

	Total potential value		Per capital value		Per unit area value	
	Trillion US$	Arrangement in the world	Billion US$	Arrangement in the world	Million US$	Arrangement in the world
United States	29.8	1	1.20	19	3.18	17
Former USSR	21.8	2	0.76	25	0.98	34
China	16.6	3	0.15	53	1.72	24
South Africa	8.9	4	2.56	10	7.29	10
Australia	6.6	5	4.14	5	0.86	39

Data source: *Mineral resources situation and policy of China in 21st century*

Table 2. per capita mineral reserve comparison between China and the world

MR	Unit	per capital of the world (Wpc)	per capital of China (Cpc)	Cpc/Wpc (%)
Raw Coal	Ton	180.8	179.6	99
Iron	Ton	18.39	2.41	13
Crude Oil	Ton	24.03	8.23	34
Cu	Kg	118.0	28.4	24
Pb	Kg	31.2	10.65	35
Sn	Kg	43.2	25.3	58
Al	Ton	5.74	0.8	14
Mn	Ton	0.29	0.053	18

Data source: *The Management of Geological Economy*

In the 21st century, the shortage of mineral resources of China will be more serious. At that time in China the primary mineral products will not be enough to satisfy the demand and some other kinds of mineral products will be in serious shortage. This situation cannot be changed by import of mineral products from international markets, because the import of mineral products will be restricted by Chain's foreign exchange reserve. Even if China has enough money, international market will not be able to meet such a great demand. The price of mineral products in international market may increase by many times because of the large amount of demand from China and the rest of the world. Maybe with the development of modern science and technology, we will find new resources to replace mineral resource. For example, maybe the solar energy will be used widely as the source of energy instead of oil or coal. And maybe we will improve the techniques of production in the future, by that time, a little consumption of mineral resources will bring a great output. However, all these are our wishes, we don't know whether or not they will be turned into reality, and we can't put our future on wish and uncertainty. We must start now to deal with problems that we may face in future.

Table 3. The demand and production of main minerals of China in the year 2000.

Unit: million ton

	Demand (D)	Production Ability (P)	P/A
Coal	1460	2700	185%
Oil	1800	1480	82%
Iron	3190	2450	77%
Cu	9.7	7.4	79%

Data source: *The Management of Geological Economy*

The problems in exploration and utilization of mineral resources
Because of the failure of former planing economic system on the guidance of mineral resources utilization, in China there exists serious waste in the utilization of mineral resources. One of the failure of the former management system of mineral resources is that mineral resources are used freely (Qian, 1996). So in order to maximum the output value of industry, excessive mineral resources are put into production, and the utilization efficiency is very low. The increase of the output value is not dependent on the improvement of the production technology but at the cost of excessive use of exhaustible resources, and in turn the lagged productive techniques will spend more resources in the procession of production. To produce a profit of one trillion dollars we use 2,668 tons of coal, while the united states only use 50.8 tons of coal for the same profit (Table 4). China's consumption of energy, steel and copper in the production of same amount of value are about 4.79 times, 3.68 times and 3.65 times compared to that of the world average.

Table 4. The utilization efficiency of resources comparison.

	Energy 10^4ton coal/10^8\$	Steel 10^4ton/10^8\$	Copper ton/10^8\$
China	26.68	1.14	197.08
United States	5.08	0.17	36.64
Formal USSR	31.55	2.14	165.00
Japan	1.91	0.25	49.78
World average	5.57	0.31	53.97
China/world average	4.79 times	3.68 times	3.65 times

Data source: Mineral resources situation and policy of China in 21st century

For the same reason the recycling rate of used resources is very low in China (Table 5). Compared to western countries, the recycling rate of used metal of China is lower than half of that of western countries. For example, in China, the recycling rate of Al is 11.2%, while western counties is about 26.3%, and Cu is 33% in China and 50% in western countries. The recycled steel used in production of one ton of steel is about 300 kg in China, in the United States the figure is 330 kg and in Italy is 660 kg.

The damage of environment and pollution are also caused in the process of mineral resources exploitation and utilization. Main problems are loss of farmland, destruction of parries and land's desertification. For example, the solid wastes of mines have caused serious pollution to the land. When we estimate the profit of mining regard to the requirement of sustainable development, we found that due to the degradation of the environment and the depletion of the resources, the cost is too high.

Table 5, Recycling rate of China and western country (%)

	Al	Zn	Pb	Cu	Steel
China	11.2	6	14.7	33	30
Western country	26.3	50.3	29.9	50	50

Data source: Mineral resources situation and policy of China in 21st century

In one word, high input, low efficiency and high pollution are characteristics of traditional production model of China. Now when we are faced with shortage of mineral resources, degradation of environment and over-increment of population. We must change our production model to fit in with the requirement of sustainable development of society. We must realize that the exploitation of resources and protection of environment are of the same importance for our development.

RESOURCES ACCOUNTING AND SUSTAINABLE DEVELOPMENT OF CHINA

Resources are the foundation of economic development. The increase of the wealth of a country partly depends on rational exploitation of its resources. Emphasizing output value of economic production, neglecting the resources base and irrational exploitation of natural resources lead to tremendous waste and depletion of resources as well as the deterioration of environment. China, with a very fast speed of economic growth in recent years, compared to a great deal of policies that are made to promote its economic growth, the concern about natural resources and environment protection is quit little. It is necessary to pay a great attention to natural resources and environment protection. Only in this way, we can achieve the aim of sustainable development of the Chinese economy.

One way to achieve this goal is to calculate the value of natural resources both in quantities and in quality, to highlight the real value of natural resources and environment and eventually take the value of natural resources and environment into the system of national account (SNA) (Du, 1996). We should include natural and environment into economic calculations in such a way as to require economic agents to base their decisions on the production and allocation of goods on an information set which explicitly include the use of such resources (Beltratti, 1995). Ultimately, this should greatly improve the overall efficiency of the economic system, which is now threatened by the current accounting organization, which under values the contribution of nature to production. Such under valuation results in various deficiencies which prevent index like Net Domestic Product to be interpreted as welfare indicators.

To take natural resource and environment into SNA is necessary requirement for human beings that is developing under circumstance of limited space and resources of earth. Current SNA is a kind economic information system with drawbacks, for it not sufficiently take the value of natural resources and environmental function in the socio-economic activities into consideration. For example, depletion of mineral resources and forest, and degradation of air and water caused by production, all these had not been regard as cost, had not been deduct out from GDP or GNP. On the contrary, the income comes from depletion of mineral resources and degradation of air and water are regarded as profit, or the aggregation of national income. Current SNA over-evaluated the profit of economic activities and under-estimated the cost. This lead to the pursuit of facetious increase of national income at the cost of depletion of natural resources and deterioration of environment. For example, in industrial countries, about 70% of the economic growth come from economic activities that has serious pollution on environment (Sheng, 1994).

Because there are expenses for cleaning or restoring the negative effecting on environment due to economic activities, the information provide by SNA was distorted even more. For example, the produce of facilities that are used to clean the leaked oil in ocean will increase the income, therefor, in terms of accounting methodology, the more destruction of the environment, the more protection activities, thus, the greater increasing of national income. Because main indexes of SNA, i.e., GNP, are widely used by news agencies, governments and international organizations, this leads to a wrong impression that GNP is regarded as an index of welfare. So governmental policies aimed at increment of GNP, while paid little attention to other things such as the quality of commodities and services, sanity and environmental condition. Take the United Kingdom as an example, national income of the country was increased 230% from 1950 to 1990, but the calculation according to the cost caused by the increasing of the distance between homes and work, worse of safety and the aggregation of the deterioration of the environment, the social profit has decreased. The cost of polluted water, air and noise is more than 22 trillion pounds, which is about 6% of the UK's national income of that periods. Such situation is also existing in other industrial countries such as the United State, German and the Netherlands etc..

In order to account for depletion and degradation of natural resources in the national account, the macro economical aggregates have to be modified in the following way (the UN, 1993). For Gross Domestic Product, GDP, which is equal to the sum of gross value added of all the resident institutional unit, when depreciation of man-made capital is deducted out we obtain Net Domestic Product, NDP. When costs of depletion of natural assets are also taken into account, it is possible to obtain the Environment Adjusted Net Domestic Product, or EDP1. After estimated cost of degradation are subtracted from EDP1, we get EDP2. In the same way, we can compute the Environmentally Adjusted Net income, ENI. ENI includes the damages to the environment that are related to production activities like government and household's expenditures on environment protection. Only in this way the index of SNA can be regard as indexes of welfare.

Here is an example of the estimation of EDP and ENI. In Mexico for 1985, the gross domestic investigation is about 22% of the GDP. After adjusted by the depreciation of man-made capital, the figure is 12% of the GDP, and after subtracted out the depletion of natural resources, the figure is 5% of the GDP, and after deducted out the cost of degradation of environment, the final figure is -2% of the GDP. That is to say, with the costs of natural resources and environment taken into account, the gross domestic investigation of Mexico in that year had caused an opposite effecting on the welfare of the country. Other developed countries such as Norway, the United States, France, the Netherlands and Australia have also done similar work on environment and resources accounting.

Since mineral resource is a kind of non-renewable resources and economic production is highly dependent on mineral resources, so Mineral Resource Accounting is an important part of resource accounting,. Though the output value of mineral industry only accounts for a little part of GNP, more than 70% of social production is supported by mineral products. The establishment of resource accounting can help us to evaluate the quality of development, and to forecast future development. Resource accounting will also provide us information to enhance of resources management of the country and help us to set up laws and regulations to manage the resources scientifically, and to utilize the resources efficiently.

In the 21 century, the most serious problem that we will be faced with is sustainable development of society. In the prerequisite of sustainable development, the UN proposed that natural resources should be taken into the system of national economic accounting (SNA), and brought up the model of the System of Environmental and Economic Accounting (SEEA) (Sheng, 1995). With natural resources and environment taken into account, SEEA is an integrated accounting system of environment and economy. The proposal of the UN was approved by most countries. The UN drew the conclusion that SEEA will gradually take the place of SNA to fit in with the requirement of sustainable development of human society.

It is necessary to point out that the study of resource accounting in China is still in its early development. The operation of resource accounting and taking resource accounting into the SNA of China are determined by the efforts of the Chinese government. The earlier resource accounting is established and taken into the SNA, the earlier the social economic development will be put into the healthy trail of sustainable development.

Acknowledgments

The research of mineral resource accounting of China is sponsored by the State Science and Technology commission of China in project 305. And thanks Dr. Andrea Beltratti of Torino University of Italy, for his notes and lecture on natural resources accounting. Also thanks Miss Fan Yin of Peking University of China, for her help to make this paper more readable in English.

REFERENCES

1. *Earth's Summit '92*, 1992, The United Nations.
2. China's agenda 21, 1994, China Environment Press
3. Li Jinchang, 1991, *the discussion of resources accounting*, Ocean Press, Beijing, China (in Chinese)
4. Qian Kuo, Du Donghai, 1996, Some points in natural resource assets management, in press
5. Du Donghai, Wang Zhixiong, 1996, Natural resources accounting: the management of sustainable development. *the Compacting of Science on Society*, No. 1. (in Chinese)
6. Du Donghai, Wang Zhixiong, 1996, A Case of Mineral Resources Aaccounting in Xinjiang, China , in press
7. Andrea Beltratti, 1995, From the theory of environment accounting to the system of environment and economic accounting, *FEEM's Newsletter.*
8. Fulai Sheng, 1995, *Real value for Nature: An over view of global efforts to achieve true measure of economic progress*, WWF International, Gland, Switzerland.
9. Andrea Beltratti, 1995, The notes of environment economics

Proc. 30th Int'l. Geol. Congr., Vol. 23, pp. 367-373
Wang Sijing and P. Marinos (Eds)
© VSP 1997

Urban Environmental Geological Problems
----Analysis of urban hazardous environmental effects in China

LIU YUHAI
Xi'an College of Geology, Xi'an, China, 710054, China

Abstract

Environmental geological problems in 200 large or medium-sized cities in China are discussed here.The total city number in China is 640(1995).Statistics shows that more than 50 cities are under threat and influence of active faults;22 cities have conditions of liquefaction; since the middle of the next century, coastal cities such as Shanghai,Tianjin,Ningbo,Guangchow would firstly face the threats of crustal subsidence and rise of the sea level; ground fissure, as a special kind of urban geological hazard, has done huge damages to civil buildings and engineering in Xi'an, Datong,Handan and Yanchow; 70 cities, including those in the upper or the middle reaches of the Yangtze River or the Loess Plateau such as Chongqing and Lanchow,are in dangers of landslide and debris flow ; ground subsidence is found in more than 50 cities due to excessive groundwater abstraction, and in some cities as Shanghai, Tianjin and Xi'an, each of the subsidence amount has exceeded 2000mm; the number of cities effectd by collapse event is increasing gradually, more than 40 cities are lying in karst regions. All of above are prominent hazardous environmental geological problems in urbanization development in China.

Keywords: City, Environmental Geology, China

INTRODUCTION

China, a country with vast territory and geological complexity, its urban environmental geological problems are becoming increasingly striking with high-speed urbanization and human engineering-economical construction in recent years.

URBAN ENVIRONMENTAL EFFECTS CAUSED BY FAULT ACTIVITY

The country is located in the eastern part of Asia continent, also belongs to the sourthern edge of Eurasian plate geo-tectonically. It is surrounded by Pacific plate and Philippine plate to the east, main part of Eurasia plate to the north and Indian plate to the south-east. The interactions among the plates strikingly influence the modern crustal activity, especially fault activity and seismicity, thus extensively effect the urban construction and development. Incomplete statistics shows that 50 large or medium-sized cities, including 14 central ones(provincial or municipalities directly under the central government) and 16 coastal cities (including Taiwan), are strikingly influenced by fault activity directly or indirectly (Table.1). The regional faults control the seismicity, but the subsidiary active faults mainly appear as dislocation deformation resulted from creepslip. They affected the urban environment geology negatively through damaging civil buildings(e.g. seven ground fissures are controlled by five active faults crossing the urban area of Datong city).

Table. 1 Statistics of The Large Or Medium-sized Cities Influenced By The Main Active Faults In China

Names of the active faults	The	influenced	cities
	Key cities	Common Large or medium-sized cities	
Tanlu Fault	Shenyang Hefei	Dalian Yinkou Anshan Liaoning Weifang Anqing	
Yilan--Yitong fault		Jiling Jiams Hegang	
Qingyuan--Mishan Fault		Fushun	
Yialujiang River Fault		Dandong Donggang	
Gongle--Lan'ao Fault	Fuchow	Quanchow Xiamen Shantou	
Zuowu--Heyuan Fault	Guangchow	Shengzhen	
Taihang Piedmont Fault	Beijing Shijiazhuang	Baoding Handan Xingtei Anyang Xinxiang Jiaozuo	
Cangdong Fault	Tianjin	Cangchow Tangshan	
Fenhe River Fault	Taiyuan	Datong Yichow Yuci Linfen Yuncheng	
Weihe River Fault	Xi'an	Baoji Xianyuang Weinan	
Helanshan--Liupanshan Fault	Yingchuan		
Northern Qilian--Corrider Fault		Jingchuan Zhangyue Jiuquan	
South-western Huashan Fault		Tianshiu	
Anninghe River Fault	Chengdu	Xichang	
The Little River Fault	Kunming	Dongchuan	
The North Fault of Tianshan	Urumqi		
Taiwan Longtitudinal Valley Fault		Jilong Gaoxun Taizhong	

URBAN ENVIRONMENTAL GEOLOGICAL EFFECTS CAUSED BY SEISMIC RISK

There were 1000 hazardous earthquakes in the history of China according to records.Tangshan earthquake(1976,M=7.8) was the most hazardous one occurring in urban area among them. According to Seismic Intensity Regionalization of China(1990), with 10% as the surmount probability in the coming 50 years, in the country's 200 large or medium-sized cities,there would be 22 ones with seismicity intensity>VII,including one with the intensity IX (Taipei of Taiwan), 10 with the intensity VIII (Beijing,Kunming, City Hu, Taiyuan, Xi'an, Haikou, Yinchuan, Lanchow, Urumqi and Lasha),11 cities with the intensity VII(Changchun, shengyang, Tianjin, Zhenchow, Hefei, Lanjin, Shanghai, Fuchow, Guangchow, Chengdu,Xining).

Obviously, the risk is premised 10% as the surmount probability in 50 years, so there are differences between the meaning of environmental geology of urban seismic risk and implication of urban seismic intensity regionalization. The former considers a more extensive time-space scale and involves far more probllems than the intensity.Thus accuratly understanding the environmental geologicial characteristics of urban seismic risk is more practical for urban construction and development than only knowing seismic intensity.

URBAN ENVIRONMENTAL GEOLOGICAL EFFECTS CAUSED BY EARTHQUAKE LIQUEFACTION

The basic causing conditions of earthquake liquefaction are:(1) local seismic intensity no less than VII;(2) layers of silt or silt clay with certain thickness within 20m underground; (3) most of the liquefaction strata belong to Holocene deposit. After their earthquake liquefrction being analyzed, the main cities in China with the condition 2nd can be determined as future liquescent cities, whose total number is 53 out of the 200 large or medium-sizd ones.

The main basis of earthquake liquefaction determination through engineering geology emphasizes four aspects as following: (1) morphological type—belonging to littoral delta plain, channel-mouth delta or alluvial plain, lower terrace and flood land, front edge of alluvial fan; (2) stratigraphic time and origensis—belonging to Q4 and al.,al+ml,al.+pl., or Q3 and al.; (3) depth of groundwater surface—generally 3--5m with 8m as the maximum; (4) lithological type—belonging to silt,fine-silt,silt clay or warp soil.

URBAN ENVIRONMENTAL GEOLOGICAL PROBLEMS CAUSED BY CRUSTAL SUBSIDENCE AND SEA-LEVEL RISE

It is predicted that even though the developed countries control their discharged amount of CO_2 in the coming days,the gas' globle amount would has increased two times the curreent amount by the end of 2050. But without controlling,the globle temperature would increase 3 degrees Centigrade then.Correspondingly the sea-level would rise 65cm,there by many coastal cities would be submerged. According to the tide-level observing data of China,the average rising velocity of the sea-level is 0.4/a during the recent century, but showing differences in different sea areas. Each of the Bohai Sea,the Yellow Sea, the Eastern China Sea and the Southern China Sea has risen 5.0cm, -2.0cm, 19.0cm and 2.0cm respectively. According to the topographic displacement monitorring data from the National Seismic Bureau, the Eastern Liaoning Gulf, the Bohai Gulf, and the delta plain of the Yellow River, the Huihe River, the Yangtze River, and the Jujiang River all belong to modern crustal subsidence area,with general velocity 2-3cm/a.

Based on the data above, it can be predicted that with the the rising trend of the globle sea-level,since the middle of the next century, the coastal cities such as Shanghai, Guangchow, Zhanjiang, Haikou and etc., would firstly face the threats and the influences from both the sea-leve rise and the crustal subsidence. In the case,coastal pretection projects, just as Holland has done, ought to be carried out to maitain the existence and development of the coastal cities costly.

URBAN ENVIRONMENTAL GEOLOGICAL PROBLEMS CAUSED BY GROUND FISSURE

Since 1960's,ground fissures of multi-origensis occured in 25 provinces in succession in Northern China and the middle and the lower reaches of the Yangtze River, coverring 60 kilometers or so of land. A few of them occured in urban area, but their hazardous effects are violent.The country's ground fissure-influenced cities, ordered in seriousness, are: Xi'an, Datong, Handan and Yanchow(Table. 2).The recent reseaching achievements show that since 1960's, these fissures gradually increased their activity. Today, the increasing trend is still continuing. The fissures in the cities induced by modern crustal activity show characteristics of growth faults,so they have genetic or direct relations to the activity of the under faults.The studies of the fissures in Xi'an and Datong indicaded that excessive abstraction of groundwater critically influences active amount of ground fissure. Seventy percentage of vertical differential settlement is induced by the abstraction.

Table. 2 Brief Introdution of Environmental Geological Effects Induced By Ground Fissures In Some Cities In China

Ctiy name	Total number and length	General strike	Main genesis and other influencing factors	Damades estimated (Chinese Yuan)
Xi'an	12 (70km)	NEE	tectonic (active faults) groundwater abstraction	>500000000
Datong	7 (20km)	NE --NEE	tectonic (active faults) groundwater abstraction	>300000000
Handan	5 (7.5km)	S--N	tectonic	>100000000
Yanchow	3 (5.0km)			

The directly hazardous effects caused by ground fissure are the main damages to buildings and underground lifeline engineering, and threaten lives and property of urban citizens, so belonging to negative environmental geological effects of nature---human interactive type.

URBAN ENVIRONMENTAL GEOLOGICAL PROBLEMS CAUSED BY LANDSLIDE AND DEBRIS FLOW

Landslide and debris flow compose the main hazardous problems of urban environmental geology in slope zones of cities located in mountains or river valleys, especially valleys of piedmont belts in the Loess Plateau,and slope zones along valleys of the middle or the lower reaches of the Yangtze River. Because of the poor basic environmental geology in cities lying in these regions,more over, influences of land recondition for urban building, mining, road and irrigation construction and etc., landslide and collapes can easily occur, threatenning urban building, traffic net and safty of the citizens.Perticularly in torrential season (daily rainfall >50mm), debris flow might more likely occur, making the situation more serious. Statistics indicate 70 cities or so are under threat of debris flow in China. In Chongqing, an important city located in the junction area of the Yangtze River and the Jialingjiang River,ancient and modern landslide account to 26 and potential dangerous places 80. Since 80's,events of slope deformation and landslides occurred one after another, and seriously worsenned the urban geological environment. The amount of economical damages and cost of treatment projects has been over 100 millions of Chinese Yuan. Some other cities as Lanchow, Tianshiu and Yan'an,located in the Loess Plateau,have been hit or influenced by landslide, debris flow for many times. Rainstorm is the main factor of the effects. For instance, there are five bebris flow ravines around Lanchow city, in Aug. 20th, 1964, debris flow induced by a rainfall of 150mm in 4 hours invaded in factories and dewelling, buried 3.36km of Lan—Xing railroad.

As hazardous urban environmental geological problems,landslide and debris flow could not be overlooked ,especially in those cities located in South-western China and North-western China.

URBAN ENVIRONMENTAL GEOLOGICAL PROBLEMS CAUSED BY GROUND SUBSIDENCE

In order to meet the need of the gradually-developing urbanization,especially of the increase of industrial companies, groundwater is much more excessively abstracted. Sequently induced ground subsidence---another kind of negative environmental geological effect in some large or medium-sized cities in China, doing damages to buildings, underground lifeline engineering, traffic and civil engineering on different levels. Besides these, in coastal cities, hazardous intensification of storm surge induced decline of the elevation and seawater encroachment induced by excessive abstraction of groundwater are indirect negative effects of subsidence. According to statistics, ground subsidence has occurred in no less than 50 cities in China. Some typical cities suffering the effects are:Shanghai, Tianjin,Wuxi, Changchow,Ningbo(Table 3). Most of them are located in coastal delta plains, some in inland basins or front zones of piedmont alluvial fans.
As to the whole country,its 1/4 of large or medium-sized cities are in effects or influences of ground subsidence, and the total number of effected cities is likely still increasing.

URBAN ENVIRONMENTAL GEOLOGICAL EFFECTS CAUSED BY CAVING-IN IN KARST REGION

According to incomplete statistics, there are nearly 40 of large or medium-sized cities located in karst region in China.In recent years, because of quick decline of groundwater level by excessive abstraction and increase of building loads in urban area,events of karst caving-in occured occasionally,resulting in destablizing of building foundations of factories, dwellings and schools, damaging buildings and underground lines, also causing breaking off traffic, water service and power supply. E.g. Wuhan city, since 1977, karst caving-in occured in the Steel Rolling Mill of South-middle China,Yuanjia street and Lujia street, forming several caves (or depressions), and caused millions of Chinese Yuans of damages, in Tangshan city, because 8 times of karst caving-in occured in the municipal stadium, the Tenth School, the Phonics Hill Park from 1991, the economics were worsened, and the complexity of the geological environment of the city were added.In this developing country,   except the eight prominent hazardous environmental geological problems mentioned above, there are still some others standing out gradually, such as: caving-in and destabilization induced by mining; seawater invasion induced by excessive abstraction of groundwater in coastal cities, leading underground freshwater be salted; water source pollution by waste solide stacking, waste water discharge and exploitation of underground thermal water, possible terran destablization induced by under space utilization and etc.

Table.3 Statistics of Urban Ground Subsidence In China

City	Morphological type	Strata		Subsidence		Active time	Observing methods and the controlled
		time	genesis	Accumulated amount(mm)	velocity (mm/a)	20th century	time*,other subsidence$
Shanghai	coastal delta plain	Q3-4	al+ml	2638	200(max)	21--65	bedrock,layer marks,crustal$
Tainjin	coastal delta plain	Q3-4	al+ml	2960	200(max)	59--94	bedrock,layer marks,crustal$
Xi'an	inland fault basin	Q1-3	al+l	2000	300(max)	70--94	layerm ark,crustal$
Puyang	allvial plain	Q1-2	al	835.0	73(max)	70--90	
Cangchow	coastal plain	Q3-4	al+ml	1000.6	100(max)	80--90	
Suchow	delta pllain of Yangtze River	Q2	al	1050	67.3	80's	
Changchow	delta plain of Yangtze River	Q	al	512.49	59.63	79-93	
Wuxi	delta plain of Yangtze River	Q	al	1025	31.4	80's	
Beijing	piedmont al.-pl. fan	Q2-3	al+pl	650		80's--	
Datong	inland fault basin	Q1-2	al+l	124	10--24	88-93	
Taipai	fault basin plain	N-Q	al	100		--66	regional monitorring net structure$

* the time when subsidence be controlled by human methods $ subsidence caused by other factors but included in the whole amount

REFERENCE

1.Erong Li and others. Environmental Geology.Geology Press,Beijing(1990).

2. Nation Seismic Bureau.The seismic intensity regionalization of China(1:4000000). Seismic Press, Beijing(1990).
3. Guolin Ren and others.Engineering Geological Map(1:4000000).Map Press of China,Beijing(1990).Urban Environmental Geological Problems In China

Proc. 30th Int'l. Geol. Congr., Vol. 23, pp. 375-385
Wang Sijing and P. Marinos (Eds)
© VSP 1997

Assessment of the Crustal Stability in the Qingjiang River Basin of the Western Hubei Province and Its Peripheral Area,China

WU SHUREN, HU DANGGONG, CHEN QINGXUAN
Institute of Geomechanics,CAGS,Beijing,China.

XU RUICHUN, & MEI YINGTANG
Survey Institute Of Three Gorges,The Yangtze River Water Conservancy Commission, Yichang,China.

Abstract

Assessment of the crustal stability in the Qingjiang river basin of western Hubei province and its peripheral area is made using fuzzy mathematics and information model proposed as a result of investigations on neotectonic activity and seismic activity in the region. According to data collected over the past four years,there are six major factors which affect the crustal stability in the region:earthquake activity,fault activity,deep-seated structures,tectonic stress fields,crustal deformation and rock porperties.This is followed by an assessment of the crustal stability of the region which are made by means of two models,fuzzy mathematical judgement and information calculation. As a result,a zonation of the crustal stability is achieved by comparing the results of the two models:four relatively substable subregions,seven substable subregions and ten stable subregions.It was found that the three dam sites of the mainstream of the Qingjiang river are all located in the stable subregions.

Keywords:Qingjiang river basin, crustal stability,fuzzy judgement,information calculation

INTRODUCTION

The Qingjiang river of 400 km in length is the longest tributary of the Yangtze river in the western Hubei province.Step hydroelectric power stations in the mainstream of the Qingjiang river are another large hydroelectric engineering construction to be built in the western Hubei province apart from the Three Gorges engineering of the Yangtze River(Fig.1). Regional engineering geological investigation and assessments of the regional crustal stability have been carried out for the hydroelectric power stations. The crustal stability of a given region involves tectonic stability,rock/soil mass stability and ground surface stability.Of them,the tectonic stability is considered to be the critical factor that controls the crustal stability[1-2],and should primarily be studied in the assessment. Therefore,the focus of this paper will be the evaluation of tectonic stability of the studied area.On the basis of analysis of the regional crustal structure, tectonic framework, fault activity, earthquake activity and present tectonic stress fields, assessment of the crustal stability in the Qingjiang river basin of western Hubei province and its peripheral area is made using fuzzy mathematics and information model.

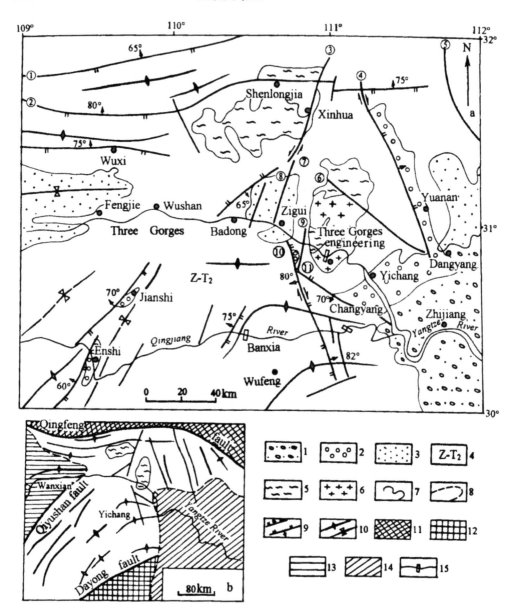

Figure 1. Tectonic framework and setting of the studied area

a. Sketch map of regional geological structures;   b. Sketch map of tectonic regionalization.

1. Neogene-Quaternay deposits; 2. Cretaceous-Eocene sandstone and conglomerates; 3. Upper Triassic-Jurassic
sandstone; 4. Sinian-middle Triassic carbonates and shales; 5. and 6. pre-Sinian metamorphic and magmatic
rocks; 7. Geological boundary; 8. Boundary of tectonic regions; 9. Normal and reverse faults; 10. Fold axis; 11.
Qinling orogenic belt; 12. Jiangnan uplift; 13. Sichuan depression; 14. Jianghan depression; 15. Dam site.

Qingfeng fault;   Yangri fault;   Xinhua fault;   Yuanan fault;   Jingmen fault;   Wuduhe fault;
Xingshan fault;   Niukou   Gongpiao fault;   Jiuwanxi fault; 10 Xiannushan fault; 11 Tianyangping fault;
12 Jianshi fault; 13 Enshi fault; 14   Wuxi fault.

QUALITATIVE ANALYSIS OF THE REGIONAL CRUSTAL STABILITY

Crustal structures and tectonics
Crustal structures. The Qingjiang river basin and its peripheral area forms a part of the Yangtze paraplatform, with a double layered rock construction consisting of pre-Sinian crystalline basement and the supracrustal Sinian-Jurassic sedimentary cover, the former having outcrops only in the core of Huangling anticline and Shennongjia anticline, and the latter widespread in all the other areas[3](Fig.1). Regional crustal structure is divided into three layers from the information of geophysical survey[4].The thickness of upper crust consisting of the crystalline basement and the sedimentary cover is 12.0-16.8km.The cover is 0-8.2km in the thickness,its seismic wave velocity is 4.6-5.7km/s.The crystalline basement with mean seismic velocity of 6.13km/s is 7.2-14.0km in the thickness.The thickness of middle crust with mean seismic velocity of 6.25km/s is 11-13km,it may composed of dioritization rock.Lower

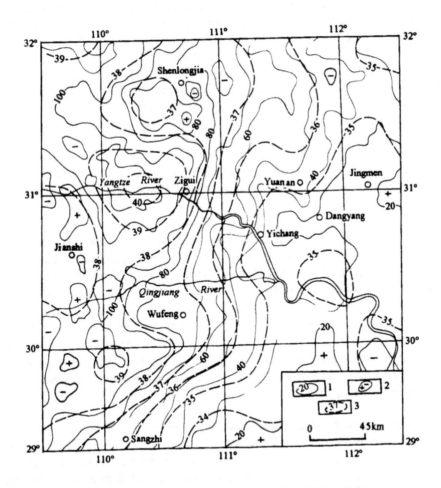

Figure 2. The regional distributive map of bouguer gravity anomaly and Moho depth
1,Negative gravity anomaly contour($\times 10^{-5}$m/s^2);2,> or <contour;3, Moho depth(km)

crust was guessed to be basaltic rock from its medial seismic velocity of 6.7km/s.The bottom surface(Moho) of lower crust is 33.8-46.8km in the depth,and the western part is the lower than the eastern part of the region.The variations of Moho depth recorded NEN-striking slope zone of the mantle uplift in the middle part of the region.The slope zone is about consistent with of regional bouguer gravity anomaly(Fig.2),and is about relationship of mirror image with present topographic form.The information shows there may not be deep fault of cutting Moho depth in the lithosphere.

Regional tectonic framework. A tripartite tectonic framework composed of E-W-, NEN-and NW-NWN-striking structural zone was formed in the major phase of the late Jurassic Yanshanian movement (Fig 1).Of them, the E-W trending Qingfeng fault belt in the northern part of the region, as the boundary between the Qinling orogenic belt and the Yangtze paraplatform, cut deeply into Moho,controlled tectonic evolution of the region since Mesozoic(Fig.1). The NEN-striking Qiyushan fault belt cut deeply into the top surface of the middle crust, as the southwestern boundary between the studied area and the Sichuan depression, controlled the formation of NEN-striking thrusts (include Jianshi fault, Enshi fault, Xianfeng fault and Qianjiang fault, *et al.*, the faults cut deeply into the top surface of crystalline basement) and fold belts in the southwestern area(Fig.1). The Baojin-Dayong fault zone separating the studied area from the Jiangnan uplift, cut deeply into the top surface of the lower crust,has a change in trend from NE to E-W. Subsequent arcuate fold zones produced in the adjoining area just follow such a course(Fig.1b).

Present tectonic activity
Regional crustal deformation. Data of repeated precise levelling serveys from 1960 to 1980 have provided a range of the variation of regional crustal deformation for different parts of the region:

Zero contour of Crustal deformation in 20 years is about parallel to the Yangtze River as a whole, subsidence in the northern part and uplifting in the southern part of the Yangtze River forms a slope dipping north with low angle(Fig.3).The largest value of uplifting is 60-80mm(3-4mm/a) near Changed city to the southeast of the region,and the largest value of subsidence is -40--50mm in the northwest of the region(Fig.3). The local distribution of present crustal deformational contour is controlled by active faults belt. For example,the crustal deformational contour in the northeast of the region is controlled by NWN-striking faults belt,and in the southwest by NEN-striking faults belt(Fig.3). The Qingjiang river basin of the center part of the region presents a uplift of 1-1.5mm/a(Fig.3).No significant change in elevation has been detected where the levelling traverses cross the faults.

Active fault and earthquake risk. The NWN-striking Xiannushan fault in the lower reaches of the Qingjiang river is one of the major active faults along which are distributed a series of landslides. Short baseline survey across the fault gives an average horizontal and vertical displacement of 0.06 and 0.066mm/a respectively. Along the fault there have occurred an earthquake of M=4.9,the largest one in the Qingjiang river basin,an earthquake of M=3.8 and ten more earthquakes of M 3 in the past 35 years.The intensity of the three hydroelectric engineering sites in the mainstream of the Qingjiang river caused by these earthquakes did not exceed V.In addition,still more active and dangerous faults and seismic zones are distributed in Zhongxiang to the northeast, Changde to the southeast and Xianfeng to the southwest of the Qingjiang river basin respectivety.they are quite far from the three dam sites(Fig.3),affecting them with an intensity of only IV. And an earthquake from the potential seismic source zones would not produce an effect greater than intensity VI.

Present stress fields. The maximum principal stress generally trends NEN-SWS from focal mechanism solutions and site stress measurements by overcoring and hydrofracturing.Some notable local variations in stress directions occur within different faults zones,e.g. the maximum principal stress axis striking NW-SE near Xianfeng fault belt(Fig.3).Present stress field of the region is achieved by material and mathematical modelling,in which the direction of loading and boundary conditions are determined from the results of the stress measurements,as well as the tectonic setting[5]. The concentration zones of maximum shear stress are distributed in Zhongxiang,Changde and Xianfeng respectivety,and are consistent with major active faults and earthquake risk zones of the region(Fig.3).

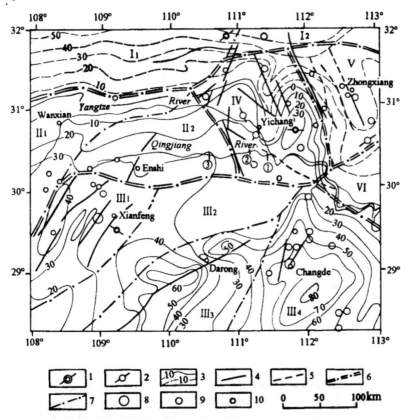

Figure 3. Today structural active subarea.
1 and 2, The maximum principal stress direction from focal mechanism solution and situ stress measurements;
3, Present crustal deformational contour(mm); 4, Fault; 5, Buried fault; 6, Zonation boundary; 7, Subregion boundary;8,6<earthquake<7;9,4.75<earthquake<6;10,3<erathquake<4.75

Zonation of tectonic activity. A zonation of tectonic activity is achieved by qualitative analysis of   space variation of the active faults, earthquake risk, present crustal deformation and tectonic stress field of the region:two sub-active regions,one weakly active region and three basically inactive regions(Fig.3). e.g. I :Shenlongjia-Baokang basically inactive region of E-W-striking slow subsidence ,with the rate of -0.5--2.5mm/a, including I_1 :basically inactive and I_2: weakly active subregion; II:eastern Sichuan-mid-upper course of the Qingjiang river basically inactive region of E-W-striking slow uplift,with the rate of 0-1.5mm/a,including two subregions of II_1 and II_2; III:southwestern Hubei-northern Hunan sub-active region of NE-striking heterogeneous uplift,with differential uplift and mid-large earthquakes, further dividing into four subregion of III_1:Xianfeng sub-active, III_2:Hefeng basically inactive, III_3: Dayong weakly active and III_4:Changde sub-active; IV: the Huangling massif and the peripheral weakly active region; V :Jinmen-Zhongxiang subjective region of NWN-striking subsidence,with differential subsidence and middle earthquakes; VI :Qianjiang-Jianli basically inactive region of slow subsidence(Fig.3).

FUZZY MATHEMATICAL JUDGEMENT ON THE CRUSTAL STABILITY

Judgement Model

The regional crustal stability can be divided into four scales of insable,relatively substable,substable and stable subregions[1] by the engineering geological investigations and mathematical modelling.Fuzzy mathematical judgement matrix is defined as:

$$V = \{instable, relativelysubstable, substable, stable\} \tag{1}$$

where the matrix element of judgement parameters(Xi) in V is

$$R_i = (r_{i1}, r_{i2}, r_{i3}, r_{i4}) \tag{2}$$

fuzzy relative matrix composed of n parameters can be expressed as:

$$R = \begin{bmatrix} r_{11} & r_{12} & r_{13} & r_{14} \\ \vdots & \vdots & \vdots & \vdots \\ \vdots & \vdots & \vdots & \vdots \\ r_{n1} & r_{n2} & r_{n3} & r_{n4} \end{bmatrix} \tag{3}$$

weight matrix of judgement parameters is

$$W = \{W_1, W_2, W_3, \cdots, W_n\} \tag{4}$$

and then,the fuzzy judgement result is showed as:

$$B = W \times R = \{b_1, b_2, b_3, b_4\} \tag{5}$$

The Determination of Judgement Parameters

There are six major factors that affect the crustal stability in the region from the above qualitative analysis:earthquake activity,fault activity,deep-seated structures,tectonic stress fields,present crustal deformation and rock properties.They are considered as the parameters of fuzzy mathematical judgement.Their geological implication and weight is described in table 1.

Calculation Results

The studied region is divided into 130 basic units of fuzzy judgement. The stability of each unit is calculated respectively using fuzzy mathematical model. The units with same value are joined together to form a distribution map(Fig.4) of the crustal stability of the region. The map shows that there are not instable subregion,relatively substable subregions are distributed near Xianfeng,Changde and Jinmen-Zhongxiang fault zones,substable subregions are distributed around relatively substable subregions and other major fault zones,stable subregions include all massifs except for fault zones(Fig.4). The result can be considered as one of major basis of the zonation of the regional crustal stability.

Table 1. The parameters and their weight of fuzzy judgement

number	parameters	geological implication	weight
f_1	earthquake activity	earthquake size, intensity and the frequency of occurrence	23
f_2	fault activity	fault size and displacement, active form and age	20
f_3	deep-seated structures	crustal structures, deep faults, Bouguer gravity anomaly	15
f_4	tectonic stress fields	tectonic stress direction,magnitude and their concentration,the evolution of Cenozoic stress fields	17
f_5	present crustal deformation	the magnitude of crustal uplift or subsidence,deformational anomaly zone	15
f_6	rock properties	the texture,strength and quality of rock mass	10

INFORMATION CALCULATION OF THE CRUSTAL STABILITY

Information Model
The quantity and quality of the information related with the crustal instability of the region can be considered in the information model. The model can be expressed by using condition probability as:

$$I_{Aj} \rightarrow B = \ln \frac{P(B/Aj)}{P(B)} \qquad (6)$$

where I_{aj} is known as the information magnitude of event B occurrence provided by parameter A in state j, $P(B/Aj)$ as probabilities of event B occurrence with parameter A and state j, $P(B)$ as probabilities of event B occurrence. From the theorem of probability multiplication,(6) can be changed into:

$$I_{Aj} \rightarrow B = \ln \frac{P(Aj/B)}{P(Aj)} \qquad (7)$$

$P(Aj/B)$ is known as probability of Aj occurrence in event B brought into existence. Total probability can be transformed into samples probability in the process of calculation,and therefrom we have

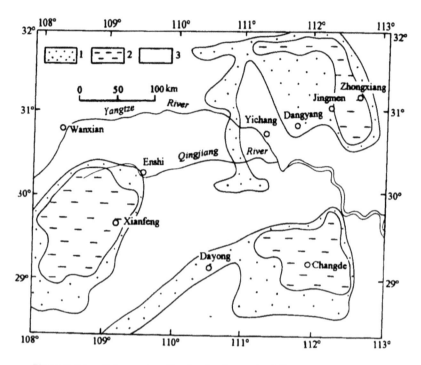

Figure 4. Fuzzy judgement results of regional crustal stability.
1, Sub-stable subregion; 2, Relatively substable subregion; 3, Stable subregion.

$$I_{Aj} \to B = \ln \frac{N_j / N}{S_j / S} = \ln \frac{S}{N} \times \frac{N_j}{S_j} \qquad (8)$$

where N_j is known as the number of the instable units with parameter Aj, N as the number of the instable units occurred in the studied region, S_j as units number with parameter Aj, S as total units number of the region.

Parameters and States of Information Calculation
There are 6 parameters and 11 states that can be considered as the parameters of information calculation of the crustal stability from the analysis of above fuzzy judgement results. The parameters and states are described in table 2.

Calculation Result Analysis
The division of information calculation units is consistent with of fuzzy judgement to compare the results of the two modles. The result of information calculation shows that there are three major factors and six states affecting the crustal stability in the region:earthquake activity,fault activity and tectonic stress fields(Table 2),and the maximum information magnitude is 11.567Nat in all units.According to the statistic frequency between the units and their

information number,and to the result of fuzzy judgement,the units with information number of more than 10Nat,5-10Nat and less than 5Nat are divided into relatively substable subregions,substable subregions and stable subregions respectively(Fig.5).The result is similar to of fuzzy mathematical judgement(comparing Fig.4 and Fig.5),and can be considered as a basis of zonation of regional crustal stability.

Table 2. Information calculation of regional crustal stability

Parameter A	State j	information calculation			Order of information magnitude
		N=6 N_j	S=130 S_j	I_{Aj} (Nat)	
Earthquake	1.Historical earthquake intensity	6	22	1.776	1
activity	2.Future earthquake intensity	6	35	1.131	4
Faults activity	1.Active faults,earthquake faults	6	23	1.732	2
	2.Fault length and displacement	6	24	1.689	3
Deep	1.Deep-seated faults	6	73	0.577	10
structure	2.Crustal structure and gravity anomaly zones	6	67	0.663	8
Tectonic	1.Precent stress state and	6	37	1.257	5
stress fields	concentration zones	6	48	0.996	6
	2.The evolution of stress fields				
Crustal	1.Anomaly zones of present	6	52	0.916	7
deformation	crustal deformation				
Rock	1.Magmatic rock	0	30	0	
properties	2.Sedimentary rock	6	68	0.648	9

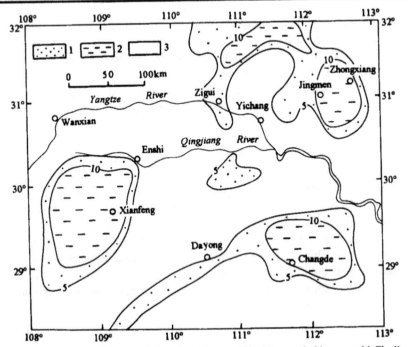

Figure 5. Information calculation result of regional crust stability(Map symbol is same with Fig 4)

ZONATION OF REGIONAL CRUSTAL STABILITY

A zonation of the crustal stability of the region is achieved by comparing the results of the two models(Fig.4 and Fig.5):four relatively substable subregions,seven sub-stable subregions and ten stable subregions(Fig.6).Of them,the area of relatively substable subregions is about 24 thousands km², and only is 13.7% of total area of the region;the area of sub-stable subregions is about 31 thousands km², and is 17.8% of the region;the area of stable subregions is about 68.5 % of the region(Fig.6).It was found that the three dam sites of the mainstream of the Qingjiang river are all located in the stable subregions,and that the relatively substable subregions are quite far from the three dam sites(Fig.6).

CONCLUSIONS

The Qingjiang river basin and its peripheral area forms a part of the Yangtze paraplatform,with a double layered rock construction consisting of pre-Sinian crystalline basement and the supracrustal Sinian-Jarassic sedimentary cover. A tripartite tectonic framework composed of E-W-,ENE-and NW-NWN-striking faults and folds was formed in the major phase of the late Jurassic Yanshanian movement.The folding is restricted within the sedimentary cover, and most of faults cut deeply into the surface of crystalline basement.

The NWN-striking Xiannushan fault in the middle-lower reaches of the Qingjiang river is one of the major active faults along which are distributed a series of landslides.Along the fault there have occurred an earthquake of M=4.9, the largest one in the Qingjiang river basin,an earthquake of M=3.8 and ten more earthquakes of M 3 in the past 35 years.The intensity of the three hydroelectric engineering sites in the mainstream of the Qingjiang river caused by these earthquakes did not exceed V. In addition,still more active and dangerous faults and seismic zones as well as stress concentration area are distributed in Zhongxiang to the northeast,Changde to the southeast and Qianjiang to the southwest of Qingjiang river basin respectively.They are quite far from the three dam sites,affecting them with an intensity of only IV.

According to data collected over the past four years,there are six major factors which affect the crustal stability in the region:earthquake activity,fault activity,deep-seated structures,tectonic stress fields,crustal deformation and rock properties.This is followed by an assessment of the crustal stability of the region which are made by means of two models,fuzzy mathematical judgement and information calculation.As a result,a zonation of the crustal stability is achieved by comparing the results of the two models:four relatively substable subregions,seven sub-stable subregions and ten stable subregions,as well as without instable subregion.It was found that the three dam sites of the mainstream of the Qingjiang river are all located in the stable subregions, and that the relatively substable subregions are quite far from the three dam sites(Fig.6).

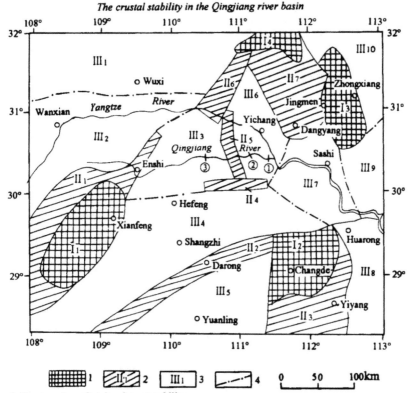

Figure 6. The zonation of regional crust stability.
1, Relatively substable subregion; 2, Sub-stable subregion; 3, Stable subregion; 4, Zonation boundary.
①, ② and ③, Dam sites.

Acknowledgments

This study was supported by China Post-doctoral Science Foundation and by the Yangtze River Water Conservancy Commision.We thank Dr. Jian Wenxing and Liu Zizhong for their helpful works in field investigation.We would like to thank the making great efforts of professor Wang Sijing and anonymous reviewers for constructive reviews of the manuscript.

REFERENCES

1.Cheng Qingxuan,Hu Haitao,Sun Ye and Tan Chengxuan.Assessment of regional crustal stability and its application to engineering geology in China,Episodes.18,69-72(1995).
2.Cheng Qingxuan,An approach to assessment of regional crustal stability:Quaternary Sciences,4 Beijing, China, (1992).
3.Wu Shuren,Chen Qingxuan,Zhao Zhizhong and Shi Ling,Structural heterogeneity and compatibility in the three Gorges area of the Yangtze River,Tectonics of China-Proceedings of the 1995 annual conference of tectonics in China,Geological Publishing House,Beijing,(1996).
4.Wu Shuren,A study of engineering geology on crustal stability in the Qingjiang river basin of western Hubei province. Press of China University of Geosciences,Wuhan,China(1995).
5.Wu Shuren,Chen Qingxuan and Mei Yingtang et al.,A study on regional structural stress fields of Mesozoic-Cenozoic era in the Qingjiang   river basin,west of Hubei province.Acta Geoscientia Sinica,35,135-151(1995).

Proc. 30th Int'l. Geol. Congr., Vol. 23, pp. 387-394
Wang Sijing and P. Marinos (Eds)
© VSP 1997

Principle and Technique of Development of GHMBS

HU RUILIN, LI XIANGQUAN, LIU CHANGLI, HE BOGAN, & GUAN GUOLIN
Institute of Hydrogeology and Engineering Geology, MGMR, China

Abstract

The geological hazard model base system (GHMBS) is a comprehensive mode base system for WINDOWS which integrates with abundant models and method modules of geological hazard forecast. It contains a lot of kinds of specialized models and forecast methods so that it can give good service to users in geological hazard forecast about subsidence, karst collapse, and seashore erosion. In development of system, the advanced techiniques of model cell generation, property-control flow and hierarchy-frame management have been introduced so that the system has characteristics of multi-task operation, ingenious queries and managment, multi-path simulation and forecast and the function for the second developer. The development of GHMBS provides an good example for the techinique of geological hazard forecast.

Keywords: Geological hazard, Simulation and forecast, Computer system

INTRODUCTION

The Geological hazards result from the comprehensive action of multiple factors, and because of interaction of these factors, complex exchangement of material and energy, and influence of human activities, the geological hazards become a much more difficult object to study. Therefore, it is an effective way to simulate and predict geological hazards by combining geological hazard with system engineering and computer technology.

The development of GHMBS (The Model Base System of Geological Hazards) is an important part of the 907th project of state key research during the period of the 8th Five-year Plan. Its task is to provide decision basis of technology for geological hazards prevention by predicting the trends of geological hazards evolution in Jing-Jin-Tang area.

THE MAIN FUNCTION OF THE SYSTEM AND ITS CHARACTERISTICS

GHMBS belongs to an applied system which has a specific objective in property. First of all, it must show clear application and advancement in the prediction and evaluation of geological hazards. On the basis of collecting a great quantity of geological hazard models, we developed the model base system which was the classified system about definitive, random and fuzzy types with the help of model base techniques in environment for WINDOWS(Figure 1). Besides of absorbing the advanced techniques at home and abroad, a great quantity of creative designs have been made and its some particular characteristics are shown as follows.

Abundant models

The system has substantial contents which contains not only many specialized models of geological hazards but also a lot of predicting mathematical methods which are divorced from geological hazard properties so that it can give good services to the users who are in deferent levels or for different objectives.

Figure 1. The platform of GHMBS

Intelligent inquirement

The system provides 6 kinds of ways for model inquirment such as directory browser, class inquirement, title inquirement, code inquirment, author inquirement and time inquirment so as to tell users to operate model effectively.

Powerful model generation

The model generation in two different levels has been developed. One is for data which is called as **data-model-generation**, and another one is for user which is called as **physical-model-generation** which is the key characteristic of the system. In the process of development, the idea of model cell generation like toy bricks has been introduced which is that compound model can be made dynamically from the main modules, model cells and C++ functions so that the work of model development has been decreased and the period of study was shortened.

Multi-paths prediction
The system can perform not only the single prediction and the united prediction but also the static simulation and dynamical simulation by setting environment.

Systematic model run
The property-control flow has been taken by GHMBS, i.e., the system running is restricted by the environment-protect file. The system can run both unsystematically and systematically because the data and environment setting was given by the property-dictionary. For this reason, GHMBS becomes more automatic and intelligent.

Servicing for different two levels
The function has been successfully developed which is both for users and for the second developers. The first is that the system gives users good running environment; and the second is that it gives the second developer a perfect model generation and management environment so as to make the function of the second model development and renewing of the system more powerful.

THE GENERAL STRUCTURE OF GHMBS

GHMBS is made up of model base, property-dictionary, model generation, model maintenance, model-run, model-inquirement, data interface and map-table-output. The general relationship of all parts is shown as follows(Figure 2):

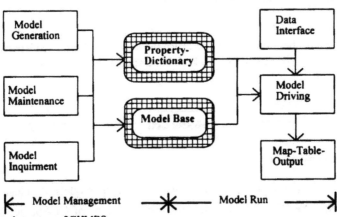

Figure 2. General structure of GHMBS

The model base is made up of model cell base and executive model base. The model cell base is used to store all kinds of basic algorithms of model generation. The executive model base is put in all kinds of specialized models and method modules.

The property-dictionary is a text file which is put in the general properties and relationship of executive models and its parameter.

The model generation system which can provide model building environment and tools is made up of model generator and model cell management.

The maintenance system includes exective model base design, model maintenance and model renew,i.e., its function is to manage models effectivelly so as to provide a favourable environment for the system running.

The model run system which can keep model running systematically is mainly made up of a series of messages of environment and operation control.

The model inquirement system which helps user to know all kinds of function index of models and parameters is made up of looking-for module and property-reader.

Data interface is the passage of data-input which mainly includes data base interface and EXECL, and also has all kinds of data preparing, data input way and the system of automatic divison of calculation elements etc..

THE MODEL DEFINITION AND GENERATION

In consideration of the function of the model development, the idea of the model cell generation like toy bricks has been suggested. It means that any kind of complex applied model can be divided into numbers of calculated elements, i.e., model cell. Numbers of model cells can composed an new compound model organically by the certain rule. In the light of the idea, we have developed as follows.

The construction of model cell
The model cells are the elements to construct the compound model. Generally, they are the general algorithms or the modules which have special function, and thire structure are shown as Figure 3.

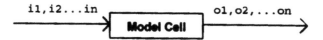

Figure 3. Structure of model cell

In which, $i_1, i_2, ..., i_n$ and $o_1, o_2, ..., o_n$ are the variables of the models. By closing the above variables and the operating to data with the structure type of CLASS in Borland C++ language, a kind of operator which can definit its object independently has been made, and its structure is shown as follows:

Class MODEL {

int   x1;   ' the variables of the model cell
float   x2;
           :

MODEL{};   ' the constructing function of model cell
void calc1(void);   ' the function of the model function
MODEL operator:=(MODEL right);   ' the function which gives data to model cell
friendistream&operator>>(istreamin,MODEL &model);   'the input function
friend ostream &operator<<(ostream &out,MODEL &model); 'the output function
};

By above rules, some common algorithms have been made into model cells, and compiled into object files which have been stored in model cell base so as to be called when model generating.

The compound model generation
In the system, the model description language which is similar to natural language have been taken to make compound models from model cells, and its process is shown as Figure 4.

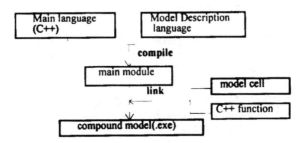

Figure 4. The process of the compound model generation

The logical relationship between compound model and model cell is like toy bricks so that the generation of the new applied model is flexible to the need of different users(Figure 5).

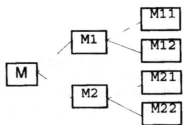

Figure 5. The relationship between compound model and model cell

THE MANAGEMENT OF MODEL BASE

The management of model base includes executive model management and model cell management which took two different ways to manage respectively according to their characteristics.

The management of executive model

The executive model management was accomplished with the help of the property-dictionary. The property-dictionary stored all general properties and relative characteristics of models and parameters which include three kinds of contents about registered information, descriptive information and the links of parameters etc. In the process of the property-dictionary construction, the relationship structure was accepted, and its formation is $R(r_1, r_2, ..., r_n)$. The R is the relationship name, and the r1,r2,...,r3 are the property variables. The file of the property-dictionary is shown as Figure 6.

Therefore, the user can operate model management according to the characteristic messages provided by property-dictionary. Its main function modules are shown as follows:

      the designer of executive model base

      the module of executive model inquirement

      the module of executive model maintenance

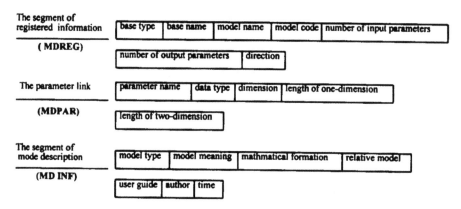

Figure 6. The form of property-dictionary

The management of model cell base

The management of model cell base is a kind of hierarchy-frame management developed according to the characteristics of model cells. It consists of two function modules which called as the editor module of the model cell trees and the editor module of the property variables of the model cells. They were shown as Figure 7 and Figure 8.

THE SYSTEMATIC MODEL RUN

The model is the core of the whole system, and the system is derived by models. The model run of GHMBS has high systematic level in which the automatic management has generally been achieved during the whole process from data-input to result-output. A systematic model run mainly experienced running environment selection, run way setting, model selection, parameter definition, data input, model calling and result output etc.

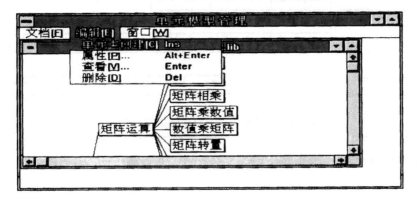

Figure 7. The editor interface of the model cell trees

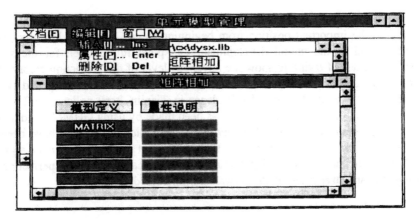

Figure 8. The editor interface of the property variables of model cell

GHMBS has two run ways of model, i.e., the single run and compound run. The former is to call a single selected model. The latter is that two relative models run according to their logical relationship in proper order in which the result of the former model provides the necessary data source for the latter model run, and it can be seen commonly in the mathematical simulation to land subsidence.

In particular, the environment-protect file has played an important role to the systematic running of the model which recorded not only a lot of essential messages of operation and control which can provide the link messages for the module running but also the present process of model running so as to provide shortcut to users to repeat. The form of environment-protect file was shown as Figure 9.

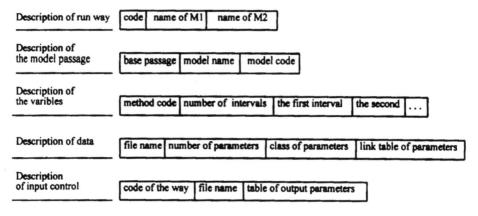

Figure 9. The form of environment-protect file

CONCLUSIONS

During the procedure of development of GHMBS, we have been putting applicability and universality on the important position from beginning to end. We have left much more leeway to expand in the model development, i.e., powerful function for the second development besides of trying to show the advanced level of the prediction theory about geological hazards at home and abroad. Therefore, GHMBS is a kind of opened system. As an example, the research work of subsidence in Tianjin, karst collapse in Tangshan and seashore erosion in Qinhuangdao area has been made by using GHMBS, and a lot of important results about the trends of the evolution of geological hazards have been got which was accepted by local departments. We believe that GHMBS will make a greater contribution to the prediction and prevetion of geological hazards by perfecting the system function and strengthening its software commercial degree.

REFERENCES

1. Wang Sen. Principle of Model Management and Design of FS-MBMS. Microcomputer, China,2,(1990).
2. Hu Ruilin, Li Xiangquan. Design of the System of Geological hazard prediction and Its Prospect of Application ,The Chinese Journal of Geological Hazard and Control,Vol.5(1994).

Proc. 30th Int'l. Geol. Congr., Vol. 23, pp. 395-407
Wang Sijing and P. Marinos (Eds)
© VSP 1997

The Special Feature of Expert System of Policy Decision Analysis for Prevention and Treatment of Geological Hazards (GHPES)

LIU YIFEN, LUO SHAOJIE, ZHANG GUANGHUI
Institute of Hydrogeology & Engineering Geology, M.G.M.R., China

WANG SONGNIAN, & YAO BAOGANG
University of FuDan, China

Abstract

The article briefly discussed the main problems in developing and study of expert system and its structure and function , the stress is put on the introduction of main characteristics of development environment of expert system of policy decision analysis for prevention and treatment of geological disasters .A breakthrough was made in solving problem about coupling of expert system, information resources database(DBASE), map and image base(GIS), method and model base in the conditions of windows. A graphical user interface for multi-objectives, -tasks and -windows was established. In the process of inference, the data can be drawn from information resources data base. The images files can be acquired from map and image base. Various prompt information can be displayed. At the same time bitmaps could be displayed on the screen. The reasoning results could be shown dynamically. This greatly widened the usability of the system. The results of retrieval from GeoRef photo disk database by the Institute of Information of Geology and Mineral Resources in China showed that up to now analogous results of study are unknown at home and abroad.

Keywords: Expert System, Structure, Function, Special Feature

INTRODUCTION

China is one of the countries having the most great varieties and the most wide distribution of geological hazards and suffering most seriously from them. The area of Beijing-Tianjin-Tangshan is the political, economic and cultural center of China, with dense population and developed economy. However, this area is also one of areas affected by geological hazards most seriously. Geological hazards exert a direct influence upon the development of national economy, bringing about enormous threats against the security of people's lives. The people of our country have accumulated rich experience and knowledge in respect of reducing, preventing and resisting geological hazards. These experiences and knowledge are valuable wealth, and their analysis, refining ,summing up, sorting out, systematizing and computerizing are of great importance.

Geological hazards owe their formation to the geological agents or superimposed geological processes induced by the anthropogenic economic activities. These geological phenomenon destroy geological environment and do harm to the development of national economy and lives and property of the peoples. With the increase of anthropogenic engineering projects, the geological hazards induced by human factors became more and more an increasingly important and nonnegligible social problem. Study on occurrence, developing, distribution regularities in space, prediction, prevention and treatment of geological hazards is a problem, over which all society shows close concern. For individual events, every kind of hazard has fortuity and regional limitations, but overall it has obvious regularities and correlations, forming a complete system of hazards. Geological hazards always occur under certain conditions of geological processes and are controlled by certain geological structures and geological elements. So far as the evolution in time and the distribution in space, they showed a peculiarity of occurrence in crowds and of regionality in distribution. Besides, all kinds of hazards have a process from a gradual change to a sudden change. Having a good grasp of these peculiarities of geological hazards, we can study their developing regularities as a system, and put forward a scientific basis for preventing and treatment of geological hazards. At present, the study on geological hazards already come from observation and analysis of single events to a step of modeling. It is used not only for mathematical, geological and physical models, but also provided conditions for establishing of knowledge processing model.

"The expert system of policy decision analysis for prevention and treatment of geological hazards in the area of Beijing-Tianjin-Tangshan" (abbreviated to GHPES) is intended to build the computer "knowledge" processing model, by use of artificial intelligence expert system, technology of computer, on the basis of exploring and modeling the thinking mode of experts when they tackle practical problems of geological hazards , by way of incorporating the strong points and selecting the essence of various schools. An expert system of Chinese-English compatible, all-Chinese interface was successfully developed by use of up-to-the-minute developing technology of soft ware in computer science in 90's, under the condition of windows 3.1, by way of programming with C and C++ languages. Successful development of it has a profound significance both in theory and in practical sense. It will promote raising the study in geological hazard science to a new level. It will open a new way for setting up an integrated pre-warning policy decision system, incorporating monitoring, prediction, prevention and treatment of geological hazards and rescue as a whole. It provides a scientific method for policy decision making and for taking part in an unified action of "international decade for natural disaster reduction". This will play an active part in raising the defense capabilities of the broad masses of people against natural disasters and their consciousness of protection of geological environment.

PRESENT SITUATION OF EXPERT SYSTEM STUDY AT HOME AND ABROAD

It is only 40 years and more past since the successful development of the first in the world digital computer(ENIAC)by Sperry & company of United States in 1946. But in such a short span of time the computer science and technology has developed at an amazing speed. Both hardware and software experienced many times great and historical technological revolution. People are accustomed to consider that the computer is able only to do a skilled and accurate numeral calculation, but in present world the great majority of problems to be dealt with have nothing to do with numeral calculation. For example, comprehension and translation of languages, show and comprehension of graph, image and sound, scientific management, policy-making analysis are not entirely the problems of mere numeral calculation. Especially the problems of government policy making and decision of important engineering projects could be correctly solved only on the basis of summarizing special experience and knowledge of leadership at all levels. Such a transformation from" numeral world" into "knowledge world", i.e. transition from data processing to symbol and knowledge manipulating is an important factor leading to the emergence of artificial intelligence expert system.

Expert system (ES) is a part of the most practical use in the study of artificial intelligence (abbreviated to AI). It is the first commercialized product among artificial intelligence. ES emerged in the mid-sixties and from then up to now it is less than 30 years past, but its development is very fast. The DENRAL expert system of interpretive type developed by American Stanford University in 1965 is reputed to be the first in the world computer expert system. In 1972 Stanford University successfully developed MYCIN system. It is a medical diagnostic system for bacterial infection of blood and meningitis. Such achievements in study are generally acknowledged as a turning point for expert system from study and experiment stage to a stage of application. In 1978 Stanford Research center of Artificial Intelligence(SRI) developed a prospector, which is a system of geological interpretation for ore-search and a well-known expert system as well. From mid-sixties to mid-eighties, more than 140 expert systems and tools were developed abroad. In the period of 15 years from 1965 to 1979 only 40 items were brought about, but in the period of five years from 1980 to 1984, nearly 100 items were developed. Of them about 20 items were commercialized. It is estimated that by 90's more commercialized and practical expert systems will be brought forth in various specialties. At the same time people will developed a lot of tool systems for building and maintenance of expert system.

The study and developing of expert system in our country begins very late, about in the end of 70's. The expert system commence its first use in medical treatment. By the beginning of 80-ies, the expert system begins to infiltrate into agriculture, communications and transportation, geological prospecting, meteorological forecast and so on. But the study and developing work is concentrated mainly in some scientific research institutes, subordinated to universities, Academy of Science and Ministries. By the mid-80's, the study of expert system expanded very fast to different applied specialties. A lot of "expert systems of practical use" come out one by one. In 1989, We developed " the expert system of geological hazard classification" using prolog language for programming. On the basis of this study from September, 1991 We began to study and develop GHPES. The study results were achieved after nearly 4 year hard work.

SYSTEM STRUCTURE OF GHPES

It provides two kinds of interactive environment: one is developing mechanism provided for knowledge engineer, geologist and specialized personnel who develop the knowledge system; the other is consultation mechanism, provided for terminal user i.e. for policy-maker in special field and system user. The system consists of following parts: knowledge gaining, knowledge base, inference machine, dynamic data base, trace debugging, interpreter, man-machine interface. It is a system in which the system programming is entirely separated from building knowledge base. Fig.1 clearly shows the system structure of GHPES:

It can be seen from the block diagram of structure that the system structure of GHPES is characterized by the following features:

(1) The expert system developed under condition of windows:
The study and development of traditional application software of computer are completed under condition of DOS(Disk Operating System). Windows were developed on the basis of DOS and are the computer platform one level higher than that of DOS. By the time of 90's of electronic brain world, it becomes an international trend to study and develop the software of computer under condition of windows. Because the conditions for developing windows have friendly, and unanimous graphical user interface, it is very simple and convenient for operation. The GDI furnished and having no bearing on the device, provides a free room for users to select needed input and output devices. Software developments need not have to take into consideration the diversity of device. This gives much conveniences to programming. It is more important that it has very effective DDE and OLE features, making the data exchange and mutual conforming of application software, improving and extending the function of application software, and making them very easy, flexible and reliable. By the aid of this option, the simplicity of operation, convenience for use, vividness and imagery of screen display and readiness for memorizing, required by policy-maker in special field and system user can be achieved. Therefore the study and development of GHPES were completed under very condition of windows. This, indeed, brought about a lot of conveniences for system users, but for the system developments, it is a brand-new style. The information concerned shows that up to date there is no ready-made, mature experiences of establishing the expert system under condition of windows which can be used for reference. Therefore the results were gained through hard work and exploration. The structure of expert system of windows style can be designed only on the basis of transformation of thinking and renewal of knowledge.

(2) GHPES was successfully developed by using C and C++ languages for programming under condition of windows. It is characterized by the following merits: modular design of C language, convenient communication interface, good adaptability, full functions of graphs, flexibility of man-computer interaction, high operation speed, and convenient for transplanting. C++ is formed by way of expansion of C language. On the basis of original C language, were added data extraction, inherited mechanism, abstract function and other facilities which improve program structure of C language, making it a flexible, high-efficient programming language which face an object and can be transplanted. Therefore, GHPES is an expert system which has style of windows and face on object.

(3) Realization of " four bases in One" system structure:
The system structure of GHPES is a "four bases in one" system structure. It solved a

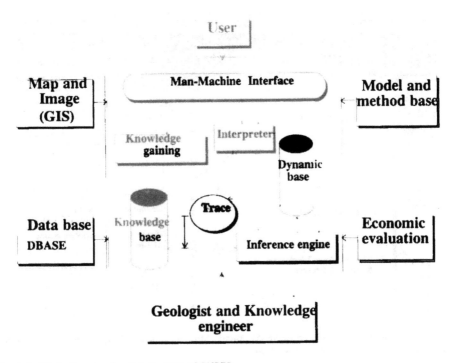

Figure 1. Block diagram of system structure of GHPES

problem about coupling of knowledge base of expert system, information resource data base, map and image base, method and model base. During knowledge reasoning DBF file can be read off information resource base (DBASE) BMP file can be read off map and image base (GIS). Besides, in the reasoning process ten bitmaps can be displayed at the same time on the screen DAT file can be read off method and model base and the conclusions of modeling prediction and economic evaluation and various prompt information's can be displayed on the screen. Multi-objective, multi-task, multi-window and imagery display of reasoning process under condition of windows was realized and reasoning results can be dynamically displayed on the screen. This extremely widened practical usability of the system.

THE FUNCTIONAL MODULE OF GHPES

The functional module of the system concretely embodies, on the one hand, the design consideration of system structure, and on the other hand directly reflects the ability of the system to solve the practical problems. The more strong and complete the functions of the system are, the more it close to the design considerations, and the higher value of application it has.

The system functions can be briefly divided into following 10 functional modules: file operation, editing and management of knowledge base, compiler knowledge base, data

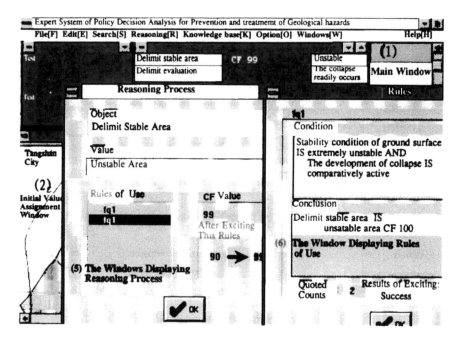

Figure 2. Indicated that the Multi-objective, Multi-task, Multi-window and Imagery Display of Reasoning process was realized under condition of windows.

and graph processing, reasoning machine, trace debugging and interpretation, reasoning, display operation option, screen display, window operation and print out. They are indicated on the following function module chart (Fig. 3)

ESTABLISHMENT OF KNOWLEDGE BASE

How to establish knowledge base is a key technical difficult point which must be solved in development of expert system. In broad line it contains three problems: knowledge acquisition, representation of knowledge and use of knowledge. The working process of expert system is a process of knowledge acquisition and use, but in computer it is necessary to have a convenient way of knowledge representation which can not only model the thinking process of human brain, but also effectively store in computer, retrieve, add, delete and change.

(1) Knowledge acquisition: Knowledge acquisition means to obtain special knowledge from experts of special field, to refine it and to transform this knowledge into the knowledge representation form in the knowledge base of expert system.

The quality of the knowledge base directly decided the ability of expert system to solve practical problems. The process of knowledge acquisition can be shown by the following chart (Figure 4):

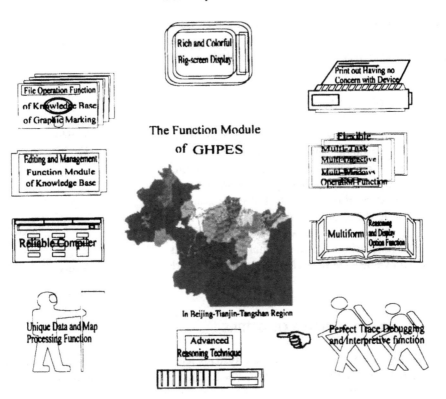

Figure 3. The function module chart of expert system of policy decision analysis for prevention and treatment of geological hazards in the region of Beijing-Tianjin-Tangshan.

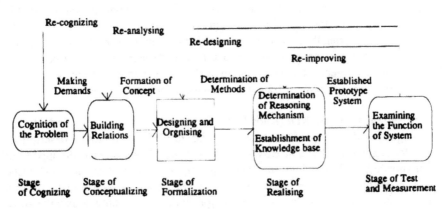

Figure 4. The Process of Knowledge Acquisition Chart

(2)Representation of knowledge

Study of the representation of knowledge is to study general method of possibility and effectiveness of representing knowledge by machine. It is an entity of data structure and control structure, considering both the knowledge store and the knowledge use. In expert system, there are a lot of knowledge representation methods which have been used. GHPES adopted production rule to represent knowledge, because production rule system has many advantages; modularity, state of nature, structuralising and universality.

REASONING MECHANISM OF GHPES

GHPES provides two kinds of reasoning mechanism: Backward reasoning: Users can give at most 20 targets in knowledge descriptive file. Reasoning machine with search knowledge base in order to obtain the values of those targets. Forward reasoning: The users need not provide any targets. They need offer only the values of known objects. This process continues unceasingly until no more new objects could gain values, see flow chart (figure 5 and figure 6).

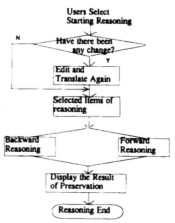

Figure 5. Flow Chard of Reasoning of GHPES

THE CHARACTERISTICS OF GHPES

To sum up, compared with the expert systems and tools published at home and abroad, for example MI, INSIGHT2, GURU, KML, CLIPS etc. GHPES is distinct not only in system structure and functional modules but also in developing environment of soft- and hardware.

(1) As to the developing environment GHPES system was developed under condition of windows. The program was compiled with C and C++ languages and a multiobject, multi-task, multi-window graphic user interface was established. Therefore, it is characterized by a friendly user interface, good systematicness, high-speed operation and readiness for transplanting. This will reduce by a wide margin the time period spent on study and development of expert system and save economic imput, man-power and material resources. At the same time it is convenient for popularizing.

(2) The system provides KDL(Knowledge Descriptive Language) of perfect functions. This language adopted productive and regular form for knowledge description. It is simple for regular writing and easy to learn and use. The system is equipped with all-screen knowledge editor, which can check up morphological and grammatical mistakes. It is very convenient in knowledge base to do operations of supplement, deletion and revision. Therefore it can help knowledge engineers to renew continuously the knowledge and establish high-level knowledge base. Up to now we have already built the expert knowledge base of policy decision analysis for preventing and treatment of "ground subsidence in Tianjin" and "Karst collapse in Tangshan".

(3)The system provides the knowledge base with dynamic linking function, which allows to make large scale knowledge base reasoning of more than 600 item rules. It also provides a management function of knowledge base, which can display the static and dynamic behavior of knowledge base.

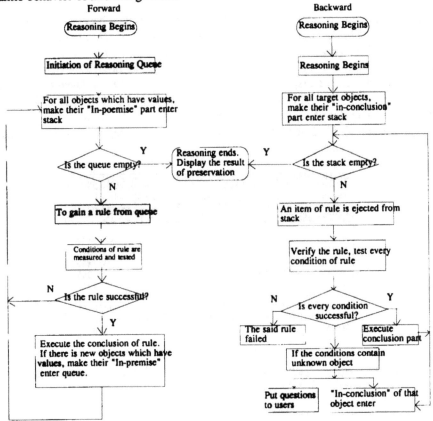

Figure 6. The chart of comparison of Forward and Backward reasoning process of GHPES

(4)The system offered two reasoning mechanisms: forward reasoning and backward reasoning. The user can carry on conveniently the cut-over between two reasoning modes under condition of menu and carry out bi-directional mixed reasoning.

(5)The system provided the interpretive function, which can interpret "why" and "how".

(6)The system has interactive debugging environment, mechanism of reasoning tracing, including a maintenance program of interactive knowledge base which temporarily revises knowledge base, an audit program of knowledge base which can preserve reasoning track and conclusion for checking afterwards, a browsing program of knowledge base which can list all facts and rules containing special expression. This greatly enhanced the ability of knowledge base system to debug, check and find out mistakes.

(7)The system adopted all-Chinese interface and can process knowledge interpretive method of two languages-Chinese and English. It is very convenient for operation and use.

(8)The system can be coupled with information resource data base (DBASE), map and image base (GIS) and model method base. In the process of reasoning, the data can be read off data base (Fig. 7), Map and image files can be acquired from Map and image base (Fig. 8), From model base got the results of model prediction (Fig. 9), various prompt massages can be displayed (Fig. 10), 10 bitmaps can be displayed on the screen at the same time (Fig. 11). This greatly widened the usability of the system. The results of retrieval from GeoRef photo disk data base by the Institute of information of Geology and Mineral Resources in China showed that up to now analogous results of study are unknown yet at home and a broad.

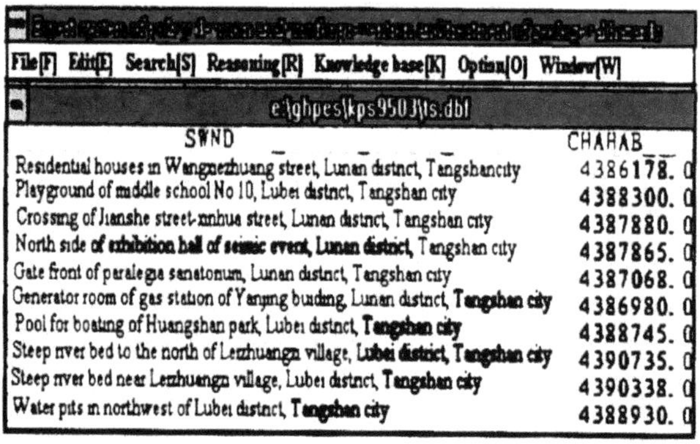

File[F] Edit[E] Search[S] Reasoning[R] Knowledge base[K] Option[O] Window[W]		
e:\ghpes\kps950J\ts.dbt	
SWND		CHAHAB
Residential houses in Wengnezhuang street, Lunan district, Tangshancity		4386178.0
Playground of middle school No 10, Lubei district, Tangshan city		4388300.0
Crossing of Jianshe street-mnhua street, Lunan district, Tangshan city		4387880.0
North side of exhibition hall of seismic event, Lunan district, Tangshan city		4387865.0
Gate front of paralegis sanatonum, Lunan district, Tangshan city		4387068.0
Generator room of gas station of Yanying buiding, Lunan district, Tangshan city		4386980.0
Pool for boating of Huangshan park, Lubei district, Tangshan city		4388745.0
Steep river bed to the north of Lezhuangn village, Lubei district, Tangshan city		4390735.0
Steep river bed near Lezhuangn village, Lubei district, Tangshan city		4390338.0
Water pits in northwest of Lubei district, Tangshan city		4388930.0

Figure 7. Part of data read off parameter descriptive data base file TS.DBF of Karst collapse spots in Tangshan in the process of reasoning.

CONCLUDING REMARKS

The important contribution of expert system is to tap the specialized knowledge out of expert's brains and to get from this large economic benefit in practice. GHPES is concerned with science of geological hazards, hydrogeology and engineering geology, environmental science, cognitive science, system engineering and computer science, and is formed due to infiltration into each other and intersection with each other among sciences. It is a set of widely usable, simple for operation and applicable in practice expert system with friendly user interface and full functions. It is established by way of exploring and modeling the thinking mode with which the human experts solve practical problems, summarizing the features of expert knowledge in geological hazard special field and by using techniques of artificial intelligence expert system. It is built under condition of windows and by using of C and C++ language for programming and its system structure achieved "four bases in one". This enabled qualitative analysis and quantitative analysis to complement each other, offering a figurative static and dynamic display function as the result of special analysis of graphs and images. After successful

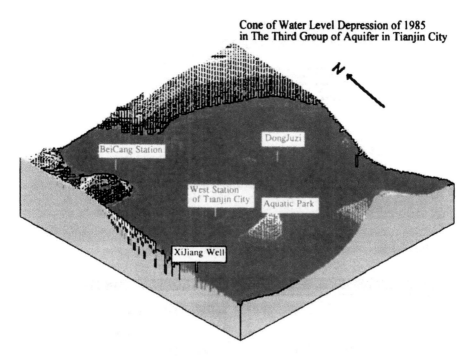

Figure 8. Image file read off from Map and image base(GIS)

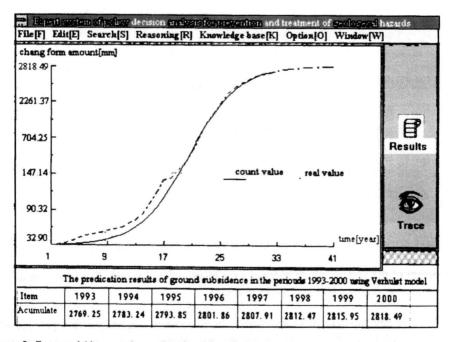

The predication results of ground subsidence in the periouds 1993-2000 using Verhulst model

Item	1993	1994	1995	1996	1997	1998	1999	2000
Acumulate	2769. 25	2783. 24	2793. 85	2801. 86	2807. 91	2812. 47	2815. 95	2818. 49

Figure 9. From model base got the results of model prediction

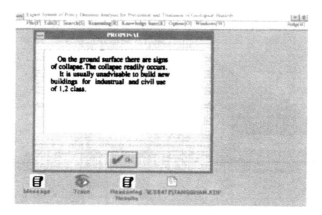

Figure 10. One of various prompt massages of reasoning process

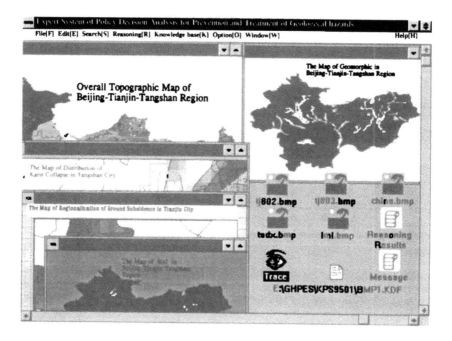

Figure 11. 11 bit maps can be displayed at the same time on screen in process of reasoning.

development of the system, all its indexes were measured and tested. The results showed that the system operates stable and works normally.

The study and development of expert system must undergo a spirally rising process of understanding of many times renewing recognition, analysis, improvement, design and practice. At present we make predictive analysis and evaluation in connection with the problem of stability of karst collapse in Tangshan, using GHPES.

The results correspond to that gained by using "grey system" evaluation method of No.01 subject. The problem of ground subsidence in Tianjin was also analyzed and evaluated in a comprehensive way. The results of analysis can serve for policy decision of planning and production departments.

The practice showed that this system not only can be directly used in Beijing-Tianjin-Tangshan region, but also can extend the scope of application to other areas or other kinds of hazards, therefore it has a good prospect of wide usage and its practical importance will become clear day by day.

REFERENCES

1. Shi chun-yi, Huang chang-ning, Wang Jia-ao, "Principles of artificial intelligence" Publishing House of Qing Hua University, October , 1993.
2. Tian sheng-feng et al, "Principle and application of artificial intelligence" Publishing House of Beijing Polytechnic University, August,1993.
3. Feng Bo-qin, "Practical expert system" publishing House of electronic industry, October, 1992.
4. Chen shi-ming et al, "Knowledge engineering language and its application", Publishing House of Nanjing University, February, 1989.
5. Shi zhong-zhi "Knowledge engineering" publishing House of Qing Hua University. February,1988.
6. Translated by Wen Du et al,"Programming windows 3.1" publishing House of Oceanography,April,1993.
7. Huang Guang, Cheng Si-qian, Li Jin,"usage manual of Borland C++ 4.0" Xue yuan publishing House, July,1994.
8. Liu Yi-fen, Hu Rui-lin,"An introduction to the progress of study and development of auxiliary computer policy- decision system for prevention and treatment of geological hazards in the area of Beijing-Tangjin-Tanshan",China Geology,1995,No,5.
9. Liu Yi-fen "Discusion about the expert system of policy decision analysis for prevention and treatment of geological hazards", China Geology,1994,No.6.
10. Liang Hong-guang "An article of required reading-The activities of 'International Decade for Natural disaster Reduction'(1990-2000)", publishing House of seismology, July,1990.
11. Avron Barr and B.A.Feigenbaum, The Handbook of Artifical Intelligence,Vol. ,Vol. ,Vol. ,Heuristech press,Stanford, California,1992.
12. Negoita,C.V.(Cinstantin virgil) 1984.Expert system and fuzzy system. The Benjamin/cummings publishing company,Inc.
13.Davi G.Goodenough, Senior Member, IEEE,Daniel Charleboel,Stan Matwin,and M.Robson, Automating Reuse of Software for Expert system Analysis of Remote Sensing Data,IEEE transactions on geoscience and remote sensing.Vol.32.no.3.MAY 1994.

Proc. 30th Int'l. Geol. Congr., Vol. 23, pp. 409-422
Wang Sijing and P. Marinos (Eds)
© VSP 1997

Liquefaction and the Associated Lateral Migration of the Hanshin District in the 1995 Hyogoken-Nanbu Earthquake (Kobe Earthquake), Central Japan

YOSHINORI MIYACHI*, KATSUMI KIMURA*, & AKIRA SANGAWA**
* *Geology Department, Geological Survey of Japan, 1-1-3 Higashi TSUKUBA, 305 Japan*
** *Osaka Center, Geological Survey of Japan, 4-1-67, Otemae, OSAKA. 540 Japan*

Abstract

The Hyogoken-Nanbu earthquake caused severe damages of buildings due to liquefaction in the Hanshin and Awaji island district. It is stressed that liquefaction took place concentrated in the filling-up ground of the sea, the river and the pond. The filling-up ground is widely distributed along the Osaka bay.

We recognized different types of deformation structures associated with liquefaction such as normal faulting, graven, compression ridges, quick sands and lateral movement of the ground, in the filling-up ground of the Kobe University of Mercantile Marine, which faces the Osaka Bay.

The vast occurrence of liquefaction resulted from the characters of the underground soils, i.g. the well-sorted sand and the overlying impermeable layer, which are appropriate for liquefaction. The lateral movement associated with liquefaction diveded the sports ground in the University into several rectangular sheets, bounded by N-S and E-W directed normal faults and gravens. These faults and gravens were generated by both the location of the same directions of underground drainage pipes and the two-directed extention stress due to sliding toward the sea (southward) and to the river (eastward). The buildings in the university were barriers against the lateral movement of the ground surface, because these foundation reached the base of the underground liquefied layer. The distance of permanent ground displacement tended to increase to the sea and reached up to one meter along N-S dierected section. It possibly reflects that the underground liquefied layer became thicker to the sea.

Keywords: 1995 Hyogoken-Nanbu Earthquake, Liquefaction, Lateral Movement, Japan

INTRODUCTION

The 1995 Hyogoken-Nanbu earthquake caused severe damages in the Awaji island and Hanshin district (Fig. 1). The epicenter of the earthquake was in the Akashi Straight. It occurred on 17th Jan., 1995. Many type of disasters took place in this region, for example, deformation of buildings and highways, landslide, fires and so on. One of the most characteristic feature of the disaster is lateral movement of the ground associate with liquefaction.

The Hanshin district is located in the E-W directed narrow plain which is bounded by the Rokko mountains consists of the granite rocks to the north. And faces the Osaka Bay to

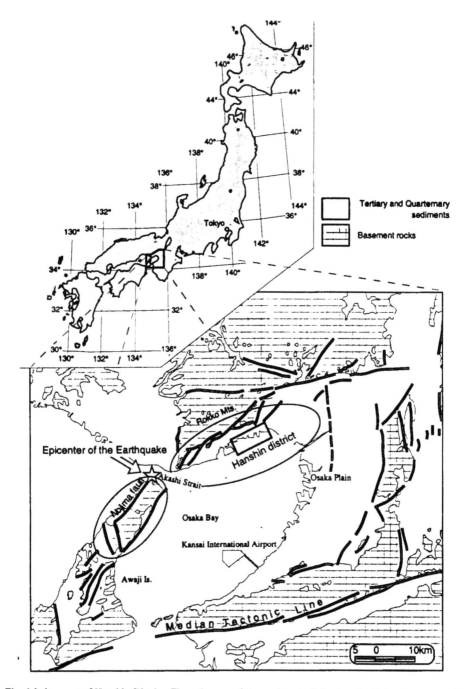

Fig. 1 Index map of Hanshin District. The epicenter of the earthquake is in the Akashi Straight. Meshed part shows basement rocks, consisting of granite, metamorphic rocks and Mesozoic sediments. White part indicates the Tertiary and Quaternary sediments. Thick solid line means a fault.

the south. Some rivers run from north to south in the plain. The district has wide filling-up ground along the Osaka Bay. Liquefaction took place regionally in the filling-up ground, while scanty in the natural ground (1)(Fig. 2). The Kobe University of Mercantile Marine is located on the filling-up ground. The disaster associate with the liquefaction were introduced in some papers (2,3). Granitic rocks are mainly used for the filling-up soils.

Some papers said that the characters of the disaster locally depend on the constructing method of the filling-up soils (4, 5). This is to say, the shore-protection reclaimed by dredge, suffered the hard disaster, while by the earthquake-proof shore-protection suffered few damages. In the case of the dredge, the shore-protection moved laterally about a few meters and the ground went subsidence in backside of the protection. The Kobe University of Mercantile Marine is considered to be reclaimed by dredge, because of the disasters.

We will discuss about the correlation with the lateral movement and the construction. As a typical example, We will describe the lateral movement and the associated deformation features recognized in the Kobe University of Mercantile Marine. These recognition are intended to serve as a prevention of the liquefied disasters.

OCCURRENCE OF LIQUEFACTION AND ASSOCIATED STRUCTURES

The map shows the distribution of liquefaction and the associated deformation structures in the Kobe University of Mercantile Marine (Fig. 3). The University faces the Osaka Bay to the south and the River Takahashi-gawa to the east. The location of paleo-shoreline can be drawn along a E-W trending road (Fig. 3), based on the air photograph taken by the U. S. Military in 1940's and geographic map of the Geographical Survey of Japan in 1885. Both data record location of the shoreline at that time. The part to be south of the paleo-shoreline is underlain by the filling-up ground, and the northern part is underlain by the natural ground, which formed the beach formerly. When the earthquake occurred, many cracks and quick sands associated with the liquefaction occurred remarkably in the filling-up ground, while they occurred poorly in the natural ground.

The southern part of the University was subdivided into the western zone with lots of buildings and the eastern zone covered by the sports ground with flat surface (Fig. 3). The ground suffered many cracks and quick sands associate with liquefaction in the eastern zone(Photo a). The southern margin of the ground were amazed most severely. In the western zone, thrusts and compression ridge distributed around the NE part of the building and open crack and quick sands around the opposite side.

EASTERN ZONE OF THE UNIVERSITY

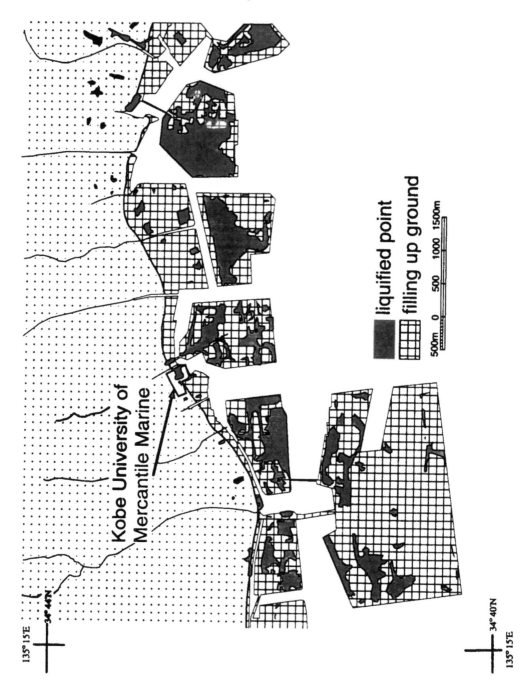

Fig. 2 Map showing the distribution of the liquefied area in the Hanshin district at the time of the 1995 Hyogoken-Nanbu earthquake (after (1)). It is noted that the liquefied area occurs concentratedly in the filling-up ground, while rarely in the natural ground.

Several N-S and E-W directed major gravens and normal faults truncated the ground into rectangular shaped blocks (Photo a, Fig. 3). Each block shifted laterally toward the bay or the river. Many minor normal faults and extension fractures were also developed inside these blocks. Most of them occured in left-hand echelon arrangement indicating right-lateral shearing near E-W directed faults, while right-hand echelon arrangement indicating left-lateral shearing near N-S directed faults. Quick sand with gravel mud were erupted along those major and minor fractures (Fig. 3 and Photo a). The ground consists mainly of a thick ballast layer and the overlying sand in shallow level as shown in walls of graven and faults (Photo c). The ballast layer has been underlain by an impermeable layer for drainage. The source layer of quick sand is estimated to be just under the impermeabile layer. The shore-protection with concrete blocks had been constructed at the southern margin of the ground (Fig. 3). Volumenous rearranged sand has been put as the foundation of the concrete blocks for shore-protection.

Gravens were about 1m wide and 30 cm deep in maximum (Photo b). The formation of the faults and gravens resulted from lateral movement of the sheets. A 20 cm-high pressure ridge occurred only at one place (Photo c), rightly under which an E-W trending drainpipe was set up. It is noted that numbers of gravels were erupted through the fracture on the crest. Fig. 4 shows three N-S trending geographic cross sections across the pressure ridge. We made these sections using both a hand level and a tape measure. The ground rised along the pressure ridge, and fell down backside of the sheets. Many tension cracks were developed around there. Voluminous quick sand was erupted through the open cracks and formed maximum 8 cm thick apron.

LATERAL MOVEMENT OF THE SPORTS GROUND IN THE EASTERN ZONE

The cumulative distance of the N-S directed lateral movement of the ground is shown in Fig. 5. It was calculated based on displacement of N-S trending ditch of concrete blocks, which separated the sports ground from 2 m-wide narrow garden with some woods and the adjacent asphalt road (Fig. 3). Each block is U-shaped with the width of 20 cm and the longness of 30 cm. Only one drainage hole was located at the connection between the ditch line and E-W trending main drainpipe. The lateral displacement caused some gap or overlapping between adjacent blocks without internal deformation of each block except the draingage hole.

The gap between blocks, showing the lateral displacement of sheets, occurred near normal faults and gravens in the ground (b, c, e and f in Figs. 3, 5) and divided the ground surface into 4 or 5 sheets of blocks (a-b, b-c, c-d, e-f in Fig. 5). The pressure ridge (d in Figs.3, 5) occurred over a main drainage pipe and the drainage hole on the ditch line there was broken and overlapped by the adjacent block. The overlapping of blocks demonstrated the shortening of 20 cm long (Fig. 5). The shortening distance was almost equal to the extension distance accumulated in the area from point a to d in Fig. 5. It indicates that the lateral movement of the ground was protected by the underground main drainage pipe in the ground. The distance of the lateral movement generally tended to increase toward the sea (Fig.5). At the southern margin of the ground the gap of concrete blocks for shore-

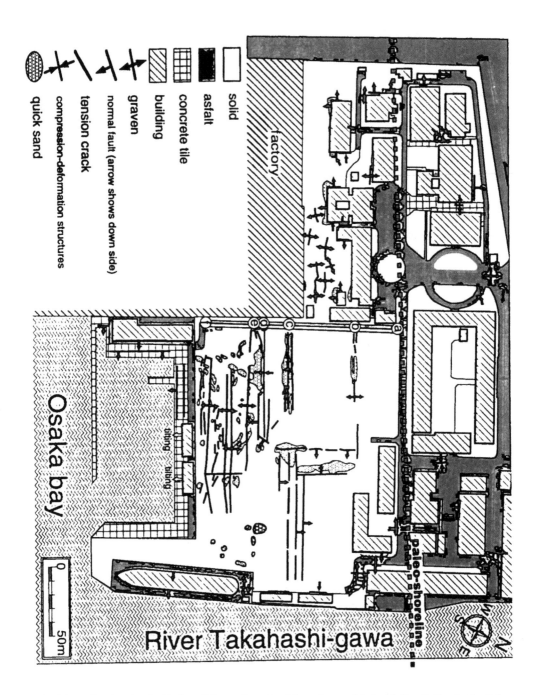

Fig. 3 Map showing the distribution of the cracks and quick sands associated with liquefaction in the Kobe University of Mercantile Marine.

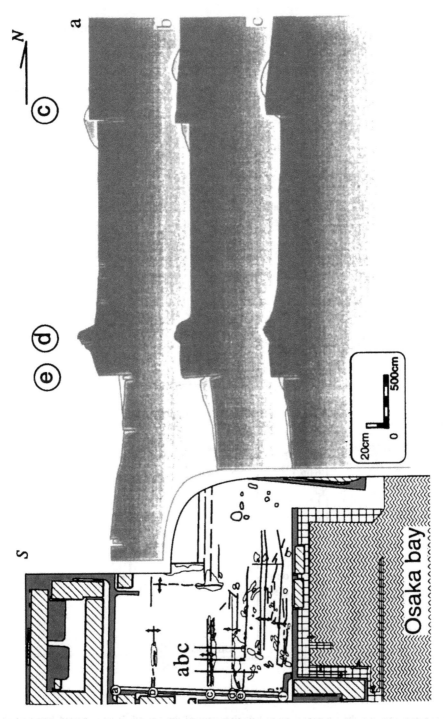

Fig. 4 N-S trending three geographic profiles along the lines (a, b and c in Fig. 3) . The ground rises on the pressure ridge. The ground falls down backside the pressure ridge.

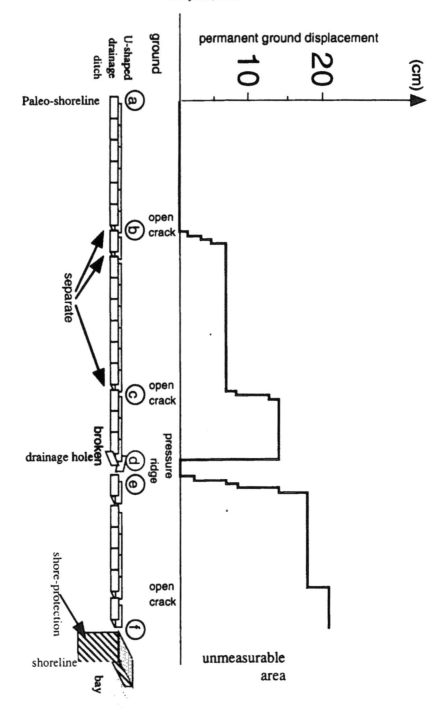

Fig. 5 Diagram showing the permanent ground displacement of the ground. It was counted based on the displacement of the ditch along the line (a) - (f) in Fig. 3.

protection reached up to one meter. The buildings there tilted seaward. This suggests the foundation of the building doesn't reach the base of the liquefied layer.

WESTERN ZONE OF THE UNIVERSITY

Many school buildings stand here and occupy 50% of the surface area. The remaining part is almost covered by asphalt. Many cracks occurred concentratedly on the asphalt surface around the buildings. Compression-induced deformation structures such as pressure ridge and thrusts were developed to the north and east sides of the building (Photo d, e), while tension cracks with the eruption of quick sand took place to the south and the west sides (Photo d). There are many tension cracks occurred in the southeastern corner of the western zone. The reason is that we estimate an factory is located to the south.

DISCUSSION

As shown in Fig. 2 and 3, liquefied area at the time of the earthquake is rare in the natural ground, but occupied the very wide area of the filling-up ground. We described in detail lateral movement and different type of cracks caused by liquefaction in reclaimed land of the Kobe University of Mercantile Marine. The features of the deformation structures are different between the eastern and western zones of the University, although both zones have been possibly constructed almost with the similar constructing method. These cases are good examples to research how deformation structures associated with liquefaction are controlled by man-made objects. The lateral movement caused by the liquefaction is explained as that a underground sediment layer is liquefied, so that the overlying non-liquefied layer could move laterally (e.g. 7). First of all, we discuss the lithology and condition of liquefied layer.

The quick sands erupted in the University consists mainly of well-sorted medium sand, and in the eastern zone, includes a few pebbles. The sand is very appropriate in grain size and sorting for liquefaction (8). In addition, it can be expected that the artificial impermeable layer covering the source layer of liquefaction may strongly promote the high pore water pressure inside the source layer. The impermeable layer had been used for drainage. That is to say, the reclaimed land in the University had the characters to introduce easily liquefaction and the associated disasters.

The eastern zone demonstrates a characteristic case of lateral movement associated with liquefaction (Fig. 3); the ground surface was divided into several sheets by N-S and E-W directed faults and gravens. The rectangular sheets were laterally displaced southward and eastward. Why are the faults and gravens N-S and E-W directed? Each main fault was located just over underground drainage pipes. It should be stressed that a main drainage pipe had been buried just under both the pressure ridge and main a N-S directed fault (Fig.3). Therefore, the location of drainage pipes controlled that of extension faults and ridge. The formation of the two-directed faults also reflected the generation of the two-directed extension stresses, which were caused by both seaward and riverward sliding of the sheets. The development of echelon arrangement of cracks inside each sheet (Fig.3) also results from two-directed extension stress.

We discuss more about the characters of the liquefied layers in the eastern zone, based on correct measurement of N-S directed permanent ground displacement (Fig. 5). The permanent ground displacement tends to correlate with two components, i.e. the thickness of the liquefied layer (H) and maximum incline between the ground surface and the bottom surface of the liquefied layer(θ) (9). In the case of the University, as the incline of ground surface is horizontal, θ is equal to the incline of bottom surface of the liquefied layer. If θ increases, the thickness of the liquefied layer increases. Therefore, the permanent ground displacement depends on the incline of the bottom surface of the liquefied layer. Figure 5 indicates that the permanent ground displacement generally increases seaward. The general trend of the displacement suggests that the bottom surface of the liquefied layer tilts southward and the thickness of the liquefied layer becomes thicker toward the sea (Fig. 6a). Such thick liquefied layer at the southern margin of the eastern zone possibly caused the seaward tilting of buildings there (Fig.3). The lateral movement of the non-liquefied layer stopped because of the main drainpipe (d in Fig. 5). It may be drawn in Fig. 6b.

In the western zone of the University, the distribution of cracks and quick sand indicates that the lateral movement were directed southwestward and buildings were barriers against the movement (Fig. 6c). It implies the foundation of the school building has reach the base of the liquefied layer.

CONCLUSION

The disasters associated with liquefaction in the filling-up ground of Kobe University of Mercantile Marine occurred widely at the time of the 1995 Hyogoken-Nanbu earthquake. The vast occurrence of liquefaction resulted from the underground soils, i.g. the well-sorted sand and the overlying impermeable layer, which are appropriate for liquefaction. The lateral movement associated with liquefaction diveded the sports ground in the University into several rectangular sheets, bounded by N-S and E-W directed normal faults and gravens. The location of these faults and gravens were controlled by that of the same directions of underground drainage pipes. The buildings in the university were barriers against the lateral movement of the ground surface, because these foundation reach the base of the underground liquefied layer. The man-made objections effected strongly the mode of the deformation structures associated with liquefaction. The seaward-increasing displacement of the lateral movement possibly reflects that the underground liquefied layer became thicker to the seaward.

Acknowledgments

We thank Dr. H. Sugita and the colleague for their helpful to our survey in the Kobe

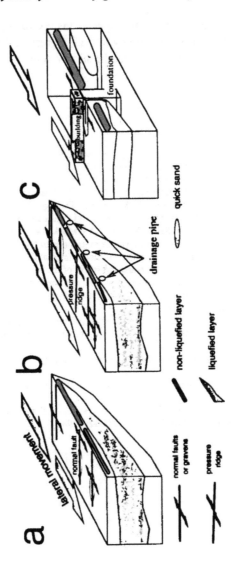

Fig. 6 Schematic model showing three types of lateral movement associated with liquefaction which are controlled by man-made objections. a: if it is flat surface and has no underground constructions, liquefied layer grow thick seaward and ground moves laterally toward the sea. b: If it has some underground constructions, the liquefied layer disappear around the constructions and the lateral movement stop along the construction. c: If it has some building, liquefied layer don't grow and pressure ridge develops foreside and tension crack backside of the buildings.

Photos

a: Sports ground in the eastern zone of the University. Many E-W and N-S directed cracks are developed. It is noted that quick sands (light part) and mud (dark colored part)are erupted along the cracks.

b: E-W trended graven in the sport ground. The wides of the graven is about 1 m and depth is 30 cm

c: Pressure ridge in the sport ground. The height of the ridge is about 20 cm. Many pebbles erupted through the crack. Sand and underlying ballast layers crop out along a NE-SW directed fault surface truncatly the ridge.

d: Voluminous quick sand erupted through open cracks at the southern wall of the school building.

e: asphalt broken by the compression at the north entrance of the school building.

University of Mercantile Marine., and special thanks to the citizen of the Kobe and Hanshin district for helpful to our survey in such a great deal of damage. And we would much indebted to Dr. Imura for help with the survey at the Kobe University of Mercantile Marine.

REFERENCES

1. Geographical Survey Institute, 1/10,000 damage distribution map of Hyogoken-Nanbu earthquake (2), Technical Report of Geographical Survey Institute. D1-No. 322.
2. M. Mitamura, K. Nakagawa, S. Masumoto, K. Shiono, S. Yoshikawa, K. Furuyama, M. Sono, S. Hashimoto, K. Ryoki, N. Kitada, N. Inoue, T. Uchiyama, S. Konishi, C. Miyakatwa, M. Nakamura, K. Noguchi, Suresh Shrestha, Y. Tani, T. Yamaguchi and Y. Yamamoto, Damage caused by the 1995 Hyogoken-Nanbu Earthquake and the Geologic Structure of Nishinomiya and Osaka area, Quaternary Research, 35,179-188
3. H. Nirei, T. Kasuda, K. Furuno, K. Satoh, Y. Sakai, O. Kazaoka, M. Morisaki and A. Kagawa, Liquefaction phenomena of instantaneous strata-collapsed type. Proceedings of Symposium on the Great Hanshin-Awaji Earthquake and its Geo-environment, Committee of Environmental Geology, Geol. Soc. Japan, 1995
4. H. Imamoto, K. Kuroda, Y. Goto and M. Nagai,the report of the harbors, airport and river group, the urgent report of the Hanshin great earthquake disaster. Soc. civil engineering, 78-83, 1995
5. H. Tsuboi, Y. Takahashi, K. Harada and H. Nitao, Comparing Remedial Measure against soil Liquefaction on reclaimed lands. Tsuchi-to-Kiso, 43-2, 67-69,1996
6. Y. Suwa, M. Hamada and T. Tabuchi, The disasters of the constructions along the water front by the Hyogoken-Nanbu earthquake, the meeting of the prompt record of the 1995.1.17 Hyogoken-Nanbu earthquake, 75-80, 1995
7. K. Ishihara, stability of Natural deposits during earthquakes, 11th Int. Conf. On S.M.F. E., vol. 1 321-376, 1985
8. Y. Sawada and I. Onodera, Report on Soil Liquefaction during the Hyogoken-Nanbu Earthquake. Tsuchi-to-Kiso, 44-2, 51-53,1996
9. M. Hamada, S. Yasuda, R. Isoyama, K. Emoto, Observation of Permanent Ground Displacements Induced by Soil Liquefaction, Proceedings of JSCE, 376, 211-220, 1986

Plate 1.

Y. Miyachi, et al.

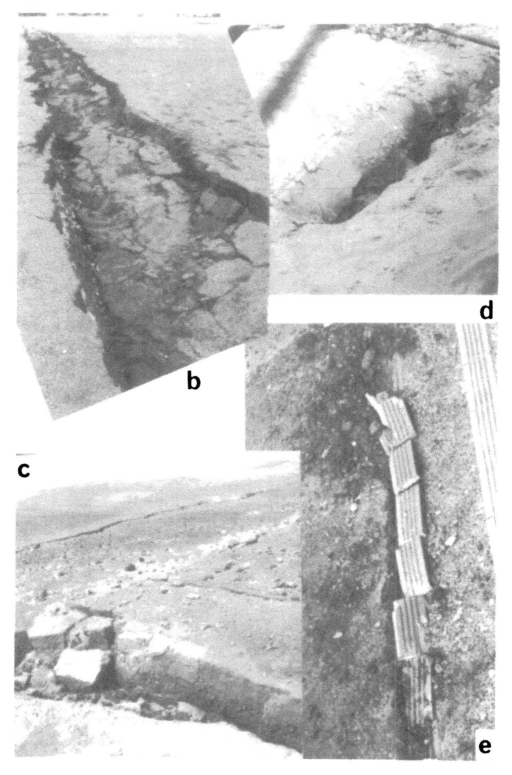

Plate 2.

Proc. 30th Int'l. Geol. Congr., Vol. 23, pp. 423-435
Wang Sijing and P. Marinos (Eds)
© VSP 1997

Common Mechanism Of Landslide Creation Along The Under Construction Egnatia Highway, In Pindos Mountain Range (W. Greece)

CHRISTARAS B., ZOUROS N., MAKEDON TH. & DIMITRIOU AN.
Lab. of Engineering Geology & hyrodeology, School of Geology, Aristotle University of Thessaloniki, GR-54006, Thessaloniki, Greece

Abstract

The part of the under construction Egnatia highway, that crosses Pindos mountain-range (W. Greece), can be characterized as one of the most difficult parts for the construction, because of the high relief, the mountain slopes are steep and the flysch, which is the dominating geological formation in the area causes important landslides, occurred almost from the beginning of the road construction.

According to our investigation the more important landslides occurred in the area are due:
1. to the geometry and the activity of the discontinuities (faults and important joints).
2. to the nature of the tectonic formation that lies under the Pindos nappes overthrusting the Ionian flysch (on the west of Metsovo tunnel),
3. to the nature of the tectonic formation that lies under the ophiolites overlying the Pindos flysch (on the East of Metsovo tunnel).

These "tectonic formations" consist of silt-stones and pelites, containing detached blocks of limestones and deep sea sediments.

Mechanically the above materials behave differently in dry and in wet conditions. In dry conditions they behave like a rock, while in wet conditions they loose repidly their cohesion and their original structure behaving like a sutured soil causing landslides along an important part of the road.

Keywords: Landslides, Slope stability, Egnatia highway, flysch, Pindos zone.

INTRODUCTION

The Egnatia highway is the most significant traffic artery in Greece. It is also the main artery linking trade and commerce from Western and Central Europe to the Middle East, as part of the traffic network of the European Union (Fig.1). The Egnatia Highway in Western Greece is crossing Pindos mountain range and is considered to be one of the most difficult parts for its construction. It includes complex geological formations that have undergone multiple tectonic deformations. The rock mass, under these conditions, is highly anisotropic and in association with the morphology (high relief and steep slopes), it presents serious geotechnical problems in the construction of high cut slopes, tunnels and bridges connecting the different sections of the highway.

The landslides are abundant along the road and are also occurred at sites where the geometry of the discontinuities in the flysch is not favourable for sliding. The landslides present similar characteristics, making necessary an investigation regarding a common mechanical behaviour of the geological formations along the road.

The geomechanical investigations and detailed studies, along with the field data and geological mapping and included in the research project entitled <<Geological - Engineering Geological Research for the Egnatia Highway>>.

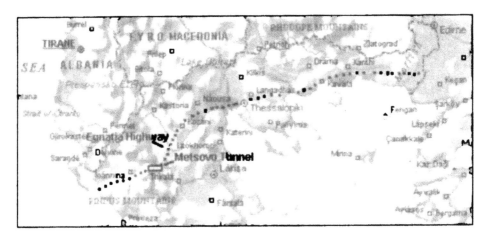

Figure 1. Egnatia highway and Metsovo tunnel area, in Pindos mountain range

GEOLOGICAL SETTINGS

The study area is located in Northern Greece, in the Pindos mountain range. Two distinct geotectonic units dominate in the area, the Pindos zone nappe and the Pindos ophiolite nappe.

Pindos zone represents the passive margin of the Neo - Tethyan ocean, composed of Mesozoic carbonate and silisiclastic rocks and the Tertiary Pindos flysch, which forms the main outcrops of the Pindos zone in the area. Pindos zone consists of a sequence of Tertiary thrusts including the Pindos nappe which over-thrusts towards WSW the flysch of Ionian and Cavrovo zones [10].

Three lithostratigraphic groups of flysch sediments have been distinguished in the study area. Politses group is very well exposed in the area, represents the nappe of the Pindos flysch and overthrusts the Zagori group sediments. The last one represents the younger sediments of the flysch of the Ionian zone. The Metsovon group appears as a tectonic window under the thrust-sheets of the Pindos Flysch nappe.

Pindos flysch (Politses group) is divided into four formations, from base to top these are: the "red flysch", alternation of red shales, pelites and sandstones with maximum thickness 100 m, the second formation comprises thin grey micaceous sandstones alternating with grey shales and marls with an average thickness 70 m, the third formation comprises thick massive sandstones and interbedded grey shales and marls with a maximum thickness 350 m and the last one characterized as "wild flysch" composed of strongly tectonized grey siltstones and sandstons.

Beneath the Pindos nappe, along the thrust front appears a "tectonic formation" consisting a melange of strongly tectonized rocks, pelites, sandstones and blocks of limestones [11]. The Pindos ophiolite complex represents fragments of the Neo-Tethyan oceanic lithosphere which was emplaced initially on the western margin of the Pelagonian zone (Cimerian micro-continent) during Late Jurassic-Early Cretaceous and subsequently over the Pindos flysch during Tertiary [1, 12].

Pindos ophiolite consists of the mafic and ultra-mafic rocks (upper mantle peridotites partly serpentinised, gabbros, mafic and ultra-mafic cumulates, sheeted dikes, massive laves, pillow laves and basic brecias), metamorphic rocks parts of the sole (amphibolites, schists and meat-sediments)as well as deep sea sediments and turbidites (pelagic limestones, sandstones, sclcarenites and micro-brecias, siltstones, green and red ribbon and nodular radiolarites) [1, 5, 9].

A tectonic formation containing blocks of all the above mentioned lithologies occurs along the tectonic contact between the Pindos ophiolite nappe and the Pindos flysch [11, 9]. In the northern part of the study area, molassic type sediments of the Meso-Hellenic Trough, were deposited during Oligocene-Early Miocene over the ophiolites and the Pindos zone sediments.

The general attitude of the contact between the Pindos ophiolite nappe and the Pindos flysch seems to be horizontal to slightly eastward dipping, as the large number of tectonic windows appearing in the study area, including the large semi-window of Malakasi confirms.

TECTONICS

Although several studies have been carried out and different explanations have been given on the emplacement of the Pindos ophiolites over the Pindos flysch, we believe that it took place during an important early Oligocene extentional tectonic event that caused a re-deformation of the tectonic melange along the ophiolite-flysch contact.

The tectonic evolution of the area is complicated. Structural analysis carried out in the area [9] show that several tectonic events took place.

Successive tectonic events arise from the structural analysis in the area. The sense of movement was established by using shear criteria and kinematic indicators. Using the methods of quantitative analysis it was possible to provide a quantitative interpretation in terms of strain from the striations observed on the fault planes[10].

Tertiary evolution started in Late Eocene times with a D0 compressional event (maximum stress 1 axes ENE-WSW) which caused detachment, folding and thrusting of the Pindos flysch before the emplacement of the ophiolite over the flysch.

D_0 compressional event caused NW-SE to NNW-SSe trending inverse faults which are the dominant tectonic features in the area and bound the tectonic slices with a movement direction towards SW. Strike slip faults with remarkable displacements of the deformation front of the Pindos nappe along them, are closely related with the above mentioned compressional features. These faults are either dextral or sinistral. The largest exists along Metsovitikos river. It is a major transverse fracture zone, known as Kastaniotiko fault [8] that interrupts the continuation of the Pindos zone.

D_0 event was followed by an important D_1 extensional event (minimum 3 axes ENE-WSW) in Early Oligocene times, which caused a seim-ductile to brittle deformation in the area and the emplacement of the ophiolites over the Pindos flysch.

Two younger successive compressional events D_2 and D_3 are responsible for the refolding, imbrication and final shape of the Pindos nappe, with the maximum stress axes trending E-W and N-S respectively, took place during the Middle-Late Miocene (the second probably evolutionary to the first).

THE TECTONIC FORMATIONS

The Tectonic formation between Pindos and Ionian flysch (to the east of Motsovo tunnel, Fig.2)

Beneath the Pindos napple, along the thrust front appears a "tectonic formation"consisting a melange of strongly tectonized rocks, pelites, sandstones and blocke of limestones [11]. Different thrusting planes have been distinguished along the thrust from of he Pindos nappe, within the tectonic formation. This formation is widely extended throughout the study area. It concerns a tectonic melange having a "chaos" structure and an appearance that reminds a "Wild-flysch" formation.

The matrix of the melange is mainly grey shales and sandstones in most cases completely sheared. Detached blocks of limestones and deep sea sediments such as thin bedded pelagic limestones, radiolarian charts, and Late Cretaceous neritic limestones with dimensions from several centimetres up to several hundred meters, are observed within the matrix. The blocks are particularly tectonized and generally fault bounded.
The tectonic formation between the opiolites and the flysch (to the west of Metsovo tunnel, Fig.3)

Figure 2. Landslide in the tectonic formation, west of Metsovo tunel

A tectonic formation containing blocks of all the above mentioned lithologies occurs along the tectonic contact between the Pindos ophiolite nappe and the Pindos flysch [11,9].

This formation resembles a tectonic melange which presents a "chaotic" structure. The matrix of the melange consists mainly of multicoloured shales, siltstone and fine grained sandstones and appears completely sheared. Detached blocks of serpentinites, basic volcanics, cherts, pelagic limestones and deep sea sediments derived from the ophiolite complex can be observed within the matrix. These blocks are strongly tectonized and fault bounded.

This tectonic formation was initially created during the Jurassic subduction-accretion evolution [5] and probable re-deformed during the tertiary emplacement of the ophiolites over the Pindos flysch [9].

GEOMECHANICAL INVESTIGATION

The region to the west of Metsovo tunnel
The more important landslides occurring in the area are not only due to the geometry of

the discontinuities of the flysch, in relation to the cut slopes directions, but mainly to the nature of the specific geological formation in which landslides are determined.

Figure 3. Landsides in the tectonic formation east of Metsovo tunnel

After our investigation, the more important geological formation considered to be responsible for landslides creation is the "tectonic formation" that lies under the Pindos nappe overthrusting the Ionian flysch; it can be observed in many places along its front [11]. The formation has a singnificant thickness of 20 to nearly 80 m, depending on the site, and extends under Pindos flysch, creating important foundation problems. The particle size destribution of the matriz of the tectonic formation is given in Fig.4.

Mechanically the material behaves differently in dry and in wet conditions. In dry conditions it behaves like a rock, having a dry density of 2.56 gr/cm3 and unaxial strength of 350 Kg/cm2. In wet conditions it loses rapidly its cohesion and its original structure and behaves like a saturated soil. It is a fine grained material of intermediate plasticity. The small plastic range between plastic (PL) and liquid limits (IL), given in Table 1 determine the ability of the material to change rapidly from the semi-soild to the liquid state, improving the significant decrease of the cohesion, angle of internal friction and bearing capacity after raining [7]. The Group Index (IG), given in Table 1, is rather high determining poor foundation conditions.

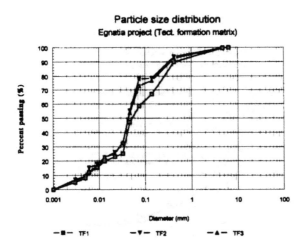

Figure 4. Particle size distribution of three samples collected from the tectonic formation

Table 1. Physical characteristics of three representative samples from the tectonic formation

Property	TF1	TF2	TF3
Liquid limit (IL, %)	36	39	38
Plastic limit (PL, %)	25	25	25
Plasticity index (PI, %)	11	14	13
Pdass No 200 sieve (%)	58	78	73
Group Index (IG)	5	11	9

One other important parameter which was investigated, was the change of the shear strength in relation to the moisture content. For this purpose three represetative samples from the area were investigated using a desk top fall cone penetrometer [6, 2]. The shear strength was measured for 10 artificially different moisture contents. The water contents in our tests covered almost the whole range of moisture, from the plastic limit to the liquid limit persentage. According to the correlation diagram of Fig. 5, an expresetative relationship can express the changes of the above properties confirming that a small quantity of water can cause a significant decrease of the shear strength.

This observation can be related to our previous observation, that a small quantity of water can change rapidly this material from the semi-soild to the liquid state, improving that in the humid conditions of Pindos mountain chain a light rain can easily create landslides on the hill-slopes.

The material should present low permeability as result of its fine grained character. So, the above described change of the mechanical behaviour should not influence the stability of the depper parts of the formation withou the presence of important faults and closedly spaced multiple open join sets, which drive the rain water dipper, in the mass of this tectonic formation.

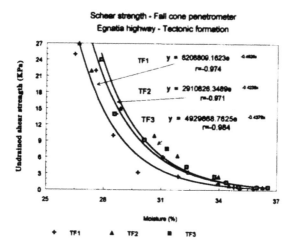

Figure 5. Correlation diagram between the shear strength and the moisture. Samples were collected from the tectonic fromation (between Pondos and Ionian flysch) and moisture changes was mad artificially.

The region to the east of Metsovo tunnel

Having already the experience of the western sida (already described), the geomechanical and general stability problems in this area, should arise mainly from the presence of the tectonic formation (lying between the ophiolited and the flysch). Furthermore the existence of large scale strike-slip and normal faults, local rock mass wedging and sliding as well as possible combinations of these features, strengthen the instability phenomena along the road.

Table 2. Physical and mechanical properties of the tectonic formation

Description	Moist. content m(%)	Uniform. coef. U	Permeab. coef.K (m/sec)	LL (%)	PL (%)	Plastic. index PI	Group index GI	Compres. index Cc	Bulk density (t/m3)
Clayey sand (SC-CH)	28	>50	10^{-7}	60	35	25	13	0,45	1,94

The tectonic formation has a singnificant horizontal extension under the ophiolitic complex and varying thickness which in most cases lies between 10 to 20 meters. The thickness decreases from the west to the east.

The particly size distribution of the matrix of the tectonic formation is presented in Fig. 6 while its physical and mechanical properties are included in Table 2.

The plasticity and compression index values show that the material presents high risk for settlemtnt and sliding [6,7].

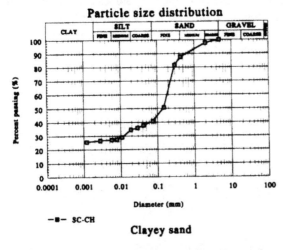

Figure 6. Particle size distribution of the tectonic formation matrix

In addition the correlation of its shear strength and moisture content presented in fig. 7, shows a very rapid decrease in shear strength with the increase of moisture content [2]. This formation and its mechanical properties are similar with the analogous tectonic formation along the thrust from of the Pindos tectonic nappe (est of Moetsovo turnnel, [3,4]).

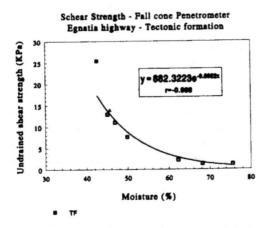

Figure 7. Correlation diagram between the shear strength and moisture contnt of the tectonic formation.

The poor mechanical properties of the tectonic formation can lead to a series of stability problems depending on the position of the formation in relation to the road design and the location of the construction (Fig 8). In some cases the road slopes cut mainly through the ophiolitic formations, Leaving the tectonic formation 15-20 m deeper from the road level. The existence of large scale normal and strike-slip faults, however, changes the position of the tectonic formation, bring it to the road level and in some cases, to the level of major constructions like tunnels and bridges.

Figure 8. Landslides at the entrance of Malakasi B tunnel constructed in the tectonic formation between the ophiolites and the flysch (east of Metsovo tunnel)

Another category of similar stability problems, mainly landslides and strongly tectonized mass movements, were encountered in cases where large scale fault zones cut through the road design. The large scale landslides activated at the slopes during the construction of the road consisted of very strongly tectonized material resembling soil formations. The zones of this material present a general E-W orientation and our investigations have shown that they are actually large scale fault zones created by the E-W strike-slip faults (Fig.9).
Combined stability problems are also related to important E-W faults, cutting the tectonic formation and collapsing probably the constructing words (tunnels etc. Fig. 10).

Finally, the sizes of the rock wedges in the most critical parts of the road were studied in detail and their safety factors were calculated. These calculations show that the formation of such rock wedges can creat serious construction problems and they should be also taken into account in the road design.

Figure 9. A representative tectonic zone, crossing perpendicularly the axis of the road. The material in the zone is broken behaving like a soil.

CONCLUSIONS

According to our investigation the more important landslides along the under construction Egnatia highway are mainly related:

1. to the nature of the tectonic formation that lies under the Pindos nappes overthrusting the Ionian flysch (west of Metsovo tunnel),
2. to the nature of the tectonic formation that lies under the ophiolites overlying the Pindos flysch (east of Metsovo tunnel).

These "tectonic formations" consist of silt-stones and pelites, containing detached blocks of limestones and deep sea sediments.

Mechanically the above materials behave defferently in dry and in west condition. In dry conditions they behave like a rock, while in wet conditions they loose rapidly their cohnesion and their original structure behaving like a sutured soil causing landslides along an important part of the road.

Furthermore, important E-W strike slip faults either menace to collapse tunnels and other constructing works or create wide tectonic zones (>100 m width), in which the rock-mass is totally broken, behaving like a soil. these zones are usually perpendicular to the road axis, causing unstable high cut-slopes.

Acknowledgments

This work has been supported by the Ministry of Environment and Public works in the frame of two engineering geological research projects that has been carried out by the Laboratory of Engineering Geology & Hydrogeology - AUTH (a) DMEO/D3b/97/1-I/29-3-94 and b) DMEO/d/1430/16-10-95).

Furthermore the authors would like to thank Prof. D. Moundrakis of the AUTH, for his many helpful suggestions during all phases of this study.

Figure 10. The tunnel is constructed in the tectonic formation, east of Metsovo tunnel. An E-W strick slip fault collapsed the left branch.

REFERENCES

1. Brunn, J.H. Contribution a l'Etude Geologique de Pinde Septentrional et de la Macedone Occidentale. Ann. Geol. Pays Hellen., 7, 1-358 (1956).
2. Christaras, B. Casagrande and Fall cone Penetrometer Methods for Liquid Limit Determination. Application

on Marls from Greta/Greece. J. Eng. Geol. Elsevier, 31, pp. 131-142 (1991).

3. Christaras, B., Zouros, N. & Makedon, The. Slope Stability Phenomena along the Egnatia Highway. The Part Ioanniana - Metsovo, in Piondos Mountain Chain, Greece. Proc. 7th Int. Congr. Iaeg, Lisboa, in Balkema, Roterdam, 3951-3958 (1994).

4. Christaras, B., Zouros, N. & Makedon, Th. Behaviour of the Votonosi Formation in Pindos Mountain (Greece). Proc. XI ECSMFE Copenhagen '95, 7, 7.23-7.28, (1995).

5. Jones, G. & Robertson, A. H. F. Tectono-Stratigraphy and Evolution of the Mesozoic Pindos Ophiolite and Related Units, Northwestern Greece. J. Geol. Soc. Lond., 148, 267-288 (1991).

6. Karlsson, R. Suggested Improvements in the Liquid limit Test with Reference to Flow Properties of Remoulded Clays. Proc. 5th ICSMFE, Paris, 1, 171-184 (1961).

7. Lambe, T.W. & Whitman, R. V. Soil Mechanics, SI Version, John Wiley & sons, New York (1979).

8. Lyberis, N., Chrowicz, J. & Papamarionopoulos, S. La paleofaille transformante du Kastaniotikos (Grece): Tecldiction, donees de terrain et geophysiques., Bull. Soc. Geol. France, 7, XXIV, 1, 73-85 (1982).

9. Mountrakis, D., Kilias, A. & Zouros, N. Kinematic Analysis and Tertiary Evolution of the Pindos-Vourions Ophiolites (Epirus-Western macedonia, Creece)., Prof. 6th Congr., Bull. Hell. Geol. Soc. (1992).

10. Zouros, N. Study of the Teconic Phenomena of Pindos Nappe Overthrust, in Epirus Area. Ph.D. Thesis, AUTH, 407p. (1993) (in Greek).

11. Zouros, N. & Mountrakis, D.. The Pindos thrust and the tectonic relation between the external geotectonic zones in the Metsovon-Eastern Zagori area (Northwestern Greece). Proc. 5yh congress, Bull. geol. Soc. Greece, XXV/1, 245-262 (1990).

12. Zouros, N., Mountrakis, D., Kilias, A. & Pavlides, S. Tertiary Thrusts and Associated Structures in the Pindos Nappe, Epirus, Nw Greece. Int. Proc. Intern. Symp. "thrust Tectonics In Albania", Tirana., Bul. Shk. Gjeol., 1, 69-79 (1991).

NEW TECHNIQUES AND APPLICATION

Proc. 30th Int'l. Geol. Congr., Vol. 23, pp. 439-448
Wang Sijing and P. Marinos (Eds)
© VSP 1997

Geophysical (Resistivity) Surveys Help Locating Rock Inhomogenities, Weak Zones And Cavities Under Land Sites Dedicated For Civil Constructions (Field Examples In Syria)

CHOUKER*, F; JARAMANI*, N; BERKTOLD**, J.
*Geological Departement, Univ. of Damascus, P.O.Box 948, Syria
** Instute f. Allg. und Ang. Geophysik, LM - University,Muenchen, Germany

Abstract

Resistivity sounding, a geophysical mean for determining electrical earth resistivity with successively increasing depth from the ground surface, becomes more and more an efficient help for civil and soil engineers investigating land sites dedicated for civil works. This paper will present cases of resistivity surveys made in Syria (in the very vicinity to Damascus) on 4 different ground pieces, close to eachother and dedicated for building of multi-etage houses. More than 500 soundings (VES-points) were measured along equidistant (every 4 m) profiles crossing the investigated sites, the sounding stations on every profile were only of 2 m distance of eachother. The electrode array used was a modified Schlumberger with the current electrode moving along OA = 1 ; 1,5 ; 2 ; 3 ; 4 ; 5 ; 7 ; 10; 10 ; 15 m, while the potential electrodes, very close to eachother, were fixed to MN = 0,5 m. The so gained resistivities were then evaluated and interpreted for every ground site. They were plotted as horizontal cross sections with the pseudo-depth (isoresistivity maps) and as vertical pseudo-cross sections along the different measuring profiles as well as a 3D-image, reflecting the structural inhomogenities in the rocks under the site, for the first 10 to 15m from the surface. The four investigated ground pieces at Dummar (urban city of Damascus), have revealed different types of inhomogenities, due to the different rock types, cutting the surface area of every site. A cavity, already cut through preparotory excavation work could be well outlined (contoured) in its extention by the resistivity image. According to the geophysical survey, soil engineer and construction engineer could locate the proper places for a representative, efficient and economical sampling (open bits or boreholes) of the earth under every ground piece, choosing the right solutions to ensure safty and stability of the planned multi-etage houses. The special array used has proved to be most valable for such engineering geological studies, because it is:
- more flexible to do measurements on small ground pieces and,
- more sensitive for vertical or steep dipping rock inhomogenities (faults, fissures, cavities...) and,
- time safing due to moving of almost 1 electrode during sounding.

Keywords: Geophysical Resistivity Survey for Civil Constructions.

INTRODUCTION

Geophysical methods are used for exploration of oil, gas, ore and mineral deposits as well as for solving engineering and environmental problems (ground water, planned and existing waste deposits, land slides, archaeology and many others).

For the growing field of engineering and environmental problems well known methods are adapted and new methods are developed. Among the most efficiently and frequently used methods are: Reflexion and refraction seismics, gravity and magnetic methods, geoelectric and electromagnetic methods including induced polarisation and georadar.

The task of our study was to investigate the rock distribution beneath four different ground pieces dedicated for construction of multi-etage houses, as an aid to find the best solution for the house foundations. For this study we used the resistivity method with a special array of electrodes convenient to solve engineering geological problems.

LOCATION AND GEOLOGY OF THE SITES

The sites of interest were located close to Dummar, NW of the city of Damascus. The area was formerly part of a local hill on the northern flank of the Qassion-anticline. It was planed by dosers and graders and is thus ready for construction of multi-etage houses. Four separate pieces of land (no. 6, 7, 9, and 10 in fig. 1), close to eachother and with a total area of more than 3000 m² were the objective of a detailed resistivity survey.

The aim was to clear rock layering and inhomogeneities, fissures, faults and cavities as well as to estimate soil parameters like stiffness beneath each piece of land. The rocks exposed at the four sites belong to Upper Cretaceous deposits (Santonian-Campanian) and they are represented by an alternation of flints, clayey and chalky limestones and marls, both massive and bedded. the layer thicknesses vary from less than one meter to about 10 meters. As part of the NW-flank of the Qassion-anticline (fig. 2) the layers dip steeply toward NW with an angle of about 35°. Locally and correlated with the hill they also dip steeply to the west in direction of the Barada valley. The dipping layers are cut at the surface areas of the four locations.

THE RESISTIVITY SURVEY

The electrical resistivity of earth material (Ohm m or Ωm resp.) depends on porosity: (pore, caverneous and crack), permeability/tortuosity, ion concentration and moisture content. It may reflect rock layering and distribution of inhomogeneities, fissures, faults and cavities in the subsurface and it may give some ideas about rock properties interesting for house foundations as well.

All four pieces of land (fig. 1) were covered with a net of measuring points along parallel profiles. As a result of test studies distance between profiles was 4 m (partly 2 m on location no. 7), and distance between measuring points was 2 m (See distribution of profiles and measuring points on the resistivity maps.).

A pole-dipole or modified Schlumberger array was used, the current electrode B being 200 to 300 m apart from the sites of measurements, the second current electrode A and the two potential electrodes M and N moving along the profiles, MN mostly being ahead. With "O" as middle of MN, the mid point of OA was attached to each measuring site. A set of 9 field studies was carried out with OA = 1, 1.5, 2, 3, 4, 5, 7, 10 and 15 m . MN was 0.5 m , except 1 m for OA = 15 m . The nine sets of field studies enabled us to map and to sound the resistivity distribution for 9 pseudo-depths . A total amount of 518 soundings were done on the four locations down to a depth of several meters.

Advantages of the pole-dipole configuration are its flexibility for soundings on very limited or small areas with hard boundaries (fences, buildings, streets) and its higher sensitivity for vertical or steep dipping features (layer boundaries, faults, cavities) and at last it is fast and time safing in acquisition of field data, due to moving of only one electrode during sounding process at every site.

ANALYSIS AND INTERPRETATION OF THE FIELD DATA

For every piece of land surveyed the apparent resistivities were calculated and plotted as :
- Apparent resistivity maps (or horizontal cross-sections) for the pseudo-depths OA =1, 3, 5, 7,10 and 15m
- Three dimensional (3D) presentations for different pseudo-depths and different angles of view.

Interpretation of these plots including geological information and borehole data led to the following conclusions:

Piece of land number 9 and 10
Site no. 10 (figs. 3, 4 and 5) shows rather homogeneous material of relatively low resistivities (10-20 Ωm) for all pseudo-depths . Contrary to location no. 10 the piece of land no. 9 shows rather inhomogeneous resistivity distribution (80-250 Ωm) with local hard rocks of apparent resistivities up to 300 Ωm. According to resistivity results and geological information the more homogeneous layer cut at site no. 10 overlies the less homogeneous layer cut at site no. 9 .

At site no. 10 with rather homogeneous resistivity distribution the rock engineer will need not more than one sampling point (an open pit) , at site no. 9 with a less homogeneous resistivity distribution two sampling points (one open pit at C-6 and one borehole down to 15 m depth at D-15) to characterize the rock parameters, are needed for foundation design of the planned multi-etage houses. This means reduced costs for drilling, excavation, sampling and analysis of rocks.

Along the northern margin of site no. 9 a small structure of increased resistivity occurs, gradually disappearing with increasing pseudo-depth . This effect may be correlated with the NW-dip of the layers. A similar effect is also observed at site no. 6.

Piece of land number 7
The most striking feature of that site (figs. 6 and 7) is a high resistivity structure in the western part, the maximum of apparent resistivity being about 3000 Ωm. This resistive feature is correlated with a cavity, that has been cut at its SW entrance during suface planing of the site. Its continuation to NNE is well contoured through the resistivity anomaly observed at the western part of site 7.

As consequence of the existence of the cavity the multi-etage house has to be buildt in the very east of site no. 7, far enough from the influence zone of the cavity.

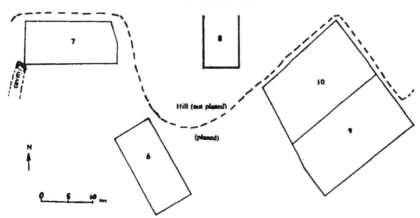

Fig. 1 Sketch map of survey site locations

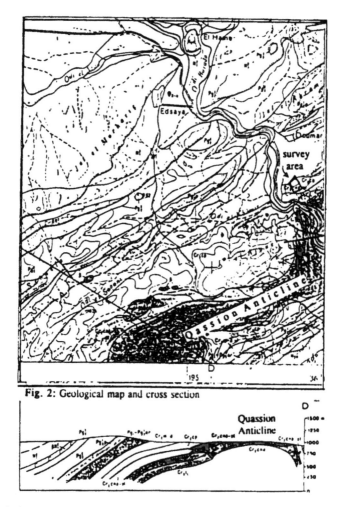

Fig. 2: Geological map and cross section

Fig. 2 Geological map and cross section

Fig. 3

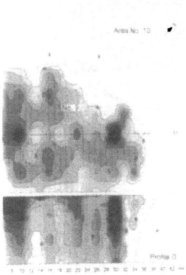

Fig. 4

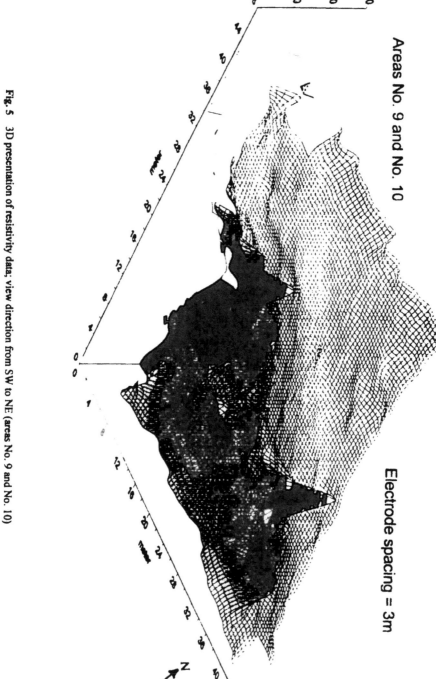

Fig. 5 3D presentation of resistivity data; view direction from SW to NE (areas No. 9 and No. 10)

Area No. 7, Electrode Spacing = 7m

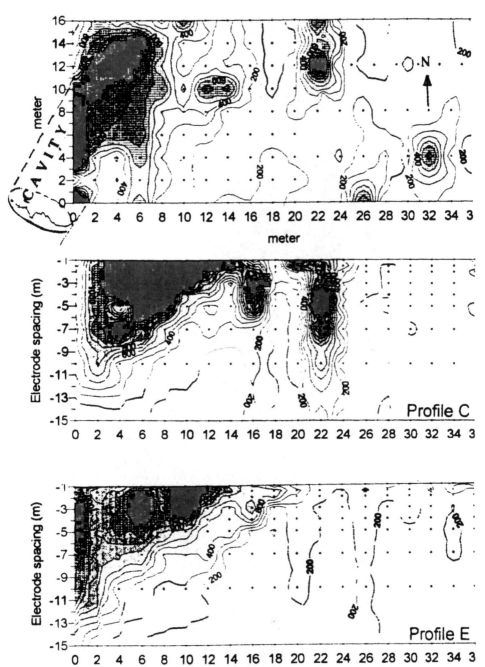

Fig. 6 Res. map (upper part); vertical cross sections (middle and lower part)

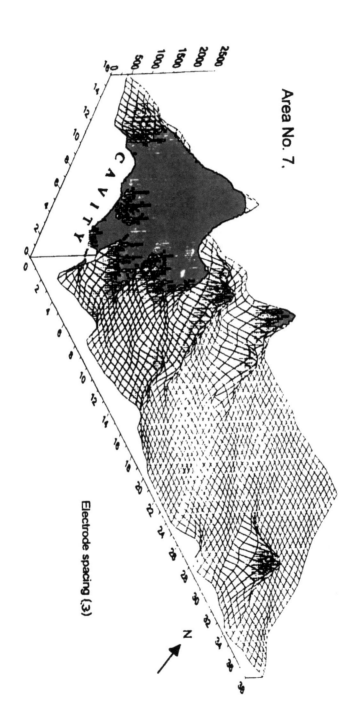

Fig. 7 3D presentation of resistivity data; view direction from SW to NE (area No.7)

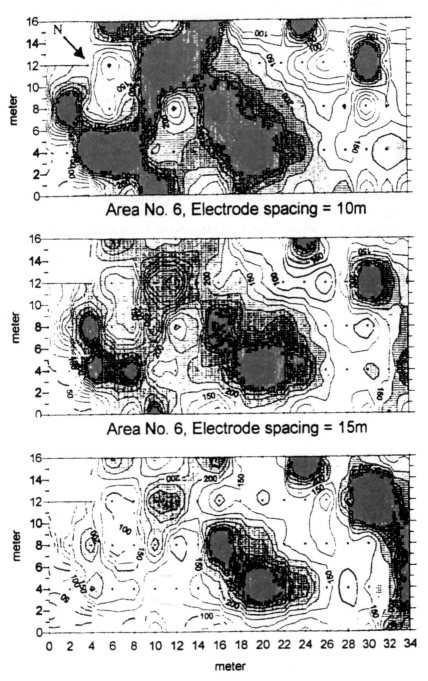

Fig. 8 Horizontal resistivity maps for different pseudo-depths

Piece of land number 6
The resistivity range on site no. 6 (fig. 8) of up to 400 Ωm and the resistivity features are similar to those of site no. 9 (figs. 3, 4 and 5), indicating that sites no. 6 and no. 9 may cut the same layer.

The apparent resistivity maps for the different pseudo-depths (fig. 8) show a tendency of three local resistivity highs shifting to NNW with increasing pseudo-depth. This result may indicate a layer dipping to NNW, fitting well with geological information of direction and angle of layer dipping and supporting a similar observation made on site no. 10.

CONCLUSIONS

The pole-dipole resistivity mapping and sounding near Dummar NW of Damascus city \was helpful for soil and construction engineers to estimate in situ rock mechanical properties, needed to design the best solutions for multi-etage house foundations. It helped to reduce costs of soil testing (reduced number of open pits, bore holes and of laboratory analysis) as well as to minimize costs for the foundation work. It added new information to knowledge of site geology (strike and dip of rock layering). Finally the strike and extension of a cavity beneath the western part of site no. 7 could be delineated with the resistivity survey. The cavity is embedded into a chalky layer already known to host many such man-made ancient cavities on the southern flank of the Qassion-anticline beneath the city of Damascus.

Proc. 30th Int'l. Geol. Congr., Vol. 23, pp. 449-460
Wang Sijing and P. Marinos (Eds)
© VSP 1997

Engineering Properties of CIPS Cemented Calcareous Sand

EDWARD KUCHARSKI, GRAHAM PRICE, & HONGYU LI
CSIRO Exploration and Mining, 39 Fairway, Nedlands Western Australia 6009, AUSTRALIA

HACKMET JOER
Department of Civil Engineering, The University of Western Australia, Nedlands, Western Australia 6009, AUSTRALIA

Abstract

An innovative new technology, known as the Calcite Insitu Precipitation System (CIPS), has been developed for improving insitu the geotechnical properties of porous sediments and rocks. CIPS is based on the crystallisation of calcite within the pore fluid and on the surfaces of constituent sand/silt grains so that the grains become strongly cemented but pores essentially remain open. This calcite cement crystallises from a proprietary solution which is permeated into the material. Because it is a non-particulate, low viscosity and water-based solution, multiple permeations are possible. Mechanical strength is significantly increased with each injection but porosity is reduced only slightly. Improvements in the mechanical properties of calcareous sands treated with CIPS have been demonstrated by a variety of laboratory tests including unconfined compressive strength (UCS), direct shear and triaxial tests. Results show that CIPS cemented calcareous sands have similar stress-strain relationships to those of natural calcarenites of similar strengths.

Keywords: calcareous sand, calcite crystal, cementation, direct shear test, grouting, mechanical strength, triaxial test, unconfined compressive strength

INTRODUCTION

Calcareous sediments and some soft porous rocks encountered offshore can pose difficulties in foundation designs for offshore structures. In some areas, such as the NW Shelf of Australia, near seafloor sediments have relatively low densities and consist of uncemented or lightly cemented bioclastic sand or silt particles. These calcareous sediments exhibit high compressibility or "pore collapse compaction" behaviour when subjected to loads which results in low skin friction on piles and large settlements beneath footings.

An obvious method of improving foundation capacity is to increase the degree of cementation within the sediment or rock. An innovative new technology known as Calcite Insitu Precipitation System (CIP System or CIPS), capable of improving insitu the geotechnical properties of these and similar materials, has been developed at CSIRO. CIPS is based on the crystallisation of calcite cement within the pores fluid and on the surfaces of constituent sand/silt grains so that the grains become strongly bonded but the pores essentially remain open. This calcite cement crystallises from a non-particulate, water-based solution of low viscosity that is injected or flushed into the sediment so that repeated

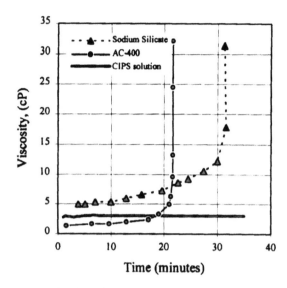

Figure 1. Viscosity versus time relationships for CIPS solution, and two common chemical grouts.

applications incrementally and significantly increase the mechanical strength of the sediment while only gradually reducing porosity.

This paper reports on the effectiveness of the CIP System in improving the mechanical properties of initially uncemented sands. An extensive laboratory testing program, including unconfined compressive strength, direct shear, triaxial and permeability tests, has been carried out on calcareous sand specimens treated with different numbers of injections of CIPS solution.

THE CIP SYSTEM

The CIP System involves injecting, or in some way permeating or flushing, the porous sediment with a specially formulated, water-based solution. The viscosity of this solution is close to that of water so it easily penetrates porous materials and can displace any existing pore fluid. Inside the pores, reactions occur within the solution over a time period which is controllable from 1 to 7 days, causing the formation of many calcite crystallites. The surfaces of the constituent sand/silt grains act as preferred nucleation sites so that the calcite crystallites grow out from those surfaces and form a coating around the pores and between the grains. This calcite coating forms a cement between the grains bonding them together in a manner similar to natural calcite mineral cement. Because the calcite cement coatings are typically thin (5-10 microns) the pores are not filled and pore throats are not blocked. Improvements in mechanical strength originate from the calcite cement which bonds together the constituent sand and silt grains, effectively converting uncemented loose sand or silt into rock.

The CIPS solution has a neutral pH and is non toxic. Commonly used chemical grouts, such as AC-400 (epoxy resin) and sodium silicate, have viscosities that increase dramatically with time (Fig. 1). In contrast, the viscosity of the CIPS solution is initially low (3 cP) and decreases slightly with time (down to 1.4 cP). This is due to the removal of calcite crystals from solution by preferential nucleation on grain boundaries. The spent CIPS fluid can be easily displaced from the pores by injections of fresh CIPS solution, thereby allowing the build up of multiple layers of calcite cement and increased bonding between the grains.

EXPERIMENTAL PROCEDURE

Calcareous sand samples with different amounts of CIPS treatment were prepared. From these, essentially reproducible specimens were taken for laboratory testing.

Material
Calcareous sand with two different size distributions in the silt to sand range (Fig. 2, Table 1) were used to prepare the samples. The calcareous sands were obtained from the seabed between Perth and Rottnest Island, Western Australia and contain approximately 96% natural carbonate shell and skeletal fragments.

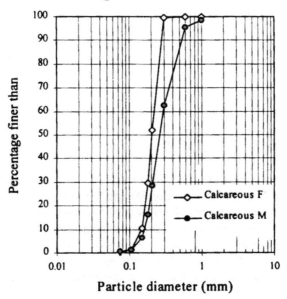

Figure 2. Grading curves for testing sands.

Table 1. Properties of sands before CIP treatment.

Sand	D_{50} (mm)	Uniform coefficient	Dry density (Mg m^{-3})	Void ratio
Calcareous F	0.21	1.43	1.40± 0.016	0.96
Calcareous M	0.27	1.82	1.47± 0.023	0.86

Sample Preparation
Approximately 170 cylindrical samples were prepared by uniformly packing dry sand into
38 mm and 63 mm ID PVC tubes of 300 mm and 425 mm length with layers of coarse clean
gravel and filter pads at both ends. Dry densities and void ratios are shown in Table 1. The
samples were initially saturated with fresh water and their permeability's measured using
the constant head method. CIPS solution was injected (flushed) from the base displacing
the fresh water in the pores. The injection system is shown schematically in Figure 3.

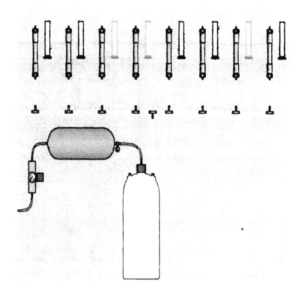

Figure 3. Schematic diagram of injection system.

The CIPS solution was prepared in an 8 litre pressurised tank connected, via control valves,
to the sample tubes. The solution was injected under pressures in the range 50-210 kPa until
double the pore volume had been displaced or until 10 minutes had elapsed. After injection
the CIPS saturated samples were left undisturbed for periods of 1, 3 or 7 days before the
next injection was applied. Each subsequent injection flushed out the spent fluid and
replaced it with fresh CIPS solution. The maximum number of injections achieved in this
study was 8. CIPS treatment was carried out at approximately constant temperature (20±1
°C).

After cementation the middle section (1/3) of each sample tube was cut out with a diamond
saw for use as a testing specimen. Constant head permeability tests were conducted in
specially constructed permeameters without removing the specimens from the PVC tubes.
These tubes were then slit longitudinally to remove the specimens.

Specimen Sizes and Test Conditions
Unconfined compressive strength (UCS) tests were chosen because they are easy to
perform, fast and cost effective. More sophisticated tests, such as triaxial, direct shear and

CPT (not reported here) were chosen to compare behaviour between the CIPS treated specimens and natural materials.

Specimens for UCS tests, which had a diameter of 38 mm and height of 114 mm (height/diameter ratio 3), were capped with high strength dental plaster to ensure uniformity of the ends. They were tested at stress rates of 1.7 MPa per minute for the weaker specimens and 4.2 MPa per minute for the stronger specimens.

To investigate the effect of moisture content on strength, both wet and dry specimens were tested. Wet specimens had been stored under water after capping and re-saturation in a vacuum desiccator. Dry specimens were exposed to air in a constant temperature room for two weeks prior to testing.

Specimens for direct shear tests were 62.5 mm in diameter and 36 mm in height. Normal stresses of 100, 200 and 500 kPa were used and the shearing speed was 0.2 mm per minute.

For the triaxial tests, specimens were isotropically consolidated at one of three effective stress levels (100, 500 or 1000 kPa) with a back pressure of 1000 kPa. They were subsequently sheared under static undrained conditions and constant total cell pressure by applying a constant rate of axial displacement of 0.2 mm per minute. During these tests the axial stress (deviator stress), axial strain and pore water pressure were monitored.

EXPERIMENTAL RESULTS

87 UCS tests, 26 direct shear tests, 6 triaxial tests and approximately 300 permeability tests were performed. Direct shear and triaxial tests were conducted on only one sand type (Calcareous M) with 3-day intervals between CIPS solution injections.

Unconfined Compressive Strength
All specimens failed at axial strains of less than 1%. UCS values ranged from 1 MPa for wet-tested Calcareous M sand with 3 CIPS treatments to 39 MPa for dry-tested Calcareous F sand with 8 CIPS treatments. Specimens with less than 3 CIP treatments were not subjected to UCS testing. Following the International Association of Engineering Geology classification [1], the CIPS treated specimens exhibit a range of strengths characteristic of very stiff soil to moderately strong rock.

Figure 4 shows a classification system developed by Deere and Miller [3] which uses both tangent modulus and unconfined compressive strength. The results of a representative selection of the UCS tests are plotted on Figure 4 together with typical ranges for concrete, limestones, sedimentary rocks and consolidated clays. The CIPS treated sands correlate with fine to coarse grained sedimentary rocks, ie. fully lithified. Presumably sand specimens treated with only 1 or 2 CIPS solutions would have lower unconfined compressive strengths.

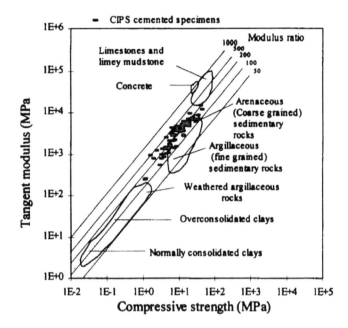

Figure 4. Engineering classification of CIPS cemented calcareous sands.

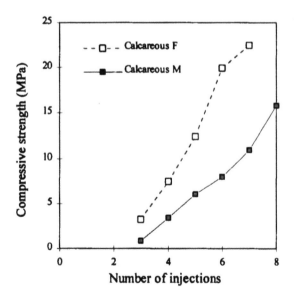

Figure 5. Compressive strength versus number of injections for wet calcareous sands.

The compressive strengths of specimens increases progressively with the number of injections of CIPS solution (Fig. 5). Strengths of over 22 MPa were achieved for CIPS treated Calcareous F sands and 16 MPa Calcareous M sands. Specimens tested after air-

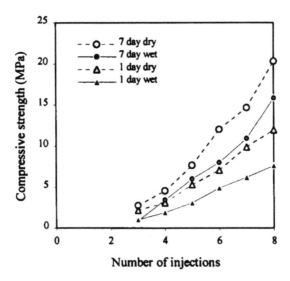

Figure 6. Compressive strengths of dry and wet Calcareous M sand with 1 and 7 day intervals between successive CIPS treatments.

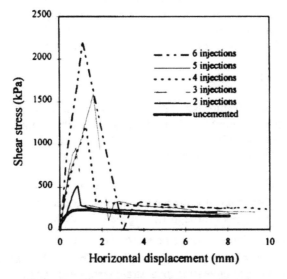

Figure 7. Shear stress versus displacement for CIPS cemented Calcareous M sand.

drying for 2 weeks have higher UCS values than wet tested specimens (Fig. 6). Also, UCS values for specimens with a 7 day interval between injections are higher than those with a 1 day interval between injections. There was no significant difference in UCS values between specimens prepared with a 3 day or 7 day interval between injections.

Direct Shear Test Results

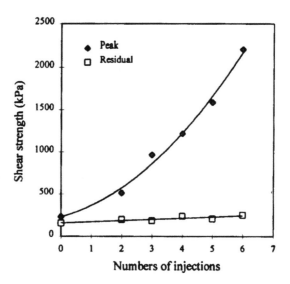

Figure 8. Shear strength (peak and residual) versus number of CIPS injections for Calcareous M sand specimens.

Relationships of shear stress versus horizontal displacement, at a normal stress of 200 kPa, are shown in Figure 7 for Calcareous M sand cemented using different numbers of CIPS injections. It shows that peak stresses develop very rapidly, generally at horizontal displacements of less than 2 mm. Peak stresses progressively increase with increasing numbers of injections. After peak stress, or rupture, shear stresses drop quickly to low residual strengths and thereafter remain constant.

Figure 8 shows the relationships between peak and residual strength for calcareous M sand specimens with different numbers of injections of the CIPS solution. The peak strengths increase with the number of CIPS injections while the residual strengths remain virtually unchanged with the number of injections.

Triaxial Test Results
Typical results of static triaxial testing under undrained conditions at constant total cell pressure are given in Figures 9 through 12. Figure 9 shows the stress-strain relationships at 3 different initial effective confining pressures (σ_3') for CIPS cemented (2 injections) Calcareous M sand. Similar behaviours are exhibited by the three specimens. The initial deviator stress responses to increasing axial strain were steep and quite linear up to peaks at axial strains of less than 1%. Obvious strength losses occurred immediately after the peaks followed by secondary increases in strength and then large axial strains at deviatoric stresses at or above the yield strengths.

A typical (selected) undrained stress-strain curve for a natural calcarenite [2] is shown in Figure 10 together with results for two CIPS cemented Calcareous M sand specimens, and an uncemented but reconstituted calcareous sand, all at initial σ_3' of 500 kPa. The CIPS cemented Calcareous M sand specimens have a similar (if slightly stronger) pattern of

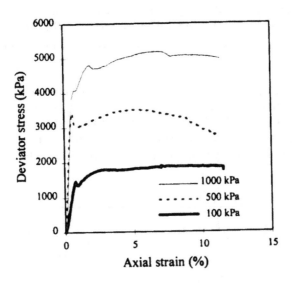

Figure 9. Stress-strain relationships in undrained compression at different initial σ_3 levels for Calcareous M sand with 2 CIPS injections.

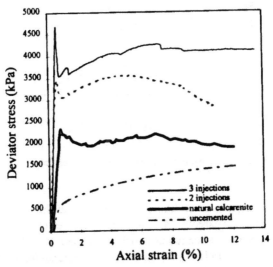

Figure 10. Undrained stress-strain relationships for one uncemented and two CIPS cemented calcareous sands and a natural calcarenite.

stress-strain behaviour to the naturally cemented calcarenite. Initial elastic responses are terminated by abrupt failures and stress drops followed by large displacements at roughly constant stress levels. The uncemented calcareous sand exhibits a much lower yield (around 500 kPa) and an overall softer behaviour.

Figure 11 shows that positive excess pore water pressures were generated early, during elastic deformation and that the magnitude of these pressures depended on the initial effective confining pressure. The higher the initial pressure, the higher the positive excess pore pressure. After yield, excess pore water pressures fell to negative values at high strains.

This pattern of pore pressure response in undrained compression for CIPS cemented calcareous sand is similar to that for natural cemented calcarenites reported by Golightly and Hyde [4] and Carter [2]. However, the fall of pore pressure is larger in the CIPS cemented sands than in the natural calcarenites and the final pore pressure was generally positive. These differences may be due to the smaller void ratio of the CIPS cemented sand specimens.

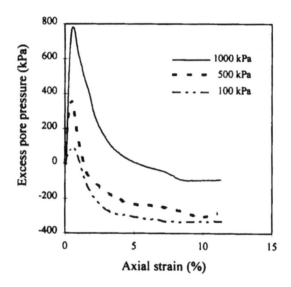

Figure 11. Pore pressure response in undrained compression at different initial σ_3' levels for calcareous M sand with 2 injections.

Figure 12 shows stress paths in undrained compression at three different initial effective confining pressures for Calcareous M sand specimens treated with 2 injections of CIPS solution. They show a type of behaviour which is characteristic of "overconsolidated" clay soil. Clearly, this "overconsolidation" was due to the cementation of the constituent grains by calcite crystals from the CIPS solution.

The best-fit peak strength line gave an effective cohesion of 133 kPa and an effective internal friction angle of 39.4° with a squared correlation coefficient of 0.995. It is anticipated that specimens with higher numbers of injections of CIPS solution will have higher values of effective cohesion but similar values of effective internal friction angle.

PERMEABILITY

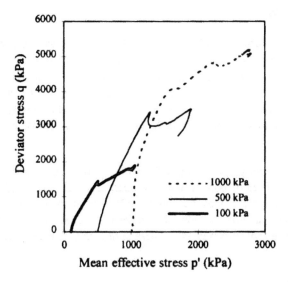

Figure 12. Stress path in undrained compression at different initial σ_3' levels for calcareous M sand with 2 injections.

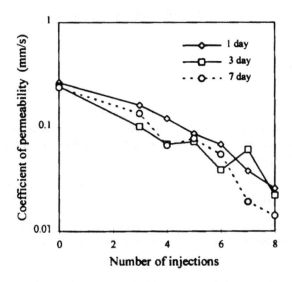

Figure 13. Variation of permeability with number of CIPS injections for Calcareous M sand.

As expected, the permeability of Calcareous M sand decreases with the number of CIPS treatments (Fig. 13). The coefficient of permeability decreases from 0.25 mm/s for uncemented Calcareous M sand to around 0.02 mm/s for specimens cemented with 8 CIPS treatments. It is notable that while permeability reduces slightly, the unconfined compressive strengths increase dramatically from zero for uncemented Calcareous M sand to around 15 MPa for the same material cemented with 8 CIPS treatments.

CONCLUSIONS

The geotechnical properties of CIPS cemented calcareous have been investigated by a series of laboratory tests. The non-particulate and low viscosity CIPS solution allows easy penetration into moderately permeable materials such as calcareous sands. Test results demonstrate dramatic increases in the mechanical strengths of calcareous sands with the number of CIPS treatments. However, permeability is reduced only slowly. The ability to apply multiple injections of the CIPS solution allows almost any desired mechanical strength to be achieved.

The CIP System has great potential for wide application to insitu improvement of the mechanical properties of porous materials such as sands. Although focussed initially on offshore sediments CIPS may be equally effective in many other applications, such as onshore deep and shallow foundations, underpinning, settlement control, slope stability, mine fill, fractured rock masses, etc. These applications are currently under investigation.

Acknowledgments

M.S. Khorshid of Advanced Geomechanics is thanked for his valuable advice on the test program. A.C.T Tan of The University of Western Australia, R. Middleton, J. Magri, D.A. Dabelstein and R. Thompson of CSIRO are thanked for their technical support.

REFERENCES

1. Anon. Rock and soil description for engineering geological mapping, *Bull. Int. Assoc. Engng. Geol.*, No.24 (1981).

2. J.P.Carter, W.S Kaggwa,. I.W Johnston,. E.A Noello, M. Fahey and G.A Chapman.. Triaxial testing of North Rankin calcarenite. *Proc. Int. Conf. Calcareous Sediments*, Perth, (eds R.J. Jewell and M.S. Khorshid), Volume 2, 515-530 (1988).

3. Deere and P.R. Miller. *Engineering classification and index properties for intact rock*, Report AFWL-TR-116, Air Force Weapons Laboratory (WLDC), Kirtland Air Force Base, New Mexico 87117 (1966).

4. Golightly and A.F.L. Hyde. Some fundamental properties of carbonate sands. *Proc. Int. Conf. Calcareous Sediments*, Perth, (eds R.J. Jewell and D.C. Andrews), Volume 1, 69-78, (1988).

Proc. 30th Int'l. Geol. Congr., Vol. 23, pp. 461–468
Wang Sijing and P. Marinos (Eds)
© VSP 1997

Some Applications of Ultrasonic BoreholeTV in Geoscience

MAO JIZHEN CHEN QUNCE QI YINGNAN
Institute of Crustal Dynamics, SSB, Beijing 100085, P.R.China

Abstract

Ultrasonic borehole TV (BHTV), by means of scanning of the ultrasonic wave can provide the instantaneous 360°
image of the borehole. With this technique, we can obtain the occurrence of stratum, joints and fissure's dip and
dip angle, the occurrence and width of fault, the geometric form of cross section of borehole, the width and depth
of borehole breakouts, size and position of cave and other important logging data. In this paper, the applications
of the BHTV test results in geoscience are discussed in detail.

Keywords: Ultrasonic borehole TV, occurrence of borehole cracks, orientating in borehole

INTRODUCTION

Ultrasonic borehole TV (BHTV), for short, is a method familiar to geoscientists. Since the
very beginning of its development, researchers of geosciences and other fields have paid
much more attention on it and deemed it a reliable tool. Its great vitality should be attributed
the fast testing speed, high precison, convenience, instanuous image showing and other
advantages. The 360° range image of the borehole can be obtained instantaneously. After
easy data analysis and processing of the photo, we can get the occurrence of stratum, the
borehole geometry and other physical parameters. The method is particularly convenient
for checking the quality of non-core drilling, counting up the number of fissures, and
discriminating rock properties. So, the system has a wide use geosciences.

ANALYZING OF ULTRASONIC BOREHOLE TV IMAGE

The principle of BHTVUltrasonic borehole TVis to scan the walls of the borehole point
point by point. When the sensor rotates across the magnetic north pole, the magnetometer
will cut the earth magnetic field and generate a instruction pulse to set the starting position.
Starting from the magnetic north pole, the scanning will repeat in the direction : East-
South-West-North. While the sensor being lifted up, the monitor on the ground will show
the helical scanning lines. staking up these scanning lines, an image of the borehole based
on magnetic north pole in the plain form will be obtained (Figure 1).

When scanning through a horizontal layer or fissure, owing to the different media, the when Bell and Gough first started the discussion of the mechanism of borehole breakout, many researchers have delved into the study. For example, Zoback and Hickman (1985), and K.A.Barton (1988) et al. have done some researches in the field and in the laboratory as well. Meanwhile, we observed borehole breakouts by BHTV in Jianchuan, Yunnan in 1985. In Zigong,Sichuan Province and Maopin, Hubei Province we also observed the phenomena. During that time we did a lot of laboratory research and tests so as to confirm that the orientation of the borehole breakout really represents the orientation of the minimum horizontal principal stress. We also discussed the theory of domain failure and compared the results of borehole breakouts with that of packer impression in the two boreholes from Jianchuan, Yunnan , and the Three Gorges area on the Changjiang River. The results agree with each other very well. From practical measurements we find out that breakouts do not appear in shallow boreholes. They usually occur below the depth of 450m. So, we can say that, based on borehole breakouts it is easy to determine the orientation of principal stress in deep boreholes.

Figure 4 shows the comparison of the results between the breakouts and packer impression. From Figure 4 we can see that the azimuths of the breakouts are N70°W and S70°E. The difference is just 180, i.e. indictive of symmetric breakouts. The orientation of the minimum principal stress is N70W, so we can obtain the maximum principal stress orientation, N20E. Compared with the results of packer impression, it can be seen that the two methods give identical results, i.e. N20E. This result agrees well with the maximum horizontal stress of western Yunnan.

Further more, for a 800 m borehole of Maopin in the Three Gorges area in Hubei, the maximum horizontal stress orientation determined by breakouts is N67W, and the orientation by hydraulic fracturing method is N69W. The results are almost the same, so, the method of determining the orientation of principal stress by borehole breakouts can be deemed as a reliable, fast and easy new technique.

Orientating the rock core
BHTV works very well in naked borehole for the orientating of the rock core. Especially in deep boreholes it shows fully its advantages. The traditional orientation method is as follows: (1) make a small hole at the expected depth in the borehole; (2) set the plinth of the orientator in to the small hole; (3) take the rock core out; (4) orientating the rock core according to the orientation of the plinth. This method is tiresome and usually the lowering down of the drilling rod will repeat at least three times. So it is very difficult to do the orientating in boreholes over 1000 m in depth. Although specific orientating instrument is available, it costs too much. However, in use of the BHTV, the rock core orientating is easy and reliable for BHTV shows high quality in the orientating in naked wells, and caneasily obtain the occurrence of the fissure and the thickness of the filling materials from the logging image. In use of this method, what we should do is to scan the expected depth, and obtain the fissure geometry and its occurrence at the end of the interval. In the Three Gorges dam area, we chose nine intervals in the Maopin 800 m borehole and operated the rock core orientation. In 1994 and 1995, in a cooperative research project with Japan, we have done the same work for five boreholes. In the latter case, for No. 5 borehole, orientated rock cores were obtained, and the results agreed well with that of BHTV.

absorption of the ultrasonic signals will change and one black horizontal line will appear on the screen and the photo (Figure 1a). In the same way, for a layer or fissure parall to the borehole axis, one black vertical line will appear. When the borehole comes across a layer or fissure with some dip angle, the image will show a sine curve with different amplitude (the difference between the crest and trough, being Δ h). The bigger the Δ h,the bigger the dip angle. The position of the wave trough indicates the trend of the fissures (Figure 1c). When there is a collapse area or a cave in the borehole wall, the orientation and the geometry of them can be directly observed(Figure 1d).

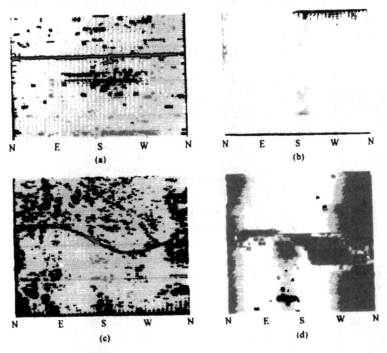

Figure 1. Some typical images
(a) Horizontal Layers or fissures   (b) Vertical joints fissures
(c) Inclined Layers or fissures   (d) Collapse, breakouts and cave on the borehole wall

Borehole breakouts take place when, with increasing depth of drilling, difference of the two principal stress reach the shearing strength. The collapse of the rock core of the borehole will occur diametrically (Figure 2a). It is almost impossible to obtain the above image by use of other logging systems.

THE APPLICATION OF BHTV IN GEOSCIENCE

The analyses of the occurrence of stratum and fissure
In use of the image of BHTV, we can analyze the stratum occurrence and the state of the fissures as well as their variation with depth in the vicinity of the borehole. Figure 2 shows some BHTV photo of Maopin 800 m deep borehole in Changjiang dam area.

After inputting into computer the data of depth, orientation indicated by the positions of crest and trough, diameter of the borehole, we can obtain the distribution of the dip and the trend of fissure (Figure 3). In the Figure 3, the x and y coordinates represent the dip and depth of fissure respectively. The indication of the oblique lines is the inclination in 360° range. Normal to the inclination dip is the trend.

The above information can provide reliable materials for the study of structural geology, engineering geology, hydrogeology, petroleum geological exploration, coal field survey and rock engineering. In use of this method, geoscientific observation will progress and, moreover, from inference to on-site measurement.

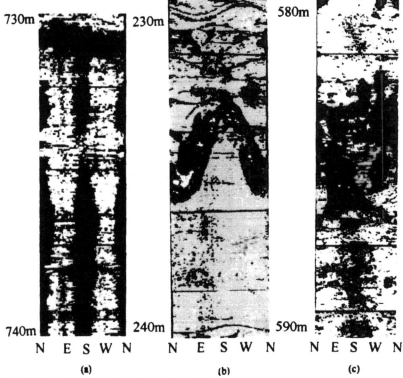

Figure 2. Some BHTV logging photographs of the 800m borehole in sanxia maoping of China
(a) Borehole breakouts in deep parts   (b) Layers of fissures   (c) Fractured zone

In addition, by using this method we can detect positions of the cave and the water-surging formation, the width of the fissure (i.e. the thickness of the filling material), and the geometry of the collapse area.

Determination of the orientation of the in-situ maximum horizontal principal stress
The orientation of the in-situ maximum horizontal principal stress can be determined by use of the information of the borehole breakouts. Borehole breakouts are the result of a kind of shear failure of rock around the borehole due to stress concentration. The phenomena of breakouts caused by stress have been paid much more attention in the world. Since 1979

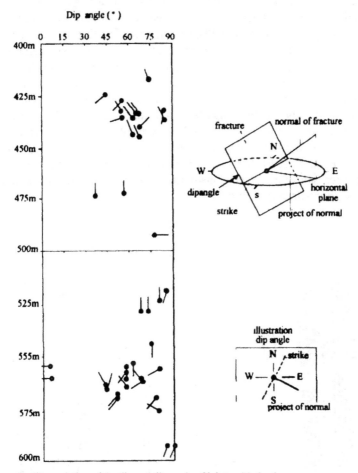

Figure 3. The variation of the dip and dip angle of joints with depth

Selection of the logging instrument position in borehole

It is well known that it is very important to select the position for the installment of measuring sensor. This is because that the measurements vary with the contact status of the sensor even with the same sensor. To obtain the satisfactory results, the first thing to be considered is the selection of the best position and environment for the sensor. Some logging sensors (such as piezomagnetic stress meter, volumetric strain meter, pressure-volumetric stress sensor, string vibration strain meter, electric resistance strain meter, packer in hydraulic fracturing, etc.) need to be settled at the place where the borehole wall is smooth, diameter of the borehole keeps constant, the rock core remains intact and no fissure and joints, no cave, and no water surging situation exist . Meanwhile, some other instruments e.g. pore pressure meterneed to be installed at the fractured zone and the aquifer position. The position of the sensor is dependent on the purpose of measurement. To satisfy the various requirements, it is necessary to detect the quality of the borehole and the status of the borehole walls. Generally, in use of previous methods, the evaluation of the above mentioned information is based on the status of the rock core during the drilling. If the rock condition is bad or the driller is not well experienced, the rock core would not be obtained and even mistakes regarding the depth will occur. We also find that, in some circumstances the rock core is complete, but the logging result of the borehole diameter changes considerably(taking the hydro-project in Thailand as an example, in which as a result of saturating of the borehole, the wall kept flaking off). So, to select the best position carefully before installing the instruments is highly recommended. Otherwise, good results can not be obtained and some loss and trouble might be produced. Recently, we have checked a few dozens of borehole for seismic stations of SSB in Xiangshan, Changpin, and selected the best fit position for monitor sensor for earthquake prediction. And these position selections were confirmed by later observations . In addition, this method have served for position selecting of packer and other logging instruments in hydraulic fracturing in-situ stress measurements with satisfactory results.

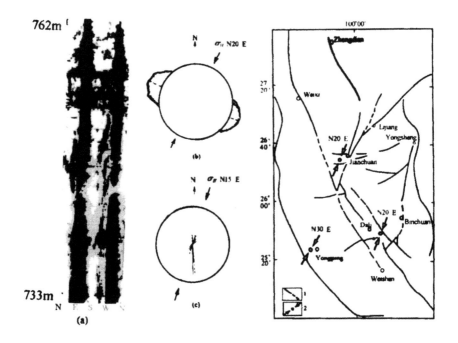

Figure 4. The orientation of principal stress determined by use of borehole breakouts
(a) Measured borehole breakouts image   (b) Hydraulic fracturing method
(c) The direction of the maximum borizontal principal steess in Yunnan of china

Exploitation of mineral resource
Oil field
The orientation of maximum horizontal principal stress can be determined by the method of

BHTV, and in use of this information, we can present more reasonable designs regarding the distribution of water well and oil well.

Coal field
By BHTV, we can get to know the distribution of the coal seam and its depth and thickness. So we can calculate the whole reserves and make reasonable exploitation scheme.

Deep mineral deposit and geothermal exploitation
At first, find the position of the mineral deposit by BHTV and determine the orientation of the maximum horizontal principal stress. Then give the master-slave drilling design conducive to the transfer of the heat and the mineral sources. In this aspect, success have been made with good benefits such as in the exploitation of the alkali deposit at the 2200 m depth of the Nan-yang oil field and the geothermal development of the old Changping City.

Checking the induced fracture
BHTV can detect induced fracture as well as the occurrence of the stratum and the geometry of natural fracture. For example, in a 300m borehole for a cooperative-operative project between China and Japan during 1994 and 1995, the detecting of the induced fracture after hydraulic fracturing test was carried out successfully. The form of the fracture of the normal fracturing test as well as the extended fracturing test were clearly detected (Figure 5). In the borehole, a few dozens of hydraulic fracturing test were carried out and almost all the induced fractures were very clear. Fast and convenient as it is to check the induced fracture with this method , but most importantly, it iscapable of detecting the form and the position of the whole fracture. Further more, by use of this method, we can obtain the distribution of the orientation of maximum horizontal principal stress varying with depth.

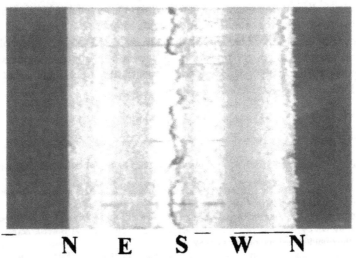

N   E   S   W   N

Figure 5. Induced fracture image after hydraulic fracturing

Above are some applications of BHTV in naked boreholes. Applications of BHTV in casing are summarized as bellow.

Some application of BHTV in casing

BHTV have good orientating function in naked boreholes and from any point of the image we can easily read out the value of the orientation, but it does'nt work in cased well. In cased well, because of the shielding effect to earth magnetic field, the orientator can not generate orientating order, so we can not determine the orientation and can give only the respective depth and geometry.

Checking of deformation and failure of casing
A 3470m borehole in east China was checked and casing deformation in many place was detected.

Checking the shot and casing deformation before and after the fracturing
In 1991, we checked the variation of the shot for an 3200m oil well in an oil field in Xinjiang before and after fracturing. We carried out two measurements, the first was after the shot, and the results showed that only shotholes of various sizes existed and no deformation or failure of the casing were detected. The second was after the fracturing, and the results indicated that the shothole was obviously enlarged, the casing was deformed or failed severely and even cleavages were observed.

The above results are helpful for the development of perforation and fracturing technique of oil fields.

CONCLUSION

From the above disscussion we get to know that, BHTV has been widely used in the field of geosciences and other geological engineering projects. A part of our practical measurements given here will be helpfull to the geoscience study and the explanation of the geophysical data. So, the method has been deemed as a reliable technique and will play a more important role in geosciences, petroleum industry, rock engineering, seismology, hydro and nuclear electricity industries, etc..

REFERENCES

1   Mao Jizhen, The application of BHTV in rock engineering, Journal of rock mechanics and engineering,
.   vol. 13 No.3, 1994
2   Mao Jizhen, Li Fangquan, et al., BHTV measurement in borehole, Reservoir-induced Earthquake Risk in
.   the Yangtze Gorges Dam Area, (edited by Li F. et al.,), Seismological Press,1992.
3   Mao Jizhen, etal, A new method for determing the direction of in-situ stress, Determing the in-situ stress
.   direction by BHTV, Institute of Crustal Dynamics, State Seismological Bureau, active tectonics and
    crustal stress field(3), Seismology press,1989.
4   Li Fangquan, In-situ stress measurements at the depth of 800m in Three Gorges area of China
.   —Comparation of the results of AE and Hydraulicfracturing in a cooperative research project between
    China and Japan, Seismology press, 1992.
5   K.A.Barton, Determing the magnitude and direction of in-situ stress in Laoton area in New Mexico
.   according to borehole breakouts, Institute of Crustal Dynamics, State Seismological Bureau, Recent of
    development of research on crustal stress, Beijing press, 1990.

Proc. 30th Int'l. Geol. Congr., Vol. 23, pp. 469-473
Wang Sijing and P. Marinos (Eds)
© VSP 1997

The Study of Mini-Penetration Test for the Liquefaction of Silt

NIU QIYING   QIU YIHUI
Department of Civil Engineering, Taiyuan University of Technology, TY 030024 P.R.C.

Abstract

Based on numerous tests of laboratory and in situ, this paper studies laboratory mini-penetration test (MPT)and various regulation of MPT blow counts; it discusses the statistical relation between the blow counts of in situ standard penetration test (SPT)and that of laboratory mini-penetration test MPTand furthermore puts forward a new approach that the laboratory MPT may replace in situ SPT for the liquefaction assessment of silt deposit.

Keywords: Liquefacient silt, Penetration blow count, Clay particecle contents, Effect of dry density.

INTRODUCTION

At present, the standard penetration test (SPT) widely used at home and abroad is the main method to determine liquefaction of silt in situ. But the SPT was affected by layer, equipment and operation. The test date is very loose. So it requires to run a few times on the same layer. Besides it requires to spend a lot of time, labour and money in doing SPT. Form 1993 to today, authors developed another new method, that is,mini-penetration test (MPT).We have applied MPT to remolded silt samples consisting of different content of clay partiecle and having different dry densites. Furthermore corresponding tests by using mini-penetrometer and standard penetrometer have been run on a lot of undisturbed silt samples in situ. Therefore authors have find various regulation of MPT blow counts. It was proved that this new method save both time and labour, and it take one person to do MPT.So it promoted the application of penetration test technology in the liquefaction assessment of the silt.

TEST METHOD

Experimental Device
Mini-penetrometer This instrument [1][2][3] was designed and made by the authors based on dynamic penetration resistance 0.86 MPa (equal to that of in situ SPT). Its index properies are as follows (Fig.1):

Standard penetration test (SPT) and boring sampling in suite are performed by SH30-2 drilling rig. The weight of standard penetrating hammers 63.5 kg.   The SPT was standardized to an energy level of 60% of the free fall energy of the hammer and the liquefaction assessment chart based on (N1)60 is now the standard used in engineering practice[4].

Sample Preparation

Silt and clayey particles used in the study were taken from the site which is located near Yifen bridge, on the west bank of Fen river in northern Taiyuan. The physical properties of soil are summarized in Table 1. Clayey mineral is kaolinite analyzed by X-ray diffraction.

Diameter of penetrating rod   1cm
Angle of penetrating cone   60
weight of penetrating hammer   400g
free fall distance of hammer   15cm
penetrating depth   25cm

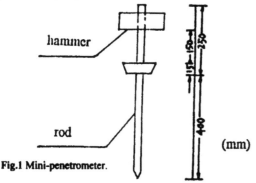

Fig.1 Mini-penetrometer.

Table 1. Silt properties

Soil	Specific gravity	Plasticity index	Medium diameter(mm)	Coefficient of uniformity	Content of clay particles(%)
Slit	2.69	7.9	0.043	2.8	0
Clayey particles		20.99			40

All the samples were made in laboratory. The content of clay particles (Pc) were 3%,6%,9%,12%,16%,and they were put in soil box size 900×400×300mm to deposit. These samples were divided into three groups according to their dry densities (γd) 14.4,15.6,16.4KN/m³ respectively. At the controlling time after the end of saturation, samples were made for test.

In Situ Test

In situ holes of SPT distribute on the bank of Fen river. Embedded depth of liquefied silt layer is 1-8 m. It is deposited by alluvial soil and diluvial soil of Fen river. Medium diameter is 0.035-0.046mm. Coefficient of uniformity is 5.8-11.25.

ANALYSIS OF TEST RESULTS

It is well known that the fine content and dry density are important factors to affect the liquefaction of silt. It order to evaluate the effect of clay particle contents and dry density to MPT blow counts, authors control time 1,3,5,7,10,20,30,60,90 days(d) for MPT. The test results were shown in Fig.2(Pc=9%).

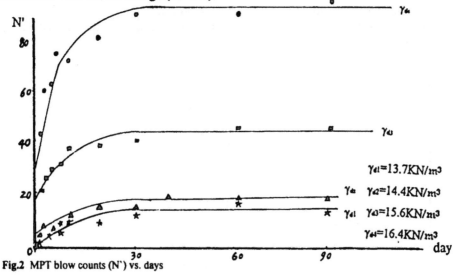

Fig.2 MPT blow counts (N') vs. days

It can be clearly seen from this figure that with the increase of time, MPT blow counts increase; but after one month, MPT blow counts is slowly increased.

Figure3 and figure4 show the results of relationship between MPT blow counts and clay particle contents or dry densitys respectively (d=90).

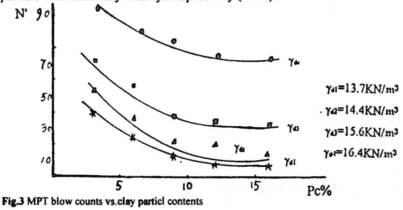

Fig.3 MPT blow counts vs.clay particl contents

It can be clearly seen from Fig.3 that with the increase of clay particle contents, MPT blow counts recreate. It can be also clearly seen from Fig.4 that with the increase of dry density ,MPT blow counts increase.

Furthermore corresponding tests by using mini-penetrometer and standard penetrometer have been run on a lot of undisturbed silt samples. Authors have revised SPT blow counts (N63.5) of different by H.J.Gibbs's equation [5].

$$N=CnN63.5$$

Cn: coefficient of revised

N63.5: SPT blow counts of any effective overburden pressure

N: SPT blow counts of the effective overburden pressure beings 100Kpa correlationship between blow counts MPT (N')and SPT (N)

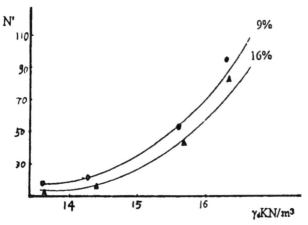

Fig.4 MPT blow counts vs. dry density(Pc=16%,Pc=9%)

The relation between forty drill holes penetration blow counts of SPT and MPT may be represented by following statistics equation.

$$N=0.077N'-0.7$$

N:penetration blow counts of SPT by the effective overburden pressure being 100Kpa.

N:pentertain blow counts of MPT .

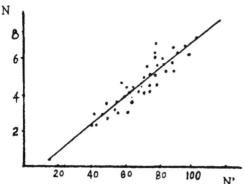

Fig.5 Correlationship between blow counts MPT(N') and SPT(N)

Correlation coefficient is 0.90.It may be considered that the relation between penetration blow count of SPT and MPT is close.

CONCLUSIONS

Based on the result of the study described above, the following general conclusions can be drawn out:
The mini-penetration blow count increases with the dry density of the silt and decreases with content of clay particles.

The relation between penetration blow counts of SPT and MPT may be represented by following statistics equation .
$N=0.077N'-0.7$

Therefore we put forward that the MPT may replace SPT for determination of liquefaction assessment of silt deposit.

REFERENCES

1. W.D.L. Finn, J.J. Emery and Y.P. Gupta. Liquefaction of large samples of saturated sand excited on a shaking table. Proceedings .Ist Canadian Conference on Earthquake Engineering, Vancouver, B.C. May, pp. 97-110 (1971).
2. J.K. Mitchell, and D.J. Tseng. Assessment of liquefaction potential by cone penetration resistance, Proceedings, H.B. Seed. Memorial Symposium, Editor, J. Michael Duncan, Vol. 2, B: Tech Press, Vancouver, B.C., pp, 335-350 (1990).
3. L.F. harder and H.B. Seed, Determination of penetration resistance for coarse-grained soils using the backer hammer drill. UBG/EERC Report No .86/06, University of California, Berkeley. California.(1986)
4. H.B. Seed, K. Tokimatsu, L.F. Harder and R.M. Chung, Influence of SPT procedures in soil liquefaction resistance evaluations. Journal of the Geotechnical Eng. Div., ASCE, Vol. 3, No.12, December(1985).

Proc. 30th Int'l. Geol. Congr., Vol. 23, pp. 475-479
Wang Sijing and P. Marinos (Eds)
© VSP 1997

The Further Study and Test of Liquefaction Resistant Characteristics of Saturated Silt

NIU QIYING & QIU YIHUI
Department of Civil Engineering, Taiyuan University of Technology, TY 030024 China.

Abstract

Presented in the paper are the results of experimental research of liquefaction characteristics of silt by varying the content of clay particles or by varying the dry density. The laboratory tests were run on remolded and undisturbed silt specimens by using cyclic dynamic triaxial device. The results show that the lowest dynamic shear strength is at 9% of content of clay particle,and the dynamic shear strength increases with the dry density of the silt. It lays the foundation of the study on silt liquefied mechanism in the future.

Keywords: Cyclic stress ratio, Clay particle content, Earthquake resistant behaviour,Effect of dry density, Time effect.

INTRODUCTION

Many earthquake disaster examples have proved that saturated silt subsoil could suffer liquefaction damage. Wang[1], Shi[2] and Zhong[3], in their investigation of the Tangshan earthquake in 1976,showed that soil deposits containing a containing amount of silt liquefied.

Now the phenomenon and assessment of the liquefaction of silt have been studied deeply and extensively by researchers.The result of cyclic load testing of silts carried out by Shi[2], Zhong[3] and Qiu[4] et al. were examined. However this is just a beginning.In order to further study liquefaction mechanism and resistant characteristics of silt, the authors run on remolded and undisturbed silt samples of varying the dry density(γd) and varying the content of clay particles (Pc) by using cyclic dynamic triaxial device. The laboratory test took us as more as two years to finish, finally drawn out the general regulation of liquefaction behaviour of silt.

TEST METHOD

Experimental Device
Cyclic Triaxial Device   The instrument used is DSD-200 model cycle triaxial device, stress

controlled,designed by china. Its index properties are as follows:

Frequency	0.01HZ-50HZ
Maximum vertical pressure	25KN
Maximum confining pressure	1.0MPa
Height of sample	10cm
Diameter of sample	5.0cm

<u>X-Rag Diffraction</u> The X-rag diffraction device is of Y-4QX type made in china (EDX).Its conditions are as follows:

Target material	Cu
Voltage	35kv
Electric current	15μA
Degree of an angle	3-45

Test Material
Undisturbed silt samples were taken from the site of which is located on the bank of Fen river,Taiyuan university of technology water treatment engineering , living quarter and scientific building site. Soil properties are as follows:

Depth of embedment(m)	4-8
Medium diameter(mm)	0.030-0.046
Coefficient of uniformity	5.8-11.25
Content of clay particles (%)	2-11.5
Dry density (KN/m^3)	14.5-15.8

Silt and clayey particles of preparative silt samples used is this study were taken from the site which is located near Yifen bridge,on the west bank of Fen river in northern Taiyuan. The soil were deposited and air dried on the bank Fen river. The physical properties of soil are summarized is Table 1. Clayey mineral is kaolinite analyzed by X-rag diffraction. All the remolded samples were made in laboratory. The content of clay particles were 3,6,9,12,16%,and they were put in soil box size 900×400×300mm to deposit. These samples were divided into three groups according to their dry densities 14.4,15.6,16.4KN/m^{3}respectively. At the time of 2 months after the end of saturation and deposit, samples were made for test.

Table 1. Silt properties

Soil	Specific gravity	Plasticity index	Medium diameter(mm)	Coefficient of uniformity	Content of clay particles(%)
Silt	2.69	7.9	0.043	2.8	0
Clayey particles		20.99			40

Samples for X-rag diffraction were air dried firstly and then rubbed with a glass bar unitl they can be cut into thin slices.

Failure Criterion
The failure criterion adopted in this study is that the strain of double amplitude is equal to 5%.

ANALYSIS OF TEST RESULTS

It is well known that the fine content is a significant factior on silt liquefied. In order to evaluate the effect of clay particle contents to cyclic stress ratio ($\sigma d/2\sigma 0$)quantitatively, the authors performed laboratory dynamic triaxial tests on all samples with different clay particle contents and different dry densities. The test results were shown in Fig.1,2,3,4.

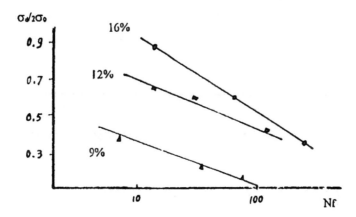

Fig.1 Cyclic stress ratio vs. number of cycles for remolded samples of different clay particle contents ($\gamma d=14.4KN/m^3$)

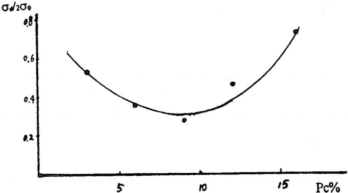

Fig.2   Relationship between cyclic stress ratio and clay particle contents (remolded samples)

It can be clearly seen from these figures (Fig. 2, Fig. 4) that with the increase of clay particle contents, the cyclic shear ratio is not monotonously increased and achieves a minimum value at Pc=9%, which samples are remolding or intact.

Figure 5 shows the result of the cyclic triaxial test. Their dry densities are 14.4, 15.6,16.4KN/m³ respectively. It can be seen that the dynamic shear strength increases linearly with the dry density.

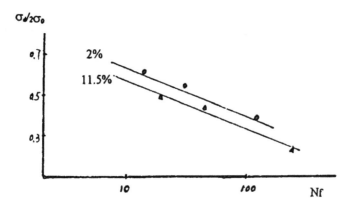

Fig.3 Cyclic stress ratio vs. number of cycles for undisturbed samples of different clay particle contents (γd=14.5KN/m³)

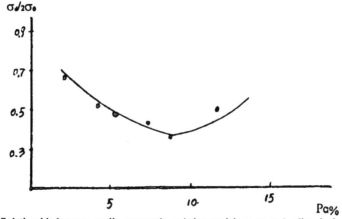

Fig.4 Relationship between cyclic stress ratio and clay particle contents (undisturbed samples)

CONCLUSION

Based on the results of the study described above, the following generial conclusions can be drawn out:

Vary content of clay particles is an important factor effecting liquefaction of silt.

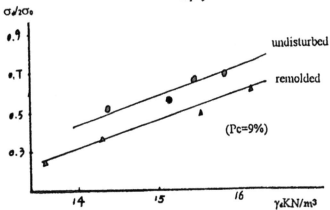

Fig.5 Correlationship between σd/2σ0 and γd(Pc=9%)

The cyclic stress ratio of remolded or undisturbed silt changes with content of clay particles in a curve of parabola. The lowest shear strength is at 9% of content of clay particles.

The dynamic shear strength increases with the dry density of the silt.

REFERENCES

1. W.S.Wang. Some findings in soil liquefaction. Water Conservancy and Hydroelectric Power Scientific Research Institute. 8,Beijin China (1979)
2. Z.J.Shi and S.S.Yu. Liquefaction behaviour and in situ identifying of sand loam. Hydrogeology and Engineering Geology. 3,pp.14-18 (1982)
3. L.H.Zhong, Analysis for evaluating liquefaction of low plasticity clays during earthquake. Chinese Journal of Geotechnical Engineering.3 (1980)
4. Y.H Qiu, S.J.Fan and W.Y.Fan. Some aspects on the liquefaction potential of dynamically compacted loesslike sandy loam. Proceedings, Ninth World Conference on Earthquake Engineering .Vol. 3. pp.225-230 (1988).

Proc. 30th Int'l. Geol. Congr., Vol. 23, pp. 481-487
Wang Sijing and P. Marinos (Eds)
© VSP 1997

In-Situ Deformation Measurements for Evaluation of Hydraulic Rock Parameters

HANS-JOACHIM KUPEL
Section Applied Geophysics, Geological Institute, University of Bonn, D-53115 Bonn, Germany

Abstract

A rarely exploited phenomenon is ground deformation due to internal pore pressure gradients. The deformation signal, although rather small, is expected from theoretical considerations: Pore pressure gradients leading to ground water flow do also deform the rock (or soil) matrix because of friction in the pore space. The gradients can be generated in a controlled form as through pumping or injection of ground water, and the deformation may be sensed through tiltmeters coupled to the ground, preferably in small boreholes. Analyses of various field experiments have shown that linear poroelasticity for saturated media appears to be appropriate to describe the dominating physical processes. This allows identification and evaluation of in-situ rock parameters not attainable through other techniques. In the full space situation, the effective hydraulic diffusivity - without neglect of the mechano-hydraulic coupling between fluid pressure and rock matrix strain - can be assessed, also the Skempton ratio and the quasi-static shear modulus of the ground volume between the pumped well and the location of the tiltmeter.

Keywords: poroelasticity, petrohydraulic rock parameters, tilt measurements

INTRODUCTION

Coupled hydraulic phenomena in saturated media have attained increasing attention amongst geoscientists and engineers during the last years [12]. Solutions of hydrogeological problems crucially depend on the knowledge of adequate rock/soil parameters. Such parameters are often insufficiently known, for example because of the scaling problem, or because even the nature of hydromechanical coupling is not yet well understood. Of particular interest are petrohydraulic rock parameters that describe the ease of fluid flow through the rock matrix as a macroscopic phenomenon, and the mechanical reaction of the matrix due to a change in pore or confining pressure. There are various methods to obtain these quantities from laboratory measurements, that is by testing rock samples; however, since the samples have been removed from their original environments and are of limited extensions, laboratory tests can only partly reveal the in-situ behaviour of rocks.

Frequently used techniques to evaluate hydraulic rock parameters under in-situ conditions are pump and slug tests, or their inverse procedures, injection and bail tests. Suchtechniques provide estimates of the hydraulic conductivity or transmissivity of aquifers and approximate the storage coefficient. A rarely considered phenomenon when pumping or injecting ground water is the deformation of the ground due to the induced pore pressure gradients. Although the deformation is small - submillimeter displacements in volumes of cubic meters -, it can be measured with adequate instruments. In previous years, the phenomenon has occasionally been observed as a noise signal in borehole tiltmeter records [2, 14]. Meanwhile, various attempts have been made to obtain such signals under controlled conditions in order to use them for assessing petrohydraulic rock parameters [3, 4, 8, 6, 7]. The tiltmeters used for sensing the deformation signal were of 0.1 microradiant resolution or better.

The paper outlines the theoretical basis of the underlying physical process, describes the type of petrohydraulic parameters that can be deduced and summarizes some field experience that has been obtained so far.

THEORETICAL BACKGROUND

Tilt signals generated by pumping activities in a well typically have a fin-like shape (Fig.1). The theory that is apt to explain the dominant petrohydraulic process appears to be linear poroelasticity. Originally elaborated by Terzaghi [11] and Biot [1], it is now mostly used in the formulation of Rice & Cleary [9].

A poroelastic medium behaves similar to a thermoelastic one in which pore pressure takes the role of temperature. Unlike in thermoelasticity, however, the two-way coupling between deformation and internal pore pressure is very strong and cannot be neglected as is often the case in thermoelasticity (heating deforms the medium, but deformation may have no significant effect on temperature). The approach is generally macroscopic, meaning the structures of pores and matrix are not specified. Application of linear poroelasticity requires almost complete saturation of the medium with a low compressible fluid (no gas), and the deformation has to be small. Only then the rock parameters may be regarded as constants.

The rheology of a poroelastic medium depends strongly on the drainage conditions, i.e. whether pore fluid can leave the considered volume or not. Therefore, two sets of poroelastic parameters need to be specified, one for drained the other for undrained conditions. In fact, four independant mechanical parameters describe the deformation behaviour of the matrix as functions of stress and pore pressure. A fifth parameter reflects the flow resistance in the matrix [5, 13].

Frequently used parameters are the shear modulus G, the Skempton ratio B, the Poisson ratios v and v_u, and the hydraulic diffusivity D. Unlike the compressibility or the Young's modulus, G is independent of drainage conditions. B denotes the change in pore pressure per unit change of confining pressure for undrained conditions and has a value between 0 and 1. v_u and v are the Poisson ratios for drained and undrained conditions, where always $0 <= v <= v_u <= 0.5$. The hydraulic diffusivity D, as measure for the flow resistance, is connected to these parameters by

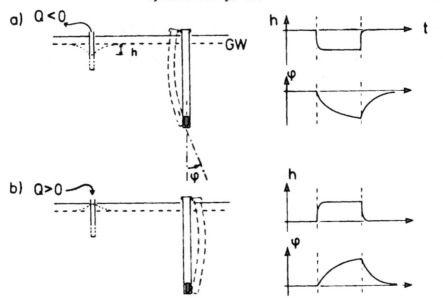

Figure 1. Schematic representation of ground deformation induced by pumping (a) or injection (b) of water from or into a nearby well. GW = ground water level, h = well level change, Q = injection rate, φ = induced tilt amplitude at bottom of deflected borehole, t = time. Modified after [6].

$$D = \frac{2}{9} \frac{(1-\nu)}{(1-\nu_u)} \frac{(1+\nu_u)^2}{(\nu_u-\nu)} \kappa \, G \, B^2 \tag{1}$$

with k denoting the Darcy conductivity, or the ratio of intrinsic permeability to dynamic viscosity of the pore fluid.

Like with linearly elastic media there exist stress-strain relations for linearly poroelastic media. Those for elastic media are simply extended by a term accounting for spatial pore pressure gradients, representing body forces. An additional equation relates spatio-temporal variations of excess pore pressure to the storage capability of the medium. To restrict the constitutive relations to a controllable number of rock parameters, the medium is usually assumed to be anisotropic and homogeneous. The full set of differential equations in 3D, or for axial symmetry or plane strain conditions, may be found in the literature [10, 5]. Note that neither intrinsic permeability nor porosity appear as individual parameters in these equations.

ROCK PARAMETERS DEDUCED FROM TILT SIGNALS

Poroelastic deformation in the far-field of a spherical pore pressure anomaly can be expressed as an analytical term. It may be used to compute the deflection of a vertical borehole (see Fig.1), assuming that the casing completely follows the strain field in its surroundings. Comparing the tilt signal in the borehole with the theoretical tilt for a known pore pressure disturbance and known geometric configuration yields in-situ rock parameters that are not accessible through conventional techniques.

In case where the pore pressure disturbance is induced through pumping with constant (negative) yield Q, starting at time $t=0$, the matrix shear strain φ in the homogeneous full space can be shown to equal

$$\varphi = c\,K\,\overline{B}\,Q/D \qquad (2)$$

[4], where c denotes a dimensionless function of effective radial distance r to the well axis, effective vertical distance z to the centre of the well screen and of D and t, namely

$$c\,(r,\,z,\,t;\,D) = (3J+E^{-})/(16\,\pi) \qquad (3)$$

with

$$E^{-} = 1 - erf(\mathscr{R})$$
$$J = [erf(\mathscr{R})/\mathscr{R} - 2exp(-\mathscr{R}^{2})/\sqrt{\pi}]/2\,\mathscr{R}$$
$$\mathscr{R} = R/\sqrt{4Dt}$$

$erf(.)$ is the error function and $R= \sqrt{r^2+z^2}$. The far-field condition is fairly fulfilled when R is larger than three times the well screen's extension. The tilt sensor has to be sufficiently deep in order to interpret tilt signals as shear strain (see below).

The amplitude of the function c steadily increases with duration t of pumping, approaching the value 0.017... (Fig. 2). If $R<=1$ or $t>=R^2/4D$, aquifer parameters may be estimated from the maximum observed shear strain value, where c may be readily replaced by 0.017 [4]. In general, it is sufficient to exploit the t_{90}-time amplitude of the tilt signal, that is the amplitude when 90% of the maximum tilt signal has evolved. Then, relation $t_{90} = R^2/4D$ together with $c=0.017$ may be used with less than 10% error. Alternatively, a type curve matching can be applied, analogous to the interpretation of conventional well tests.

The other parameters in eq.(2) are

$$K = 2rz/R^3, \qquad\qquad \text{a geometric coefficient, and}$$
$$\overline{B} = B/3\,(1+v_u)/(1-v_u); \quad \text{obviously,} \quad B/3<=\overline{B}<=B.$$

The assumption that tilt signals represent mostly shear strains in the subsurface is not valid near stress-free boundaries like the earth's surface. Here, rotational movements are more dominant and superimpose on shear signals. The relevance of near surface effects is subject to current investigations [7].

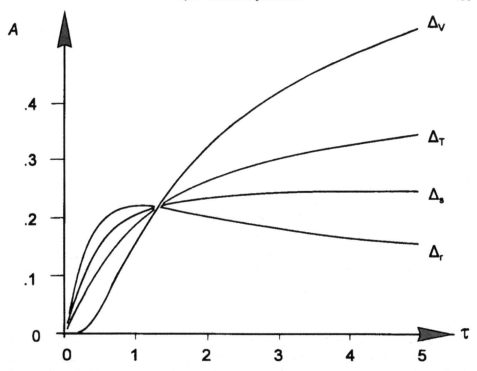

Figure 2. Theoretical shear strain φ (=Δ_s), volume strain Δ_V, tangential strain Δ_T, and radial strain Δ_r in full space poroelastic vicinity of a suddenly activated, point-like pore pressure disturbance of constant injection rate Q at distance R. Amplitudes A are normalized to $QB/(4\pi RD)$; here, function c (eq.3) is scaled as $A/4\pi$. Transient build-ups of strain signals are represented as functions of dimensionless time $\tau=4Dt/R^2$. Modified from [6].

CASE STUDIES

Experience with pumping induced rock deformation has been gained in several field studies. In the cases listed here, the deformation was sensed by tiltmeters of 100 nrad ($\cong$20mseca) resolution or better. Individual experiments, though, differed considerably from each other. Important aspects are summarized below.

Case 1:
Site Medelby, almost flat grassland and forest region in Schleswig-Holstein, Northern Germany, with sandy ground. Two biaxial borehole tiltmeters of 1 nrad resolution, installed at 30 m depth. Ground water level at 2.5 m below the surface in unconfined aquifer extending to about 60 m depth. Pump and injection tests with Q values up to ±4 m³/h at distances 10 to 120 m have induced tilt amplitudes up to 400 nrad; t_{90}-times around 20 minutes. The hydraulic diffusivity was found to be around 0.4 m²/s, one order of magnitude less than estimated from conventional pump tests; $B \cong 0.65$ [4].

Case 2:

Underground rock laboratory of the Swiss NAGRA, mostly in granodiorite. One biaxial borehole tiltmeter of 1 nrad resolution, installed at 17 m below the floor of the major gallery. Intermittent closure of naturally draining, 100 m long, horizontal drill hole with contact to an extended fissure system lead to variations of 4 MPa in excess fluid pressure which induced tilt signals up to 2 urad; t_{90}-times around 20 days. The derived hydraulic diffusivity was around $8 \cdot 10^{-4}$ m^2/s, one order of magnitude higher than estimated from laboratory tests on intact rock samples; $B \cong 0.85$ [4].

Case 3:
Test area near Kirchheim-Teck, Rheinland-Pfalz, Southern Germany; grassland over loamy to sandy ground, $\cong$ 1% slope. Various tiltmeter positions in 0.6 m deep pits, instrumental resolution 100 nrad. Unconfined aquifer from 0.5 to 4.5 m depth. Pumping activities of 2 hours duration with yields from -0.35 to -0.7 m^3/h induced significant tilt signals up to 20 m distance. In the near-field (3 to 6 m), tilt amplitudes reached values of 40 urad within t_{90}-times from 10 minutes to 2 hours. Average hydraulic diffusivity around 10^{-2}m^2/s agrees for hydrologic and poroelastic pump tests [8].

Case 4:
Site Nagycenk, Western Hungary; vicinity of well which is used for water supply, with screen sections around 58 and 70 m depth; flat topography, alternating gravel, sand and clay/loam layers; ground water level at 3 m depth. Two biaxial borehole tiltmeters of 100 nrad resolution installed at 5 m depth, 7 and 8 m distance to the well at different azimuth. Water yields from -30 to -40 m^3/h induce tilt signals of 3 rad amplitude with t_{90}-times around 2.5 hours. Here, the signs of the tilt signals are opposite than expected from the full space solution. It is speculated that this is due to surface subsidence [7], which is currently under investigation.

CONCLUSIONS

Observation of pump induced deformation in the subsurface appears to be a mean to quantify poroelastic rock/soil parameters under in-situ conditions. The Skempton ratio B and the hydraulic diffusivity D (without neglect of matrix deformation) are parameters not directly obtainable through traditional techniques. It seems that rock values estimated from pump induced tilt signals reflect integrated properties of the effective volume, that is some volume between the pore pressure disturbance (the well screen) and the location of the tilt sensor. This may explain the discrepancies between the hydrologically and the poroelastically deduced hydraulic diffusivities in cases 1 and 2. Cases 3 and 4 show that rather standard tilt sensors may suffice to observe pump induced ground strain. Two other aspects of poroelastic ground tests are: (1) Poroelastic (tilt) deformation signals seem to be measurable at greater horizontal distance than the pore pressure disturbance (observable as well level subsidence). The former are also visible outside the tested aquifer, in more shallow or deeper layers, where the pore pressure may not be affected. (2) When the azimuth of the maximum tilt effect deviates from the direction towards the pumped well, anisotropic fluid flow is the reason and may thus be recognized. Cases 1 and 2 have evidenced that phenomenon [4]. Clearly, the greater relevance of rotational tilt over shear strain expected at shallow depths needs further investigation. A tilt sensor fixed to the ground cannot distinguish between these two signal types.

Application of poroelastic testing in hydrology will certainly be restricted to the study of special problems, because conventional hydraulic tests can be carried out more easily and often provide sufficient information. Yet, many problems *are* special and may justify to do poroelastic pump tests, for example reconnaissance of: delimitation of regions used for drinking water supply, tightness of waste sites, effective fluid diffusivity in host rocks, direction of underground fluid contaminations, optimized locations of producing wells in oil fields, characteristics of fluid flow in geothermal fields.

REFERENCES

1. M.A. BIOT. General theory of three-dimensional consolidation. J. Appl. Phys. 12, 155-164 (1941).
2. H.-J. KUPEL. Neigungsmessungen zwischen Hydrologie und Ozeanographie. PhD thes. Univ. Kiel, 165 p. (1982).
3. H.-J. KUPEL. Gesteinsverformungen durch Porendruckgefale. Kurzberichte aus der Bau-forschung, Fraunhofer Ges., Ber.Nr.39, 145-146 (1988).
4. H.-J. KUPEL. Verformungen in der Umgebung von Brunnen. Habil. thes. Univ. Kiel, 198 p. (1989).
5. H.-J. KUPEL. Poroelasticity: parameters reviewed. Geophys. J. Int. 105, 783-799 (1991).
6. H.-J. KUPEL. In-situ Deformationsmessungen zur Bestimmung hydraulischer Bodenkennwerte. In: B.MERKEL, P.G.DIETRICH, W.STRUCKMEIER & E.P.LONERT (eds.), Grundwasser und Rohstoffgewinnung, GeoCongress 2, Sven von Loga, Koln, 291-296 (1996).
7. H.-J. KUPEL, P. VARGA, K. LEHMANN and Gy. MENTES. Ground tilt induced by pumping - Preliminary results from the Nagycenk test site, Hungary. Acta Geod. Geoph. Hung, 31(1-2), 67-79 (1996).
8. R. MAIER and S. MAYR. Bestimmung von Aquifereigenschaften durch Messung von Neigungen bei Pumpversuchen.- In: DGG-Mitteilungen, spec. vol. 3. DGG-Seminar Umweltgeophysik (1996), in press
9. J.R. RICE and M. CLEARY. Some basic stress solutions for fluid-saturated media with compressible constituents. Rev. Geophys. Space Phys. 14, 227-241 (1976).
10. S.ROJSTACZER and D.C.AGNEW. The influence of formation material properties on the response of water levels in wells to Earth tides and atmospheric loading. J.Geophys.Res. 94, 12.403-12.411 (1989).
11. K.TERZAGHI. Die Berechnung der Durchlasigkeitsziffer des Tons aus dem Verlauf der hydrodynamischen Spannungserscheinungen. Sitzungsber. Akad. D. Wiss. Wien, Math.-Naturw. Kl., Abt.IIA 132, 125-138 (1923).
12. Th. TOGERSEN. Modeling and testing coupled hydrological processes. EOS Transactions 75, 73-78 (1994).
13. H. WANG. Quasi-static poroelastic parameters in rock and their geophysical applications. Pure Appl. Geophys. 141, 269-286 (1993).
14. A. WEISE. Neigungsmessungen in der Geodynamik: Ergebnisse von der 3-Komponenten-Station Metsaovi.- PhD. thes. Techn. Univ. Clausthal (1992).

Proc. 30th Int'l. Geol. Congr., Vol. 23, pp. 489-499
Wang Sijing and P. Marinos (Eds)
© VSP 1997

Infiltration of Rainwater from Slopes of Pyroclastic Flow Deposits Using Automated Electrical Prospecting

SHUICHIRO YOKOTA*, TETSUYA FUKUDA**, AKIRA IWAMATSU**,
SHIN'ICHI UDA***, TAKUYA WADA***, & HIDEKAZU MASAKI****
* *Dep. Geosci., Fac. Sci. and Eng., Shimane Univ. ,1060 Nishikawatsu, Matsue 690, Japan.*
** *Inst. Earth Sci., Fac. Sci, Kagoshima Univ.,1-21-35,Korimoto, Kagoshima 890, Japan,*
*** *Osaka branch of CTI Engineering Co., 1-2-15 Otemae, Chuo-ku, Osaka 540, Japan.*
*****Fukuoka branch of CTI Engineering Co., 2-1-10 Watanabedori, Chuo-ku, Fukuoka 810, Japan*

Abstract

Rainfall and its infiltration from the slope surface may be one of the most important factors for slope failures in monsoon regions such as South and East Asia. To obtain the process of the infiltration of rainwater, temporal changes of the apparent resistivity inside of the slope have been measured using automated electrical prospecting in southern Kyushu, Japan. A target slope is underlain by the Quaternary pyroclastic flow deposits characterized by soft and permeable properties. Results of measurements after a heavy rainfall indicate decrease in apparent resistivity value and its inward extension. This suggests that the inward infiltration of rainwater have proceeded from not only top surface but also slope surface. The decreasing of the resistivity continues for longer than a day in deeper portion, and consequently, the effect of the infiltration was recognized extending over 20 meters from the surface. On the other hand, the resistivity value tends to change rapidly between wet and dry in response to weather condition in shallower portions, while it changes very slowly in deeper portion.

Keywords: Slope Failure, Rainfall, Resistivity, Electric Prospecting, Pyroclastic Flow Deposits, Quaternary, Japan

INTRODUCTION

Slope failures during a rainy season are annual occurrences in southern Kyushu, where most of slopes are underlain by the Quaternary pyroclastic flow deposits. They are homogeneous non-welded dacitic tuff, and are soft and permeable. They are sometimes called as "Shirasu", a local name of "whitish sand"[1]. Many studies have been made to understand the mechanism of slope failures and to estimate the recurrence period of failures in the region (for example,[2]). As a result, soft and permeable properties of the deposits and rapid deterioration (weakening) of slope surface have been believed to be primary cause for slope failures apart from steep slopes observed in this region[3]. On the other hand, triggering of failures may be associated with heavy rainfall in most cases similar to the mechanism obtained in the monsoon regions.

Wada *et al.*[4]attempted to measure temporal changes of the apparent resistivity beneath the pyroclastic plateau using automated electrical prospecting[5, 6]. Downward infiltration from the top of the plateau has been recognized by the resistivity measurements. However, to discuss slope failures, such prospecting should be made directly on steep slopes. To achieve this, the authors have attempted to make similar electrical prospecting on a steeper slope in the same region.

TOPOGRAPHICAL AND GEOLOGICAL OUTLINE

Topographical and Geological Outlines around Measurement Profile
Fig.1 shows the location of the study area. The Quaternary pyroclastic flow deposits is widely distributed in southern Kyushu including the study area[7, 8]. As shown in topographic map (Fig.2), the location is situated on the rim of a pyroclastic flow plateau whose elevation is about two hundred twenty meters. Deep erosion observed at rim of the plateau results in steep slopes. Measuring profile for prospecting was set on one such a slope.

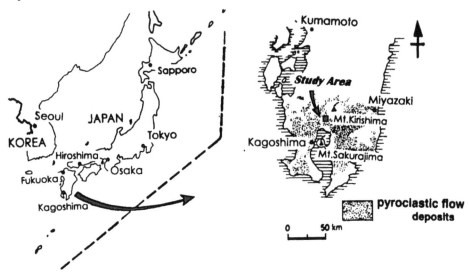

Figure 1. Location of the study area. Dotted part in right figure shows the extents of the Quaternary pyroclastic flow deposits.

Fig.3 shows the topographical and geological section of a line A-A'. While slope dips 40 to 50 degrees as a whole, sometimes it dips more than 70 degrees. It is mostly composed of soft pyroclastic flow deposits, that date back to about 25,000 years[8]. Welded tuff forming a jointed rock mass and sedimentary rocks mainly composed of silt stone and sandstone underlain by these pyroclastic flow deposits. While the overlying pyroclastic flow deposits are highly permeable, the underlying rocks have very low permeability. On the other hand, top of the plateau is covered with thick volcanic ash fall with pumice and black fumes soil of poor permeable properties[4]. The surface of the slope along the measuring profile is mostly covered with vegetation. However, difference in age of vegetation and existence of three concave portions on the slope may indicate three stages of past failures since the past hundred years or so.

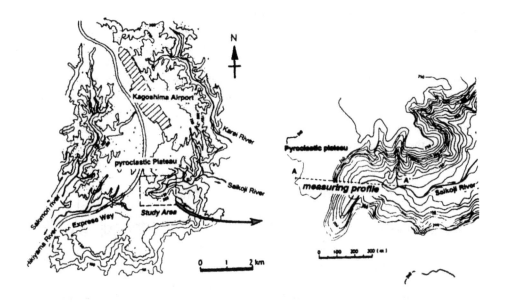

Figure 2. Pyroclastic plateau of the study area and topographical outline around measuring profile. Dotted part in the left shows the portion higher than 200 meters in elevation. Line A- A' is shown in Figure.3

pyroclastic plateau

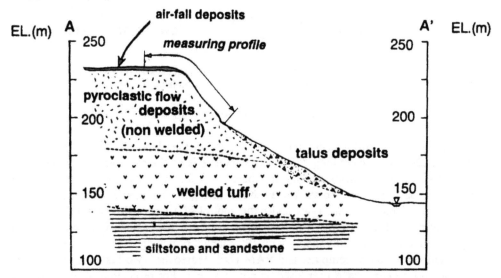

Figure 3. Topographical and geological section of A-A' (see Figure.2). Measuring profile covers from flat part on the top to steep slope.

PRINCIPLE OF MEASUREMENT

Infiltration of rainwater and apparent resistivity

The principle to obtain infiltration process by electrical prospecting is shown in Fig.4. During the rainy days, rain water may infiltrate inward from the slope surface and some portion inside of the slope may have wet condition. This may be revealed as the decreasing of the resistivity and its inward extension. On the other hand, on a fine weather day, surface may become dry due to evaporation. This may bring about an increasing of the resistivity value. If many electrodes are distributed on both the slope and top, and measuring is repeated cyclically, wetting or drying process may be revealed as temporal changes of the apparent resistivity.

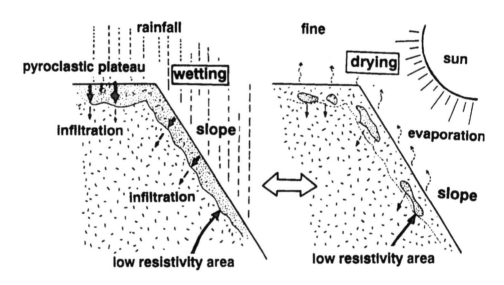

Figure 4.   Principle of automatic electrical prospecting. Decreasing of apparent resisitivity along the slope and its inward extension are revealed during rainfall.

Measuring Profile and Automated Measurement of Apparent Resistivity by Electrical Prospecting

Measuring Profile and distribution of electrodes on the slope are shown in Fig.5. the profile is about 65 meters long. Of which, 45meters is on the slope and 20 meters on the top of the plateau. 98 electrodes are distributed with 0.5 meters spacing on the slope and 2.0meters spacing on the top surface as shown in Fig.5. Measurement was automatically made with a three hours intervals under the control of a personal computer. To achieve this, apparatus described by Wada et al.[9] has been used. Resistivity meter has the capacity of 15mA and 400 voltages with 100 channel switching. Control and data storage were done by personal computer and RAM card. Moreover, Two batteries (12 voltages) and two solar panels have been used as power source. For protection of electrodes and cable against rodents, they are mostly covered with flex pipe. Rainfall have also been recorded during the same period.

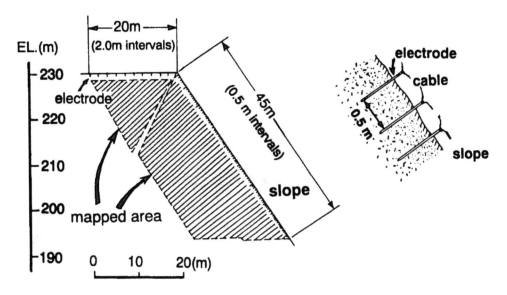

Figure 5. Distribution of electrodes and their intervals.

MEASURED RESULTS AND CHARACTERISTICS OF INFILTRATION OF RAINWATER

Apparent Resistivity and its temporal changes
Measurement has been continued during six months from June to November 1995 except for August and September. Fig.6 shows examples of apparent resistivity obtained at June 3rd and 27th. Two diagrams in left side of Fig.6 are the distribution of the raw apparent resistivity values expressed in ohm-meters, and on the right the ratio of resistivity between each days is expressed in percent[10]. The ratio gives a temporal change during the period. Consequently, we obtained many such recordings during different period with records of rainfall.

Wetting process and drying process
Fig.7 shows a change from 0:00 hours of June 3rd to 0:00 hours of June 4th indicating a wetting process after a heavy rain. More than 150mm rainfall was recorded during this 24 hours. Based on these diagrams, it seems that rainwater may have infiltrated from the slope surface, and it may have gradually proceeded inward. After 24 hours, the front of 5percent lower zone approaches the depth of 7 or 8 meters from the surface. These evidence of the infiltration was finally recognized extending over 20 meters from the surface.

On the other hand, Fig.8 shows a change from June 3rd to June 8th. Only 15mm rainfall was recorded during these 5 days. Base stage used for calculating ratio in these diagrams are same as that in Fig.7. Considering that little rainfall was recorded during this period,

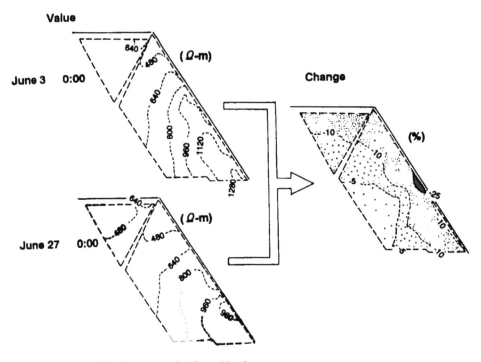

Figure 6. An example of apparent resistivity and its changes

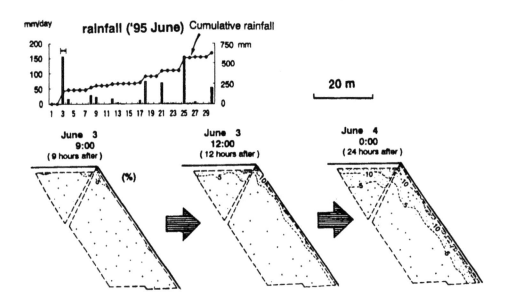

Figure.7. An example of wetting process (from June 3rd to June 4th).

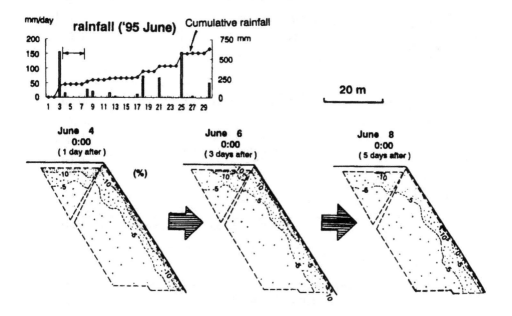

Figure 8. An example of drying process (from June 4th to June 8th).

these diagrams may show a drying process within the slope. During the period, configuration of the low resistivity zone becomes relatively small, and high resistivity zone appears partially within outermost portion of the slope. This means that outermost portion of the slope becomes rapidly dry by evaporation. However, most of deeper portion may not change during the same period.

According to these various changes of resistivity with rainfall, following characteristics on infiltration process are pointed out;
   (1) Rainfall may infiltrate from the surface of slope, and it gradually proceeds in ward.
   (2) Relationship between temporal changes of resistivity and rainfall indicates the existence of distinct time lag between them.
   (3) Resistivity seems to change rapidly between wet and dry in response to weather condition in outermost portion. On the contrary, the changes may be very small and slow in the deeper portion.

A ratio of the area of low resistivity zone to total analyzed area and its changes with time may be an effective index for quantifying the amount of infiltration and its time lag during rainfall. Such ratio was calculated in two dimensional section using the distribution diagrams at many stages. Fig.9 shows such temporal changes of the ratio of the area lower than -5 percents to total area in three depth zone. Here, three division of the zone are; (A) 1-2m, (B) 2-5m, and (C) 5-20m respectively. On the other hand, (D) in Fig.9 is the daily

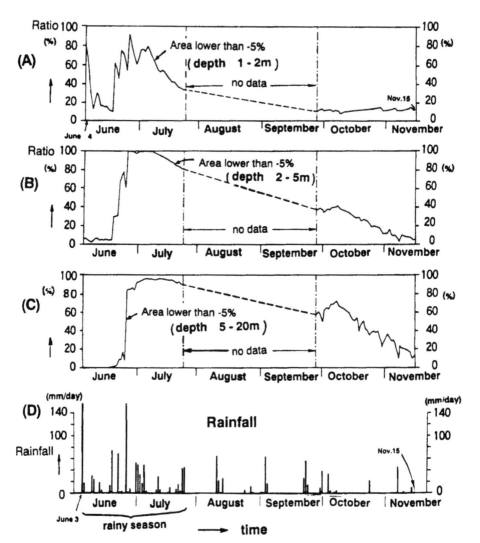

Figure 9. Changes of ratio of low resistivity to total area (A) 1-2m, (B) 2-5m, (C) 5-20m, (D) daily rainfall.

rainfall during same period. Difference in curves (A), (B) and (C) indicates the time lag with depth. The time lag seems to become larger with the depth. Especially, it approaches about three months after a rainy season in the case of (C).

The resistivity distribution curves resemble those of the Antecedent Precipitation Index (API)[11], well known as effective rainfall derived from record of daily rainfall, therefore we can compare these curves with those of ATP.

Apparent Resistivity and Antecedent Precipitation Index (API)
The index API is widely used to evaluate the effects of rainfall for slope failure in general, and is defined as follows[11];

$$Dn = a^{n-1}r_1 + a^{n-2}r_2 + \ldots + a^1 r_{n-1} + r_n \tag{1}$$

where D_n : Antecedent Precipitation Index (API) at n-th time,

r_n : precipitation at n-th time,

a : coefficient expressing reducing rate, and take a value between zero and one, that is $0 <= a <= 1$.

Fig.10 shows examples of API curves. They are the cases that a = 0.0, 0.5, 0.9 and 1.0 calculated on a simple rainfall respectively. The case when a =1.0 is special on which API is equivalent to the cumulative rainfall. On the other hand, the case of a = 0.0 is that of daily rainfall.

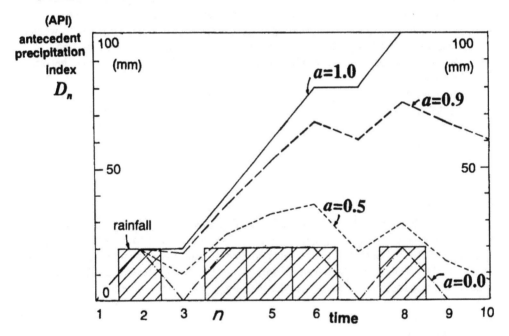

Figure 10. Antecedent precipitation index Dn. While the graph of a =1.0 is equivalent to cumulative precipitation, that of a = 0.0 corresponds daily precipitation.

Comparing each curves in Fig.9 with those in Fig.10, we can estimate coefficients "a" fitting each curves. Calculated coefficients " a" are as follows:

a = 0.90 for depth 1 - 2 m,
a = 0.97 for depth 2 - 5 m,
a = 0.98 for depth 5 - 20m.

Difference in coefficients mentioned above means that the deeper the infiltration proceed, the larger the time lag becomes. Understanding of practical relationship between resistivity values and water contents within the deposits composing the slope may be next step for the study on slope failures. If such relation is obtained in future, it may become possible to predict individual slope failures by using these coefficients and records of rainfall. Moreover, these may contribute to construct regional hazard map for slope failures[12]with studies on various primary causes.

CONCLUDING REMARKS

As a concluding remarks;
(1) Changes in apparent resistivity which indicates infiltration of rain water into the slope was recognized after a heavy rain by using automated electrical prospecting.
(2) Infiltration may begin from the surface of slope, and gradually proceeds inward.
(3) The infiltration was recognized extending over 20 meters below the surface.
(4) While resistivity tends to change rapidly between wet and dry in response to weather condition in shallow portion, it changes very slowly within the deeper portion.
(5) Distinct time-lag is recognized between rainfall and infiltration, and it may become larger with its depth.
(6) The ratio of a temporal changes of area of low resistivity zone to total area resemble the API (Antecedent Precipitation Index) curves. Therefore, we can quantify time-lag of wet condition to rainfall by coefficients "a" in API, and evaluate the unstability of the slope.

Acknowledgments

These project have been done under the financial support of the River Information Center of Japan and the CTI Engineering Co. Some engineering support have been made by Mr. Ichikawa of the CTI Science Co. and Mr. Furudate of   the Geo Consultant Co. We would thanks to Dr. Venkatesh Raghavan of the Media Center of the Osaka City Univ. for reading the manuscript, in addition for discussion and valuable comments.

REFERENCES

1. Iwamatsu, A., Fukushige,Y. and Koriyama S. Applied geology of so-called "Shirasu", non-welded ignimbrite. Japanese Jour. Geography, 98, 379-400 (in Japanese with English abstract) (1989).
2. Shimokawa, E., Jitozono, T. and Takano, S. Periodicity of shallow landslide on Shirasu (Ito Pyroclastic Flow Deposits) steep slopes and prediction of potential landslide sites. Transactions, Japanese Geomorphological Union, 10, 267-284 (in Japanese with English abstract) (1989)
3. Yokota, S. Deterioration of "Shirasu" and slope failures. Report on 1993 Kagoshima heavy rain disaster, Part 2, 63-72 (in Japanese) (1995).
4. Wada, T., Inoue, M., Yokota S. and Iwamatsu, A. Downward infiltration of rainwater by using continuous electric prospecting within pyroclastic plateau (Shirasu-Daichi) in Southern Kyushu, Japan. Jour. Japanese Soc. Eng. Geol., 36, 349-358 (in Japanese with English abstract) (1995).
5. Inoue, M. Survey of underground water by the expression of resistivity cross section (in Japanese with English abstract). Jour. Japanese Soc. Eng. Geol., 20, 12-19 (1979).
6. White, P. A. Measurement of ground-water parameters using salt-water injection and surface resistivity, Ground Water, 26, 179-186 (1988).

7. Aramaki, S. Formation of the Aira caldera, southern kyushu, ` 22,000 years ago, Jour. Geophys. Res., 89, 8485-8501 (1984).
8. Nagaoka, S. The late Quaternary tephra layers from the caldera volcanoes in and around Kagoshima Bay, southern Kyushu, Japan. Geogr. Repts. Tokyo Metrop. Univ., 23, 47-122 (1988)
9. Wada, T. ,Masaki, H., Nishiyanagi, R., Fukuda, T., Iwamatsu, A. and Yokota S. Successive measuring of resistivity on steep slope of "Shirasu" (Part 1)-Measuring method and system setup. Proc. '96 annual meeting of the Japanese Soc. Eng. Geol., 237-240 (in Japanese) (1996).
10. Fukuda, T., Iwamatsu, A., Yokota, S., Wada T. and Masaki, H. Successive measuring of 1. resistivity on steep slope of "Shirasu" (Part 2) - Relationship between resistivity and rainfall. Proc. '96 annual meeting of the Japanese Soc. Eng. Geol., 241-244 (in Japanese) (1996).
11. Kobashi, S. ed. Sanchi-hozen-gaku, Bun'eido Publishing Co. Tokyo, 280pp.(in Japanese)   (1993).
12. Yokota,S. Multi-purpose digital hazard map for slope failures. Geoinformatics, 7, 51-59 (1996).

CONSTRUCTION MATERIAL

Proc. 30th Int'l. Geol. Congr., Vol. 23, pp. 503-513
Wang Sijing and P. Marinos (Eds)
© VSP 1997

Evaluation of Soil-Tire Mixtures as a Lightweight Fill Material

CHIEN-JEN CHU & ABDUL SHAKOOR
Department of Geology, Kent State University, Kent, Ohio, 44242, USA

Abstract

Mixtures of two different soil types and shredded scrap tire containing 0%-100% shredded tire, by weight, at 10% intervals, were tested for moisture-density relationships, permeability, shear strength parameters, unconfined compressive strength, and consolidation behavior. The soils used included a non-plastic silt (ML) and a low plasticity clay (CL). For each soil type, three different size ranges of shredded tire (7mm-13mm, 13mm-25mm, 25mm-38mm) were used to prepare mixtures. The results indicate that the density of soil-tire mixtures decreases and permeability increases with increasing shredded tire content for both soils and all three shredded tire sizes. The dry density values of soil-tire mixtures are reduced to 2/3 of the density values for soils alone (a requirement for lightweight fill) with the addition of 60% shredded tire of any of the three size ranges used. At 60% tire content, the permeability increases by 5-6 orders of magnitude for silt and 6-7 orders of magnitude for clay, the increase being greater for larger sizes. The addition of all three shredded tire sizes improves the friction angle of both silt and clay but has little to no effect on soil cohesion. Unconfined compressive strength deceases with increasing shredded tire content for both soils and all three shredded tire sizes. The chemical analyses of pure tire material, and of the leachate from soil-tire mixtures, indicate the presence of minor amounts of Zn, Fe, and Mn but their concentrations are less than the maximum allowable contaminant levels as specified by the U.S. Environmental Protection Agency.

Based on the preliminary data, it is believed that soil-tire mixtures can be used in engineering applications where reduction in density or improvement in the drainage characteristics of low permeability soils is desired. Such applications may include development of football fields and playgrounds in areas of silty and clayey soils, reconstruction of unstable slopes where a light-weight permeable soil may be desirable, and as a general lightweight structural fill.

Keywords: Shredded tire, soil-tire mixtures, lightweight fill, leachate analysis

INTRODUCTION

In the United States, over 279 million scrap tires are generated each year (House Bill S2462, 1990). Of these, nearly 85% are landfilled, stockpiled, or illegally dumped. Since whole tires do not compact well, they tend to rise to the surface of the landfill where they disrupt the landfill cap and allow water to infiltrate the landfill. When stockpiled, scrap tires provide an ideal breeding space for rats, mosquitoes, and other disease vectors, causing a public health hazard (House Bill S2462, 1990; Hudson and Lake, 1977). Scrap-tire fires are extremely difficult to extinguish and pose a serious threat to health and environment due to liquid and gaseous emissions.

In order to reduce the environmental and health hazards associated with scrap tires, it is essential to drastically reduce the landfilling and stockpiling of scrap tires by finding alternative uses. A large-scale potential use of scrap tires can be in soil stabilization. Tire-stabilized soils can be used as a lightweight, or semi-lightweight, fill material for embankments and for reconstruction of potentially unstable or failed slopes in highway engineering as well as in situations where improvement in drainage characteristics is required. Previous studies (Ahmad, 1992; Bosscher et al., 1992; Lamb, 1992; Upton and Machan, 1993; Black and Shakoor, 1994) have indicated the potential for such applications but additional research is needed to quantify the effect of increasing tire content on engineering properties and chemical characteristics of soil-tire mixtures.

OBJECTIVE

The objective of the research presented in this paper was to investigate the engineering properties of soil-tire mixtures containing 0% to 100% shredded tire by weight, at 10% intervals, so that the optimum amount and size of shredded tire needed for improving the desired soil properties could be determined, find the chemical characteristics of the leachate generated by the interaction of water and various soil-tire mixtures, and evaluate the potential applications of tire-stabilized soils.

METHODOLOGY

Two different soil types, a silt and a clay, were used for this study. Bulk samples of the two soils, weighing about 1000 lbs (450 kg) each, were collected from local borrow areas and oven dried at 105 C for 24 hours before subjecting them to various tests. Samples of shredded tire were obtained from Continental Turf Systems, Inc., Continental, Ohio. The samples consisted of three different size ranges: 1/4"-1/2" (7mm-13mm), 1/2"-1" (13mm-25mm), and 1"-11/2"(25mm-38mm). Approximately 500 lbs (227 kg) of shredded tire was obtained in each size range for the lab tests.

The two soil types were tested in the lab to determine Atterberg limits, moisture-density relationships, permeability, unconfined compressive strength, consolidation behavior, and shear strength parameters. All three tire sizes were tested for their dry densities and permeabilities. The soil-tire mixtures, containing 0% to 100% shredded tire by weight, at 10% interval, were tested for moisture-density relationships, permeability, unconfined compressive strength, consolidation behavior, and shear strength parameters. The soil-tire mixtures were prepared using both soil types and each of the three sizes of shredded tire. The permeability, unconfined compressive strength, consolidation behavior, and shear strength parameters were determined for samples compacted to at least 95% of the maximum dry density and within *2% of the optimum water content. All engineering property tests, except the consolidation test, were performed according to the methods specified by the American Society for Testing and Materials (ASTM) (ASTM, 1993). The consolidation characteristics were evaluated from the volume change data obtained during the consolidation stage of the triaxial testing of soil-tire mixtures containing 1/4"- 1/2" (7mm-13mm) size shredded tire. The observed volumetric decrease, converted to axial strain, was taken as a measure of the compressibility of soil-tire mixtures.

In order to investigate the environmental impact of the usage of soil-tire mixtures, the leachate generated in the field by the interaction of rain water and all three sizes of shredded tire material, as well as soil-tire mixtures, was analyzed for iron, barium, cadmium, copper, chromium, lead, zinc, and cobalt.

ENGINEERING PROPERTIES

The engineering properties of the two soil types used in this study are presented in Table 1 whereas Table 2 provides the dry density and permeability values for the three tire sizes. Based on Atterberg limits, and using the Unified Soil Classification System (USCS), the silt can be classified as a nonplastic silt (ML) and the clay as a low plasticity clay (CL). The dry density and permeability values for the three tire sizes are quite similar to each other. The variation of engineering properties with increasing shredded tire content of 1/4"-1/2" (7mm-13mm) for soil-tire mixtures is shown in Figures 1 through 8. The trends for the other two sizes of shredded tire were found to be similar.

Figures 1 and 2 show that the maximum dry density decreases linearly with increasing amounts of shredded tire material for both silt and clay and for all three size ranges of shredded tire material. The optimum water content decreases only slightly up to 60% tire content, beyond which it shows a rapid decrease from over 14% to less than 3% at 100% tire content (Figures 3 and 4).

Table 1. Engineering properties of the soils used.

Property	Soil Used	
	Silt	Clay
Liquid Limit(LL)	26.9	31.5
Plastic Limit(PL)	24.2	20.3
Plasticity Index(PI)	2.7	11.2
Optimum Water Content(%)	15.7	16.8
Maximum Dry Density (pcf)*	106.8	102.5
Permeability (cm/sec)	4.48E-07	4.32E-08
Compressive Strength (psf)**	4077	6600
Soil Classification (USCS)	ML	CL

* 1pcf = 0.016Mg/m^3; ** 1psf = 0.048KN/m^2

Table 2. Dry density and permeability of shredded tire.

Property	Shredded Tire Size		
	1/4"-1/2"	1/2"-1"	1"-1 1/2"
Dry Density (pcf)*	43.2	43.5	43.6
Permeability (cm/sec)	0.16	0.18	0.18

* 1pcf = 0.016Mg/m³

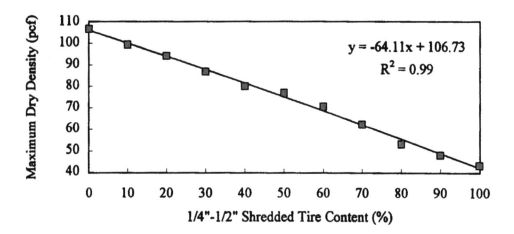

Figure 1: Maximum Dry Density vs shredded tire content for silt-tire mixtures.

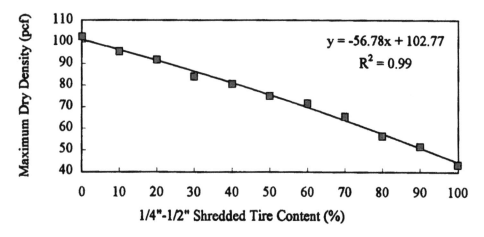

Figure 2: Maximum Dry Density vs shredded tire content for clay-tire mixtures.

The permeability of soil-tire mixtures was found to increase with increasing shredded tire content for both soil types and all three tire sizes (Figures 5 and 6), with the maximum increase in permeability (six orders of magnitude for both soil types) occurring at 40% shredded tire content, beyond which the permeability continues to increase but at a much slower rate.

The plots of unconfined compressive strength versus shredded tire content for both soil types are shown in Figures 7 and 8. The unconfined compressive strength decreases exponentially with increasing tire content. The shear strength characteristics of soil-tire mixtures were determined by performing the triaxial test. Table 3 shows that the friction angle increases and cohesion decreases with increasing tire content for silt-tire mixtures.

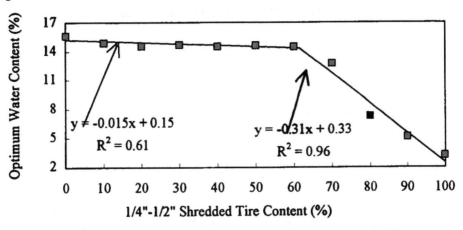

Figure 3: Optimum water content vs shredded tire content for silt-tire mixtures.

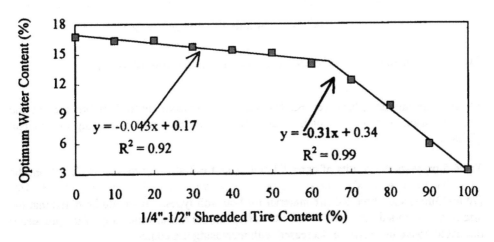

Figure 4: Optimum water content vs shredded tire content for clay-tire mixtures.

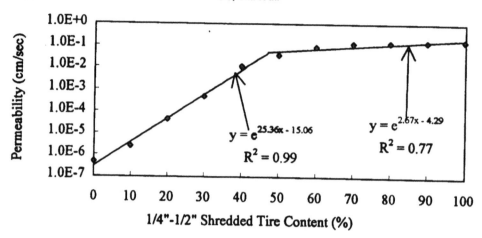

Figure 5: Permeability vs shredded tire content for silt-tire mixtures.

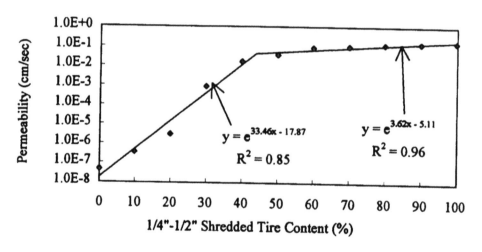

Figure 6: Permeability vs shredded tire content for clay-tire mixtures.

For clay-tire mixtures, however, the friction angle increases up to 20% tire content and then decreases, whereas the cohesion continues to decrease with increasing tire content (Table 4).

Table 5 presents the results of consolidation tests for both soil types. It is clear from the table that the compression index values increase with increasing amounts of 1/4"-1/2" (7mm-13mm) size shredded tire material for both soil types. It should be noted that the size of the shredded tire material had little effect on the rate at which any of the properties discussed above increased or decreased with increasing tire content.

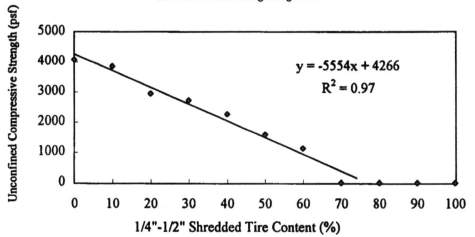

Figure 7: Unconfined compressive strength vs shredded tire content for silt-tire mixtures.

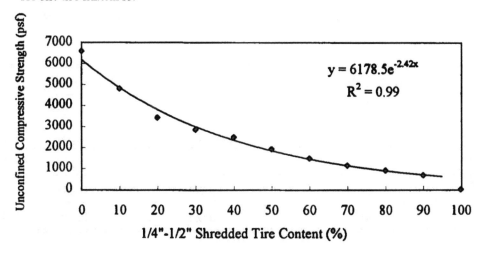

Figure 8: Unconfined compressive strength vs shredded tire content for clay-tire mixtures.

Table 3: Shear strength parameters for silt and silt-tire mixtures (1psf = 0.048KN/M²).

1/4"-1/2" Shredded Tire	Friction Angle (°)	Cohesion (psf)
0	30	1656
10	32	1498
20	34	1122
30	36	985

Table 4:   Shear strength parameters for clay and clay-tire mixtures.

1/4"-1/2" Shredded Tire	Friction Angle (°)	Cohesion (psf)
0	35	2635
10	36	1829
20	38	1289
30	32	1200

* 1psf = 0.048KN/m^2

Table 5: Compression index values for silt, clay, silt-tire and clay-tire mixtures containing 1/4"- 1/2" (7mm-13mm) size shredded tire material.

Shredded Tire Content (%)	Compression index (Cc)	
	Silt-Tire	Clay-Tire
0	0.051	0.101
10	0.070	0.108
20	0.089	0.138
30	0.099	0.219

LEACHATE ANALYSIS

The leachate was collected periodically from all three size ranges of shredded tire as well as from selected soil-tire mixtures for a period of one year.   All leachate samples were analyzed for important trace metals (aluminum, barium, cadmium, cobalt, chromium, copper, iron, manganese, lead, and zinc) using ICP techniques. The results, presented in Table 6, show that concentrations of Cu, Cd, Cr and Pb are less than the Maximum Contaminant Levels (MCLs) established by U.S. environmental Protection Agency (EPA, 1992) for all three tire sizes and all three soil-tire mixtures tested. The concentration of Ba is close to the MCL for all groups except silt and clay mixtures with 30% shredded tire of

1/4"-1/2" (7mm-13mm) size.

Table 6-a: Results of leachate analysis for pure tire.

Ions	Pure Tire (1/4"-1/2")	Pure Tire (1/2"-1")	Pure Tire (1"-1.5")
Al	0.44	0.31	0.44
Ba	2.28	2.09	2.69
Cd	0.002	0.003	0.003
Co	0.05	0.06	0.08
Cr	0.02	0.03	0.02
Cu	0.03	0.01	0.01
Fe	1.17	4.65	3.11
Mn	0.73	1.52	3.38
Pb	0.01	0.003	0.005
Zn	67.09	42.50	32.73

Table 6-b: Results of leachate analysis for soil-tire mixtures.

Ions	70% silt with 30% Tire (1/4"-1/2")	70% clay with 30% Tire (1/2"-1")	40% clay with 60% Tire (1"-1.5")
Al	0.11	0.01	0.14
Ba	0.94	0.03	2.05
Cd	0.002	0	0.001
Co	0.04	0	0.04
Cr	0.04	0.04	0.03
Cu	0.02	0	0.01
Fe	10.41	50	2.31
Mn	84.11	15	9
Pb	0.003	0	0.009
Zn	19.00	6.00	27.60

ENGINEERING APPLICATIONS AS LIGHTWEIGHT FILL MATERIAL

The requirements of a lightweight fill material are that it should have low density, high shear strength, and good drainage characteristics. Since tire-stabilized soils meet these requirements, they can be used to construct highway embankments on soft ground, such as peat and clay, to reconstruct already failed or potentially unstable slopes, to develop play grounds, and to fill low lying areas.

CONCLUSIONS

The conclusion of this study can be summarized as follows:

1. The maximum dry density, optimum water content, and unconfined compressive strength of soil-tire mixtures decrease, and the permeability increases, with increasing shredded tire content for both soil types and all three tire sizes used in this study.

2. The concentrations of leachate for all three tire sizes and selected soil-tire mixtures are significantly below the maximum allowable contaminant levels as specified by the U.S. Environmental Protection Agency for Cu, Cd, and Cr and are close to the maximum allowable limit for Ba.

3. The silt and clay soils stabilized with metal-free shredded tire material, ranging in size from 1/4"-1/2" (7mm-13mm), can be used as a lightweight fill material without any detrimental effects on the environment.

4. The optimum amount of shredded tire material needed to convert a silt or a clay into a fill is 60% at which the dry density of soil-tire mixtures is reduced to 2/3 of the density for soil alone.

Acknowledgments

The authors would like to thank the Ohio Department of Transportation for providing the funding for this research project. Thanks are also due to Karen Smith for typing and formatting the manuscript.

REFERENCES

Ahmad, I., 1992, Laboratory Study on Properties of Rubber Soils: Report No. FHWA/IN/JHRP-91/3, School of Civil Engineering, Purdue University, West Lafayette, Indiana.

American Society for Testing and Materials, 1993, Soil and Rock, Building Stones, Geotextiles: Annual Book of ASTM Standards, Vol. 4.08; Philadelphia, Pennsylvania.

Black, B.A. and Shakoor, A., 1994, A Geotechnical Investigation of Soil-Tire Mixtures for Engineering Applications: Proceedings 1st International Congress on Geotechnics of Waste Materials, Edmonton, Canada, pp. 617-623.

Bosscher, P.J., Edit, T.B., and Eldin, N., 1992, Construction and Performance of a Shredded Waste-Tire Embankment: Department of Civil and Environmental Engineering, University of Wisconsin, Madison, Wisconsin.

Environmental Protection Agency, 1992, Federal Register, 57 FR, No. 246, Washington, D.C.

House Bill S. 2462, 1990, Tire Recycling Incentive Act of 1990: 101st U.S. Congress, 2nd Session, April 19, 1990, Washington, D.C.

Hudson, J.F. and Lake, E.E., 1977, A Planning Bibliography on Tire Refuse and Disposal: EPA Report 68-01-4362, Office of Solid Waste Management Programs, U.S. Environmental Protection Agency, Washington, D.C.

Lamb, R., 1992, Using Shredded Tires as Lightweight Fill Material for Road Subgrades: Draft Report, Materials and Research Laboratory, Minnesota Department of Transportation, Maplewood, Minnesota.

Upton, R.J., and Machan, G.M., 1993, Use of Shredded Tires for Lightweight Fill: Oregon Department of Transportation, Salem, Oregon.

.

Proc. 30th Int'l. Geol. Congr., Vol. 23, pp. 515-517
Wang Sijing and P. Marinos (Eds)
© VSP 1997

Natural Stone Slab Combined with Wood - an Interesting New Building Material

PEKKA IHALAINEN
Tampere University of Technology, Engineering Geology, P.O.Box 600, Tampere Finland

Abstract

During the year 1995 a new type of composition material was introduced by the Finnish stone producer Ericstone Ltd. The idea of this innovation has been to combine wood with stone so that the best properties of the both materials are achieved. The result of the development work is that now there exists a light and flexible building material with durable stone surface. This combination of properties opens many possibilities for the material to be used for example in room fittings and furniture production. The special manufacturing process of this Ericstone slab is also under controll and protected by a patent.

Keywords: Building stone, Composition material

INTRODUCTION

In some cases the high specific density as well as the rather low durability against tensional and bending loads have been the most problematical properties of natural stone. A new composition type of material was introduced in 1995 by the stone producer Ericstone Ltd. in Finland. The goal of the development work has been to combine the desirable properties of both the natural stone and wood. This new stone product is named as the Ericstone slab.

METHOD OF MANUFACTURING AND THE STRUCTURE OF THE SLAB

The structure of the Ericstone slab can be characterised as a typical sandwich type with a thin slab of natural stone laminated together with a thicker slice of balsa wood. The principal problem in manufacturing is to get the covering stone slice thin enough. Otherwise the specific density of the material becomes too high and therefore the most significant advantages of the material are lost. In practice the manufacturing process goes like that: Slabs of wood, in this case balsa, are laminated to the both sides of a calibrated stone slab using fibreglass and epoxy glue. Furthermore the same kind of fibreglass-epoxy lamination is made on the outer faces of the both wood slabs. It is essential that the grains of the wood slabs come exactly perpendicularly to the large surfaces of the slabs. After lamination the wood-stone-wood slab is sawn in two identical pieces through the stone component. After sawing the surface of the stone is polished. The type of the wood as well as the type of stone used in this manufacturing procedure can both be chosen rather freely.

TECHNICAL PROPERTIES AND METHODS OF TESTING

According to the technical properties of a 33 mm thick Ericstone slab (Table 1.) the advantages of both the stone and wood can be achieved. The stony surface of the material is hard and durable but the complete slab itself is much lighter and flexible than a corresponding massive natural stone slab. The compressive strength was tested according to German norm DIN 52 105 [1] using cubical samples. This test corresponds the strength in the edge of the slab. In practice the strength coincides with the strength of the wooden component of the material. The compressive strength in the middle of the slab was determined by compressing a large slab (300 mm x 300 mm) using a steely spacer with the dimensions of 40 mm x 40 mm. In this case the strength represents the shear strength of the covering stone slice. However, the compressive strength is an important parameter if the Ericstone slab is used as the floor material. Depending on the demands in the target the compressive strength of the slab can be passed by an optimal proportion between the thickness of the stone and the wood. Modulus of rupture was determined according to German norm DIN 52 112 [2]. The strength against bending load proved to be higher than that of the natural stone in average. In connection with bending test the amount of bending of the test beam was measured with a special electronic gauge. The Ericstone slab proved to be significantly more flexible than the corresponding massive stone slab. The high flexibility comes important where there occur torsional forces or continual vibrations of structure. Tensional strength over the joint between stone and wood was measured by pulling a small slab fastened in stone surface. In most cases the joint between wood and stone proved to be stronger than the tensional strength of either the stone or the wood itself. Thus, there is no need for the stroger joint than the epoxy glue enables. The principal technical properties of the 33 mm thick Ericstone slab are given in Table 1.

Table 1. Technical properties of an Ericstone slab; a 6 mm thick slice of metamorphosed Wasa quartzite laminated together with a 25 mm thick balsa wood

Technical property	Parameter
Compressive strength in the middle of slab	25 MPa
Compressive strength in the edge of slab	7 MPa
Modulus of rupture	24 MPa
Relative vertical bending before the rupture	1 %
Tensional strength over the joint between stone and wood	1,1 MPa
Water intake during 2 weeks in 100 RH conditions	3-4 %
Water intake during 2 weeks when soaked in water	33-43 %
Specific density	710 g/dm^3 (23,4 kg/m^2)
Maximum slab size	3 m^3

CONCLUSIONS

In many cases new innovations open new uses for the traditional materials. In the case of the Ericstone slab the laboratory tests have proved that both the good properties of stone and wood can be combined by utilising the idea of composition materials and sandwich structure. Several potential targets for the use of Ericstone slab can be pointed out. Despite of its lightness and low specific density the surface of the material is genuine stone with high durability. Its low specific density enables its use for instance in some parts of ships, lifts and other public conveyances. Because of its luxurious look and low specific density many new changes are also opened for its large use in room fittings as well as in furniture industry. Because of its wooden component the use is restricted only to dry conditions.

REFERENCES

1. DIN 52 105. Prüfung von Naturstein. Druckversuch. Deutsche Normen 1988.
2. DIN 52 112. Biegeversuch. Deutsche Normen 1988.

Proc. 30th Int'l. Geol. Congr., Vol. 23, pp. 519-531
Wang Sijing and P. Marinos (Eds)
© VSP 1997

Construction Material Investigation and Experiment Studies of Wan An Xi Concrete Faced Rockfill Dam

ZHANG YONG LIANG & LIU FU MING
Fujian Research Institute of Investigation and Design of Water Conservancy and Hydropower

Abstract

In the region of megaporphyritic-like medium to coarse grained granite, the weathered overburden of lithosome is very thick, only under the protection of fine grained granite at the top border facies of the rock mass can the magmatic body within overburden be found. The feldspar content of granite amounts to above 2/3, with a weak bond between two groups of medium cleavage, which is advantage for percussion and rotation drilling operation. Granite, having a developed joint-fissure structure, can be quarried to produce main rockfill and transition rockfill material through one-shot differential blasting method. Having been compacted by a vibrating roller, the rockfill shows adequate strength and modulus of compressibility, which are up to design requirement. Bedding material, composed of sound granitic crushed rock mixed with fully weathered granitic sand containing fine-grained soil, has fairly good physical and machanical characteristics, which meet the requirement for load transfering zone and the second seepage barrier of concrete faced rockfill dam.

Key words: Concrete faced rockfill dam, granitic rock quarry, overburden weathered stratum border facies, bedding material

INTRODUCTION

Wan An Xi hydroelectric power station is the first stage power station on Wan An Xi stream, a tributary of Bei Xi stream , the main branch of Jiu Long river in Fujian province, located on the south side of Wan An village in Long Yan city , 81km away from the city proper. The dam is of concrete faced rockfill dam type, which stands 993.8m high, with a normal storage water level of 365.0m above sea level, storage capacity $228 \times 10^6 m^3$. The project was put into operation in August 1994 and generated electric power in December of the same year.

The dam required a large amount of good quality rockfills of different gradations, which demands a quarry meeting the following specification: 1 sound rock with high water resistance; 2 quarriable rockfill meeting the required design gradation; 3 thin overburden weathered stratum with large volume deposit of usable material; 4 short hauling distance with convenient quarrying operation etc. The reconnaissance of rock material quarry was carried out in different stages of general, preliminary and detailed surveys, applying the laws of geological distribution and rock weathering process, searching for good quality granitic rockfill with abundant reserve. Rockfill test, in-situ blasting and compaction tests showed that megaporphyritic-like medium to coarse grained granite with developed joints and fissures could be extracted through one-shot blasting for rockfill and transition material, which were up to design requirement in terms of strength and modulus of deformation. A study on preparing semi-pervious bedding material, mixed with sand containing fine grained soil from wholly weathered granite was also carried out, which met the design requirement both for load transfer structure and the second barrier of seepage prevention, with a remarkable economic benefit. It thus opens up a new way for the source of bedding material in China.

RECONNAISSANCE OF ROCK QUARRY

Outcrop and Distribution of Granitic Batholite
Wan An Xi hydroelectric power station is located on a huge granitic batholite, which was formed in the third active intrusion episode of the third phase in the early Yan Shan Stage $(\gamma_5^{2(3)C})$, about 140.7~143.9 million years ago. It is called Gu Tian lithosome in the Fujian Provincial Geological Map with a scale 1:500000, having a length of 70km, a width of 5~35km and outcrop area of 1368km², as shown in figure 1. The petrographic facies of Gu Tian lithosome is relatively developed, showing rather deep degradation. The internal facies of megaporphyritic-like medium to coarse grained structure can be widely observed , and at the top and the periphery of the lithosome, the border facies of fine grained structure has been observed at many different spots.

The dam site is situated in the north eastern part of the lithosome, at a spot 5km from the eastern boundary, about 20km from the western and southern boundaries and 10km from northern boundary. A riverwise fault of about 14km long passes through the upstream of the dam site along NE direction of the river reaches, which is called Song Pe Tan fault. Tri-angle-faces fault formed of silicified mylonite can be seen at many places along the river, in regular and grandeur arrangement. The river valley is deeply cut, having steep slopes, covered with densely forest and vegetation, causing difficulty for the investigation work of rock material.

Outcrop Analysis and Site Selection for Exploration
Within 5km of the dam site area, the following outcrops have been observed : 1 megaporphyritic-like medium to coarse grained granite, the belt-shaped batholite outcrop of which can only be found scatteradly on the river bank; 2 strongly silicified mylonite of Song Pe Tan fault, which outcrops at many places along the upstream river reaches of the damsite; 3 Under the protection of silicified tectonite of Song Pe Tan fault at the bottom wall of which, such as the dam site and at some spots of steep valley, a wide area of moderately weathered batholite has been surveyed; 4 quartzose sandstone outcrop has been spotted on the residual caprock of an adjacent rock on the top of Da Wo Tou mountain on the left bank and the upstream massif of the dam site; 5 outcrop of fine-grained granite can be clearly seen where it is located, which reveals as veins or border and crest rim fine grained facies or post intruding small lithosome $(\gamma_5^{2(3)d})$. Since the border of the huge lithosome is rather far away from the damsite, the exploration of rock material has been focused on the crest

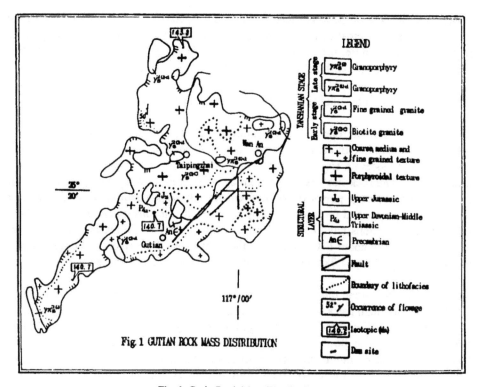

Fig 1 GUTIAN ROCK MASS DISTRIBUTION

Fig. 1. Gutin Rock Mass Distribution

border facies.

During the preliminary survey, two exploratory tunnels were dug by the side of a small path at the junction of two gullies, 2km southward of the damsite. The tunnel opening is on a bank wall of fully weathered megaporphyritic-like medium to coarse rock, which has a bank slope of 60~70 degree. The tunnel is more than 30m long, of which 14~21m is covered with fully weathered zone, followed by strongly weathered rock intercalated with little amount of moderately weathered rock block, which can be broken by hand, and the quarry rubbish can be crushed into rock debris as it falls down along the slope, with rock particle larger than 10cm being hardly seen. Several exploration sections were analysed and the result showed that the ratios of waste to usable reserve were mostly langer than 1. This not only needs high investment for unusable material, but also no palce can be found for dumping ground and if it is dumped into the river valley, a disaster can surely be expected.

A great deal of investigation data indicates that under the conditions of warm and humid climate , the megaporphyritic-like medium to coarse granitic rock is more liable to weathering than sedimentary, extrusive rocks and ordinary granite. Its porphyritic crystal is K-feldspar, having a coarse grain size of generally $1\times1.5cm^2$, with some of which $2\times3cm^2$ and few being up to $3\times6cm^2$. the porphyritic crystal content is commonly 5~25%, or up to 30~40%. Upon weathering, plagioclase is first decomposed into kaolinite, the crystal links of other mineral particles are then damaged, partly or even totally lose their cohesive forces, but most of K-feldspar and quartz particles reveal as sand and psephitic shaped mosaic structure, preserving the structure of the parent rock, forming very thick and totally weathered overburden stratum widely spread in this region, especially at the ridge of planation surfaces of different grades and at the crest of residual hill, showing thicker stratum and not suitable for rock quarrying ground.

Siliceous tectonite outcrops at 5 spots along the river, upstream of the damsite with a reserve of one hundred thousand to several hundred thousand m3. The rock is quite sound and located at moderate elevation. But a very hard rock might require a high drilling bit consumption. The rock has developed a group of fracture cleavage, dipping towards the left bank with outcrop points distributed on the left bank resulting in poor diagenetic feature of rock mass and forming a cliff by the river. It is inconvenient to quarry and most of the rock formations have limited depth and small reserve . Therefore, a rock quarry, having a reserve of about 670×10^3 m^3, on the right bank of the river, at the upstream site of Shan Jie Keng ferry, 1.7km away from the damsite was finally chosen as quarry No. 1.

The quartzose sandstone quarry on the residual caprock of an adjacent rock on the top of Da Wo Tou mountain has a small reserve, located in high elevation, it is thereby unusable. Nevertheless, this discovery has encouraged the geologist to search for the possibility of the existance of crest border facies fine-grained granite at nearly moderate elevation. Finally, fine-grained granite outcrop was discovered in a dense forest on the same crustal block in an interstream area downstream. Further exploration was immediately carried out, gradually increasing the depth and the area of the job, which result in finding out the best and the only usable quarry for the project--quarry No.15.

Genesis Analysis of Quarry No. 15
There are two reasons which cause the formation of fairly thin weathered overburden at quarry No.15, namely 1 The existance of fine-grained granite on the border facies at the crest Gu Tian lithosome . 2 Located on the western side of a rather narrow crustal block in an interstream area, and on the middle and upper parts of steeper slope. At crest elevation of 552m and 542m two boreholes, at 100m apart , have been drilled, giving the thickness of fine-grained granite 35.5m and 45.15m and the depths of fully weathered rock 16.54m and 12.42m respectively. Under the protection of fine-grained granite, the underlying medium to fine-grained granite (as transition facies) and megaporphyritic-like medium to coarse-grained granite (as internal facies) show a slower weathering process. The weathered material produced on the steep slope has been abrased and washed away so that the weathered overburden is relatively shallow, about 0~7.3m, and the bedrock is well exposed. Within the area of 40000m^2 to be quarried, the average depth of overburden is 8.7m, totaling an amount of waste 350000m^3, only 17% of the usable reserve 2040000m^3, being a good quarry rarely available in the region. Short gullies are developed below the quarry, which transfer the washed away flow into the river channel. By the sides of short gullies are residual ridges, forming a series of alternating geological structures. It has been inferred that the crest of residual ridges is the grade III of the planation surface in the region, with a depth of weathered overburden of more than 30m. Therefore, except at the upper location of the gallies, where the bed rock can still be seen, the rest below the quarry

No.15, is covered by very thick overburden, this makes the possible extension of the quarry towards the lower elevation and both upstream and downstream sides very limited.

QUARRY SELECTION AND CONSTRUCTION USE

Quarry No. 15 is situated on the left bank downstream, 0.5km away from the damsite. During the detail exploration, it was suggested the quarrying be set between elevations 360~552m, with difference in elevation of 192m, natural rock slope of 33^0 and the average quarrying depth of 50m. The quarry would use a bench stoping method, with an average back-slope of 60^0, giving the usable reserve 2040000m^3.

Construction organization design had considered to haul rock material from quarry No.1, 1.7km away from the damsite, for the lower part of the dam , while the middle and the upper part of the dam would be suppied with rock material from quarry No.15 by constructing 2~3km long of haulage way.

During dam construction, the construction unit abandoned quarry No.1 due to the fact that the rock was silicified etc., and quarry No.15 became the only quarry available for dam construction. Rockfill quarrying operation was carried out from the lowest possible elevation, i.e. to open a loading platform at elevation 348m, and gradually extend towards downstream, high elevation, lengthwise and depthwise, 90% of 1040000m^3 rockfill required for dam body construction was obtained from the quarry No.15, and the actual blasting operation was finally extended to elevation above 460m.

The rocks at the middle and the lower part of the quarry No.15 are mainly of megaporphyritic like medium to coarse granite, showing ultimate saturated compressive strength of 73.2~115.0Mpa, with an average being 86.3Mpa. It is of sound rock, with a mean dry unitweight of 26.0KN/m^3, but slightly lower than ordinary granite in strength, with a softening coefficient of 0.60~0.72 and an average of 0.70,i.e. lower than 0.80, the requirement of the relevant specification ; and the other indexes are up to the Chinese code requirement.

The lower compressive strength of rock has a relatively low competency against impact and torsion developed during percussion and rotation drilling operation, this result in a lower consumption of drilling bits. The internal causes of which are:1 Granite quarried from the middle lower part of the quarry No.15 having coarse grained structure; 2 High content of K-feldspar, with a total content of K-feldspar and plagioclaze being above 70%. The feldspar mineral has two groups of mostly perpendicular moderate cleavages, along which it is easily broken. Close observation on the splitted surfaces of megaporphyritic-like crystals containing K-feldspar, it can be seen that quite a large of them being the cleavages of the crystals. It follows that the forces of the two group cleavages in feldspar mineral of granite are much lower than those among mineral particles, and the mineral is most easily broken along clevage surfaces when disturbed by drilling impact and torsion and pulverized into rock debris and powder. While quartz, has a mineral hardness of 7 degree, harder than feldspar by 0.5~1.0 degree, it has no cleavage and shows a conchoidal fracture when being hit and splitted. Therefore, it needs more energy to crush than feldspar.

Practice has thereby proved that with a view to saving drilling bit consumption, granite has a superiority to silicified rocks, quartzite and quartzose sandstone, and megaporphyritic-like medium to coarse grained granite has a superiority to fine-grained granite.

JOINTS ,FISSURES OF GRANITE AND BLASTING TEST

The quarry No.15 neighbours with Song Pe Tan fault, the main fractures of which is striking at NEE direction, with nearly vertical air faces. At the blasting test ground, there are three fault having the same strike such as F_2 etc. and joints of the same group also develop, dipping upstream (group a) and downstream (group b). Group a develops in the hanging side of the fault F_2 , having an average interval of 36cm or 18cm at some concentration section, with quite developed micro-closed fracture cleavages. Group b and NW joints develop in the bottom wall of the fault F_2, having an average interval of 40–45cm. NW joints intersect with air faces at an angle of about 30^0, dipping into (group c) or out of (group d) the slope. At the terrain where group b,c,d of joints and fissures dissect the downstream bottom wall of the fault F_2, it is suitable for quarrying small size rockfill . Based on the above mentioned analysis, the first blasting test was carried out in the full experimental segment to quarry main rockfill material, and for transition rockfill material, it was performed at the downstream terrain. The blasting test proved that the result was fairly ideal.

Since there is no main geological structures in positive slope directions or dipping towards air faces. Therefore, the frame work of blasting operation is focused only on the direction of the air faces, without relating to structural planes. The recommended design values of grain size distribution for main rockfill and transition rockfill materials together with the results from the first blasting test are compared as shown in tables 1,2 and figure 2.

It can be seen from tables 1,2 and figure 2, the effective particle size of main rockfill material from the first blasting test result is slightly larger, while the rest are quite ideal. The main parameters and the result of the first blasting test performed at the quarry No.15 is shown in table 3.

The blasting test and construction practice has thus proved that the rockfill quarrying in granitic rock show a bright prospect.

STRESS AND STRAIN CHARACTERISTIC OF GRANITE ROCKFILL

Generally, the stress and strain relationship of CFRD rockfill reveals a remarkable nonlinearity, and the stress and strain curves obtained from the lab. triaxial test are commonly expressed by approximate mathematical function for the purpose of applying finite element calculation. The most commonly used model both home and abroad is a

Table 1   Comparison of Grain Size Distribution (Main Rockfill Material)

Characteristic particle size	Recommended design value (mm)	The first blasting test result (mm)
max. particle size D_{max}	800~600	700
controlled particle size d_{60}	280~108	127
average particle size d_{50}	200~60	88
controlled particle size d_{30}	71.0~13.2	39
effective particle size d_{10}	8.9~1.08	15

Table 2   Comparison of Grain Distribution (Transition Rockfill Material)

Characteristic particle size	Recommended design value (mm)	The first blasting test result (mm)
max. particle size D_{max}	400~300	500
controlled particle size d_{60}	137~51	100
average particle size d_{50}	100~30	73
controlled particle size d_{30}	36.5~6.9	24
effective particle size d_{10}	4.75~0.62	4.3

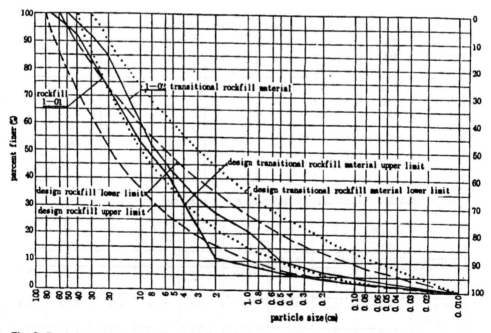

Fig. 2. Recommended design grain distribution curves for rockfill and transitional materials with curves obtained from the 1ˢᵗ blasting test at No. 15 quarring site

hyperbolic model developed by Duncan and Chang. Some engineering cases showed that the approximate model had basically given satisfactory results in representing the stress-strain characteristic of rockfill materials. The eight parameters determined from triaxial test for Duncan and Chang model are as follows,

ϕ: *internal friction angle*
C: *cohesive force*
R_f: *failure ratio*
K: *base of initial modulus of elasticity*
n: *exponent of initial Poisson's ratio*
G: *base of initial Poisson's ratio*
F: *coefficient of initial Poisson's ratio*
D: *Poisson's ratio coefficient*

This project used the method of grain size distribution similarity in preparing samples for triaxial testing. The sample size of main rockfill and transition rockfill materials is ϕ 300×750mm, and that of bedding material ϕ 200×500mm, applying the chamber pressures of 0.2, 0.4, 0.8 and 1.2Mpa and a rate of shearing 2.0~2.7 mm/min. The test results are shown in table 4.

From the eight parameters shown above, the tangent modulus and targent Poisson's ratio at any points of the hyperbolic curve can then be obtained. For convenience, Duncan replaced the tangent Poisson's ratio by tangent bulk modulus expression in 1980, and put forward the so-called E-B model, the parameters of which are shown in table 5.

Table 3   Main Parameters and Results of the First Blasting Test at Quarry No.15

material from blasting (code name)	blasting framewor area (m²)	ignition mode	charge structure	unit charge consumption (kg/m³)	ununiformity coefficien Cu	coefficien of curvature Cc	d<5mm content (%)	rate of large rockfill (%)
main rockfill material (1-01)	2.15× 2.15	in rows	continuous	0.774	8.5	0.8	9	3 (>600mm
transition rockfill (1-02)	2.15× 2.15	"V" shaped	continuous	0.779	23.3	1.3	18	7.5 (>300mm

Table 4   Rockfill Parameters of E~γ Model from Large Triaxial Test

sample material	dry unit weight g/cm³	C KPa	ϕ degree	K	n	G	F	D	R_f
bedding material	2.10	150	41.6	1080	0.45	0.41	0.09	7.60	0.67
transition rockfill	1.97	100	40.1	1160	0.36	0.44	0.21	8.26	0.96
main rockfill	1.99	100	40.1	1080	0.42	0.47	0.24	6.16	0.98

Table 5 Rockfill Parameters of E-B Model from Large Triaxial Test

sample material	φ_0	$\Delta\varphi$	K	n	B	m	R_f
bedding material	54.4	10.5	1800	0.32	1100	0.19	0.72
transition rockfill	49.5	8.1	3000	-0.26	740	-0.28	0.86
main rockfill	50.3	8.6	3000	. -0.33	780	-0.39	0.85

where φ_0:: *internal friction angle when σ_3 equals to atmospheric pressure.*
$\Delta\varphi$: *internal friction angle difference, i.e. The reduce value of φ when σ_3 increases up to 10^0.*
B: *base of bulk modulus*
m: *exponent of bulk modulus, the rest symbols are the same as previously defined.*

Generally, the values of n and m for both E-γ and E-B models are in positive number, but the above table shows the negative values, which is uncommonly seen. This indicates that the sharp edges of granitic rockfill are sheared-off during triaxial testing process, showing a shear-contraction behaviour although the strength of rock itself is fairly high. Therefore, the tangent modulus of elasticity and the tangent bulk modulus will decrease as the chamber pressure increases.

The stress and strain analysis of Wan An Xi concrete faced rockfill dam has used the above mentioned parameters for 2D and 3D finite element calculation, where the effect of construction sequence, the difference of mechanical properties between bedding material and face slab concrete etc. have been put into consideration. The main results are shown in tables 6 and 7, which are in conformity with those of similar projects both home and abroad.

Table 6 Maximum Stress and Displacement of Dam Body (2D finite element calculation)

calculation conditions / cross section	upon project completion			after reservoir impoundment		
	vertical displacement (cm)	horizontal displacement (cm)	σ_1 (MPa)	vertical displacement (cm)	horizontal displacement (cm)	σ_1 (MPa)
0-72	5.10	-0.49	0.25	6.60	1.35	0.32
0-18	27.30	2.20	0.84	33.60	16.70	0.97
0+54	5.40	0.68	0.22	6.90	2.10	0.26

Table 7   Maximum Stress and Displacement of Dam Body (3D finite element calculation)

stress and displacement / calculation conditions		upon project completion	after reservoir impoundment
max. displacement of dam body (cm)	vertical	20.6	24.5
	horizontal	1.90	11.1
max. deflection of face slab (cm)	normal	0.46	14.53

STUDY TEST ON MATERIAL SELECTION OF BEDDING MATERIAL CONTAINING DECOMPOSED GRANITIC COARSE SAND

Conventionally , bedding material are prepared by mixing fresh and hard crusher-run rockfill and sand. But the productivity of crusher-run sand sized particles at Wan An Xi CFRD project was limited, it could hardly meet the requirement of construction schedule and its unit cost was also too high. Micro-differential extrusion blasting method produced too much oversized particles, which needed further screening to meet the design requirement for grain size distribution, and added the cost as well as constructional difficulty. Although natural river sand mixed with crusher-run rockfill showed good workability, it had fineness modulus mostly of 3.4~4.30, with small percentage of fine particles (d<0.1mm), which did not meet the design requirement of semi-imperviousness material. Through a great deal of laboratory and in-situ tests, decomposed granitic coarse sand, which is plentiful nearby the damsite (located in the lower part of zone B and the upper part of zone C of the fully weathered megaporphyritic-like medium to coarse grained granite) has been used for the preparation of bedding material. Such decomposed granitic rock has a content of particle size smaller than 0.1mm about 9~20%, with d<0.005mm being 2~8%. Most of the particles is composed of sand and small peblle , having coefficients of ununiformity and curvature of 13.8 and 1.74 respectively, showing well graded grain size distribution, being most ideal for the fine particles (d<5mm) of bedding material.

Crusher-run rockfill, which is mixed with three different proportions of decomposed granitic sand, is directly placed on the upstream zone of the dam for compaction comparison test.
1 Crusher-run rockfill mixed with 25% decomposed granitic coarse sand.
2 Crusher-run rockfill mixed with 15% decomposed granitic coarse sand and 10% natural sand (coarse sand with particles finer than 0.1mm below 1%).
3 Crusher-run rockfill mixed with 25% natural coarse sand.

The in-situ compaction test used 16t smooth steel-drum vibaratory roller, having an actuating vibration impact of 50t and a running speed of about 0.3~0.5m/s. Water was added to mixed material adequately for lubrication effect on particle surfaces, with the main object to facilitate dense compaction. The compaction thickness per lift was set at 40cm, with the number of compaction being 2,4,6 and 8 roller passes. Settlement of each compaction was measured and dry density with porosity were obtained from the relevant test pits, using water injection method, so that in-situ permeability coefficients could also be obtained for different mixed proportion of bedding material at different vibrating roller passes. The test result is shown in table 8.

Table 8 Roller compacted test results of bedding material on dam face, mixed with 25% fully weathered granitic sand containing fined grained soil

item	compactio index	Number of 161 vibratory roller compaction n				
		4	5	6	7	8
physical property	percentage of d<5mm(%)	32.0		39.7~35.9 37.8		32.2~35.6 33.9
	percentage of d<0.1mm(%)	0.8		1.4~1.2 1.3		2.9-3.2 3.05
	coefficient of uniformity Cu	37.5		31.5~27.0 29.25		55.0~47.1 51.05
	coefficient of curvature Cc	0.56		0.54~0.48 0.51		0.93~0.84 0.885
	dry density Pd (t/m^3)	1.929		2.084~2.003 2.044		2.296~2.222 2.259
	porosity n(%)	32.3		19.2~22.4 20.8		11.0~14.0 12.5
mechanical property	compaction settlement h(mm)	12.0	19.8	26.8	32.4	37.0
	deformation modulus E(MPa)		22.9	25.5	31.9	38.9
	compression modulus E0(MPa)		36.8	40.9	51.2	62.4
hydraulic property	coefficient of permeability K(cm/s)	7.4×10^{-3}		5.8×10^{-3}		3.6×10^{-4}

The result shows that crusher-run rockfill mixed with 25% decomposed granitic coarse sand and compacted to 8 roller passes has the best physical and mechanical indexes.

Grain size distribution
Particles finer than 5mm hold 32.2%~35.6%, with a coefficient of ununiformity between 47.1~55, particles finer than 0.1mm occupy 3.9%~4.2%, basically in conformity with the upper zone of grain size distribution zone proposed by James L. Sherard, which is up to design requirement and conducive to minimizing particle segregation during construction.

Dry unit weigh and porosity
Mean dry unit weight is 22.59KN/m^3 with a corresponding porosity of 12.5%. Some data on dry unit weight and porosity of CFRD in China are: 21.70 KN/m^3 and 19.9% at Guang Xu dam, 23.74 KN/m^3 and 13% at Xibei Kou dam . It is thereby clear that Wan An Xi dam bedding material has a high density although it is composed of decomposed granitic coarse sand of having a lower specific gravity.

Coefficient of permeability

Water injection test carried out on site gives an average permeability of 3.6×10^{-4}cm/s, being semi-pervious material, while laboratory test shows that fine particle movement or piping does not develop as the hydraulic gradient increases up to 54. It has a stable internal structure, which is up to the requirement for the second barrier of seepage prevention within the concrete faced rockfill dam.

Compression modulus

Compression modulus measured from in-situ vibratory compacted layer (without superimposed overburden load) is 62.4 Mpa, while in the lab., within the normal pressure of 0.1~0.2Mpa, is 68.5Mpa. Zhushu Qiao CFRD in China used weathered slate as rockfill and had a compression modulus of 31Mpa~44Mpa, nevertheless, it did not cause any detrimental effect on concrete face slab and joints. The recorded compression modulus of some CFRD in foreign countries are: Segredo dam 30Mpa~60Mpa, Khao laem dam 40Mpa, Foz do Areia dam 35Mpa. Based on project construction and operation records, they are still considered satisfactory by experts in the field of earth and rockfill dams.

Shear strength

Direct shear test and triaxial test performed in the lab. show the internal friction angle of bedding material decreases as the chamber pressure or normal pressure increases. e.g. the internal friction angle (max. value) expression of the bedding material mixed with 25% decomposed granitic coarse sand is as follows,

$$\phi = 46.9^0 - 6.9^0 \lg(\sigma_3/P_a)$$

where P_a: atmospheric pressure

Based on the above physical and mechanical characteristic of bedding material mixed with decomposed granitic coarse sand, finite element seepage analysis on temporary dam section during construction has shown that almost 70% of hydraulic head is concentrated and decreased in the bedding zone and the transition zone. It will not affect the downstream slope stability. Swedish method and Simplified Bishop Method used in the maximum dam cross section stability analysis also show no possibility of planar sliding in the zone of bedding material, with factor safety of dam against sliding between 1.591~1.856, satisfying or being greater than the requirement specified in Chinese code on earth and rockfill dam design.

The fully weathered biotite granite can be classified into three zones A,B and C. Zone A is mostly composed of brownish red clayed soils, with a clay content higher than 20%. Except quartz particles, the rest mineral particles have been fully weathered into clayey mineral, such as kaolinite etc. exposed as eluvial horizon with uniform hue and good binding property. Since it contains a high percentage of clay particles, it manifests as lump of soil, which can not be separated evenly into the void of crushed rock material. Zone B has a cerise colour to powder-like white colour sandy clay. It is of sand material containing fine grained soil, with clay content of 10~20%, and a slightly large amount of silt particles. The plagioclase has been totally weathered into kaolinite, so does most of the K-feldspar. K-feldspar sand hardly exists or almost does not exist at all. The upper part of zone C is mostly greyish white fine grained sand, with a clay content less than 10%. K feldspar mineral has not been fully weathered into kaolinite. When rubbed in hand, the remained sandy particles can be felt. The lower part of zone C is mostly composed of sand and coarse sand, containing fine grained soils. The crystal between mineral particles have been totally damaged. The classification boundary with strongly weathered granite is defined by the easiness of hand-twisted breakage.

Between elevation 310~340m of monadnock, below the quarry N0. 15 , and after excavation of 1m deep in zone A and the upper part of Zone B, is sand containing fine particles in the lower part of zone B and the upper part of zone C. But in actual excavation, the material without containing brownish red clayey soil is considered the acceptable upper limit, while the lower limit is controlled by the possible normal operation of backhoe. The representative material from this zone about 2~4m below ground surface is sand containing fine grained soil, of which d<0.1mm amounts to 20.3%. The preparation of bedding material by mixing cresher-run rockfill and the above mentioned decomposed granitic sand give a particle content of smaller than 0.1mm about 5% or so, which is up to the design requirement.

REFERENCES

[1] Zhang Yong Liang, Liu You Ke, Liu Jing Hui, Investigation and Application of Rockfill in Wan An Xi concrete faced rockfill dam, Water Power, September 1994

[2] Chen Yu Nuan, Liu Fu Ming, Chen Yu Bin, Study Test on the Characteristics of Bedding Materials Containing Decomposed Granitic Coarse Sand, Water Power, May 1994.

[3] Zhang Yong Liang, Engineering Geological characteristic of Granitic Weathered Soils in Fujian Province, Developments in Engineering Geology, February 1987